D0705312

Designing Sustainable Forest Landscapes

Designing Sustainable Forest Landscapes is a definitive guide to the planning, design and management of forest landscapes, covering the theory and principles of forest design as well as providing practical guidance on methods and tools. Including a variety of international case-studies, the book focuses on ecosystem regeneration, the landscape planning of natural forests and the design of plantation forests. Using visualisation techniques, design processes and evaluation techniques, it looks at promoting landscapes that are designed to optimise the balance between human intervention and natural evolution.

A comprehensive, practical and accessible book, *Designing Sustainable Forest Landscapes* is essential reading for all those involved in forestry, natural resources and landscape professions, as well as those with a wider interest in these areas.

Simon Bell is a forester and landscape architect. He is Senior Research Fellow at Edinburgh College of Art, where he is a director of the OPENspace research centre, Associate Professor at the Estonian University of Life Sciences and also a landscape consultant. He was formerly Chief Landscape Architect to the British Forestry Commission and is involved in research, teaching and project landscape design.

Dean Apostol is a landscape architect and ecological restorationist with 30 years of experience. He teaches at the University of Oregon, and is presently active in the design of a new city (Damascus, Oregon). He was chief landscape architect for Mt Hood National Forest in Oregon, and has been active in the Society for Ecological Restoration International. He recently edited *Restoring the Pacific Northwest: the art and science of ecological restoration in Cascadia.*

Designing Sustainable Forest Landscapes

Simon Bell and Dean Apostol

Taylor & Francis
Taylor & Francis Group
LONDON AND NEW YORK

First published 2008 by Taylor & Francis
2 Park Square, Milton Park, Abingdon, Oxon OX14 4RN

Simultaneously published in the USA and Canada
by Taylor & Francis
270 Madison Avenue, New York, NY 10016

Taylor & Francis is an imprint of the Taylor & Francis Group, an informa business

© 2008 Simon Bell and Dean Apostol

Typeset in Frutiger by
Florence Production Ltd, Stoodleigh, Devon
Printed and bound in Great Britain by
The Cromwell Press, Trowbridge, Wiltshire

The publisher makes no representation, express or implied, with regard to
the accuracy of the information contained in this book and cannot accept
any legal responsibility or liability for any errors or omissions that may be
made.

British Library Cataloguing in Publication Data
A catalogue record for this book is available from the British Library

Library of Congress Cataloging in Publication Data
Bell, Simon, 1957 May 24–
     Designing sustainable forest landscapes/Simon Bell and Dean Apostol.
          p. cm.
     Includes bibliographical references.
     1. Forest landscape design.   I. Apostol, Dean.   II. Title.
     SB475.9.F67B45 2008
     634.9–dc22   2007017622

ISBN10: 0–419–25680–6
ISBN13: 978–0–419–25680–9

Contents

Acknowledgements

This book is the outcome of the experience of many years of development, research and application. While we may have written it we cannot claim that it has solely been our project, or that there have not been significant contributions from many other people. Thus our debt of gratitude is immense. We need to thank our mentors who first set us on the road to get involved in forest design. Since we each need to thank a long list of different people, Simon Bell will offer his first, followed by Dean Apostol.

I, Simon, must acknowledge people such as Dame Sylvia Crowe as the person who first started the process in the UK and also has an influence in the USA. While she had retired before my time, I met her on several occasions before her death and her writings were an inspiration. Then Duncan Campbell, who recruited me to become a forest designer, also deserves a special mention. The colleagues who worked with me in the Forestry Commission and with whom the process and practice was developed and refined over the years – Oliver Lucas, Maggie Gilvray, Alison Grant, Carol Anderson, James Swabey, Gareth Price, Sandra Hanlon, Dafydd Fryer, Russell Bailay, Richard Howe, Roger Worthington and Nicholas Shepherd – need a word of thanks. The researchers at the Forest Research Agency (FRA) who worked on the application of the landscape ecology to the restoration of native woodlands, a project initiated and commissioned by myself, were Graham Pyatt, Jonathan Humphrey, Chris Quine and Joe Hope. Also at the FRA was Max Hislop who contributed material on public participation techniques. Also in the UK I need to thank Steve Conolly of Cawdor Forestry, with whom I have worked on a number of projects for the material in the Dochfour case study and for some of the examples of computer visualisation. Thanks too to Clive Steward of the Woodland Trust, for allowing the use of the Victory Wood case study. Scottish Woodlands staff – Stuart Wilkie and Colin McNair – are also to be thanked for permission to use computer graphics from some projects I did for them.

In Canada there are a lot of people for me to thank. In British Columbia the project to introduce forest landscape design was managed by Jacques Marc, the senior landscape specialist, with help from Luc Roberge, Lloyd Davies, Ken Fairhurst, Peter Rennie and Larry Price. The time they worked on this was very productive. Also, the development and integration of landscape ecology into the design process would not have happened without Larry Price. Thanks also go to Jacques for permission to use some of the materials collected or developed for the visual landscape design training manual project. Subsequently my business association with Linnea Ferguson, when they ran Vista Design Ltd in Nelson, BC, was a very fruitful period. Also I need to thank Alex Ferguson of Slocan Forest Products and Paul Jeakins of Kokanee Forest Consulting for the opportunity to work on the many projects with them, especially the Bonanza face and for the permission to use it in this book as a case study. Across Canada, in New Brunswick and Nova Scotia, I need to thank Robert Whitney of the Maritime Forestry Technology College and Tom Murray and Julie Singleton of the Canadian Forest Service in Fredericton for the opportunity to develop, train and pilot the process over there. Thanks also to Tom and the staff at StoraEnso, including Gerry Peters, now retired, for permission to include the Sutherlands Brook project as a case study.

In Finland Minna Komulainen of the Kainuu Rural Advisory Centre and Airi Matila of Tapio, the Finnish Private Forest Owners Association, are to be thanked for the opportunity to work on the Vuokatti project and for permission to use it in this book as a case study. Baiba Rotberga of the Latvian State Forest Service and Olgerts Nikodemus of the University of Latvia are also to be thanked, both for their work in developing the forest landscape planning and design and the applied landscape ecology work for Latvia, and for permission to use it as a case study or as other material.

In Tasmania, Australia, I need to thank Bruce Chetwynd for hospitality, for showing me the forests of that beautiful island and for allowing me to use an illustration.

Finally, I need to thank my wife Jacquie for her support and help during the arduous process of completing the book. She has put up with my many absences, word processed early drafts and has been dragged around the forests of the world, patiently listening to my comments about the appearance of logging in every landscape we visited.

I, Dean, need to offer thanks to many colleagues who had a part in developing the intellectual and case study possibilities for the content of this book. First, thanks to the many former colleagues at the US Forest Service. Richard Shaffer, my mentor as head landscape architect at Mt Hood National Forest, had the courage and vision to carve out some work space for me to develop my initial ideas on forest design. Fellow landscape architects Jennifer Burns, Diana Ross, Mike Abate, Pete Heiden, Jurgen Hess, Paul Gobster and Terry Slider all shared important ideas, feedback and/or case studies along the way. Warren Bacon provided crucial moral and financial support for development of the Forest Landscape Analysis and Design publication that initiated the American interest in the subject. Nancy Molina (formerly Diaz) is a creative and thoughtful ecologist and neighbour who saw the possibilities in merging ecology and design and developed the initial analytical framework. She has also been a strong proponent and teacher of the process to others across the globe. Carol Spinos had the vision to introduce forest design to the Applegate area of Southern Oregon.

Pat Greene, landscape architect and close colleague, had a huge role in applying and improving forest design on Mt Hood National Forest. Pat passed away at too early an age and unfortunately was unable to see this book in print.

We are both most fortunate to have crossed paths with Dave Perry, Jerry Franklin, Hamish Kimmins and other pioneering forest ecologists who are changing the nature of forestry through their research, writings and patient approach to educating the rest of us about how forests work. Paul Hosten, another ecologist and artist (believe it or not) did some crucial work in reconciling design with ecology at Carberry Creek. David Andison is doing similar work in the Boreal forests of Canada, helping to shape large-scale forest harvests that closely mimic natural disturbance patterns.

Robert Ribe of the University of Oregon, who is doing important research on forest landscape aesthetics, has been a close colleague of mine. He has provided timely support over the years, including sending the occasional paying client my way. Stephen Sheppard of the University of British Columbia is engaged in pathfinding research on the linkage between forest aesthetics, ecology and economy. Peter Barz is a peripatetic, cranky and eccentric German who used to live in Scotland and is a living history book of Central European forestry, and has beaten me over the head more than once on the importance of defining sustainability before preaching it.

I want to join Simon in thanking Robert Whitney of the Maritime Forestry Technology College in New Brunswick, Canada, a most creative spirit trapped in a forester's body.

And finally, many thanks to my partner Marcia Sinclair and my son Simon James Apostol. Marcia has taught me all I know about the art of public involvement. Both have put up with an unfinished farmhouse, an often empty bank account and a day-to-day struggle to make ends meet as I pursue my stubbornly independent life that provides an irregular income and too many unfinished projects.

Foreword

Awareness of the crucial role played by spatial patterns in the dynamics of Nature is a relatively new phenomenon in science. The outlines of landscape ecology began to emerge in Europe at least 60 years ago. However, until the 1980s, foresters and most ecologists in North America focused largely on plant communities (or stands) and circumscribed habitats, discrete entities with more or less clear boundaries. Landscapes were not totally ignored; ecologists in both Europe and the USA did seminal work relating the distribution of plant communities to environmental gradients across landscapes.[1] But the concept of "landscape ecology", i.e. that landscape patterns could influence processes and functions, and that those processes and functions could in turn feed back to influence landscape patterns, came more slowly.

The situation changed in the early 1980s when Forman and Godron, influenced by work in Europe, published a paper introducing the basic elements of landscape ecology to a North American audience. Forman wrote[2]

> What do the following have in common? Dust bowl sediments from the western plains bury eastern prairies, introduced species run rampant through native ecosystems, habitat destruction upriver causes widespread flooding downriver, and acid rain originating from distant emissions wipes out Canadian fish. Or closer to home: a forest showers an adjacent pasture with seed, fire from a fire-prone ecosystem sweeps through a residential area, wetland drainage decimates nearby wildlife populations, and heat from a surrounding desert desiccates an oasis. In each case, two or more ecosystems are linked and interacting.

It was an idea whose time had come. The International Association of Landscape Ecology was formed in 1982. The journal, *Landscape Ecology*, was initiated in 1987 and served to link the more established European traditions with those newly emerging in North America. Papers and books proliferated through the 1980s and output has been accelerating since (typing "landscape ecology" into Google Scholar today returns 233,000 hits). Ecologists have now extended their spatial view across the ecological hierarchy with concepts such as metapopulations, meta-communities and metaecosystems, all explicitly recognizing the importance of spatial patterns on ecological processes and functions. Earth System Science is nothing more than landscape ecology written large and with an added temporal dimension.

The logical application of landscape ecology to landscape design also occurred earlier in Europe than in North America, driven in the former by the ubiquitous presence of humans and their various needs and desires. In contrast, conservation was the initial driver in North America, the stage being set in the 1960s by MacArthur and Wilson's immensely influential theory of island biogeography, which linked spatial patterns (island size and proximity to a mainland) to biodiversity via patterns of colonisation and extinction, and led to a thriving cottage industry centred on the theory of reserve placement, design and connection. By the end of the 1980s, a few ecologists were moving beyond conservation to address issues such as how landscape patterns influence the spread of disturbances.

Initially much of this was theoretical and had little impact on how lands were actually managed, but that changed dramatically beginning in the early 1990s. In 1994, in response to the Spotted Owl crisis in the Pacific Northwest, a large team of scientists drew on the principles of reserve design and interconnectivity to produce a regional conservation plan unprecedented (and still unsurpassed) in scope. Somewhat earlier, in 1992, landscape ecologist Nancy Diaz (now Molina) and landscape architect Dean Apostol, at that time employees of the Mt Hood National Forest in Oregon, produced a landscape ecology/design manual for land managers. The first of its kind and the reigning gospel for foresters until the publication of the work at hand, Diaz and Apostol made the crucial leap of treating landscapes as wholes rather than just background for reserves and corridors. Moreover, they began the process of teaching managers how to envision flows and design management with an eye towards those flows. During this time, Simon Bell began to teach principles of landscape design to foresters in Britain and Canada. By the late 1990s, innovative educators such as Robert Whitney at the Maritime School of Forest Technology began offering short courses in landscape ecology and design to foresters.

Landscape design, as envisioned by Apostol and Bell, adds a third ingredient to ecology and architecture: people. Forests are vital to the economy of many rural areas, but they also provide crucial ecosystem services and, for many, evoke ancient, deep connections with the natural world. In old Ireland, marriages, coronations and other important ceremonies were performed beneath ancient trees, effectively the elders who served as witnesses. For the Hawaiians, the mountain forests were "Wao Akua", the realm of the gods, and the Romans struck at the spiritual heart of Celtic Britain by cutting down the forests. In his classic 1898 text on the roots of religion, Frazer wrote of a widespread belief in which "trees considered as animate beings are credited with the power of making the rain to fall, the sun to shine, flocks and herds to multiply, and women to bring forth easily."[3] Submerged for a time by the materialistic exuberance of the industrial age, during the 1960s these old feelings began to re-emerge with force at various places throughout the world and led to unprecedented confrontation between different views of how forests should be used.

The conflicts have not gone away, but they have entered a new era in which protagonists increasingly sit down together and work towards a common vision. Landscape design has a central role to play in this. The dominant cultures in today's world have an unfortunate tendency towards a generic fundamentalism that sees the world in terms of "ors" rather than "ands": good or evil, extraction or preservation, economy or environment, my way or the highway. I often show a slide of a zebra to illustrate the point that if we back up and look at wholes rather than parts, we sometimes find that things can be both black and white. This is not intended as a Pollyanna-ish view that everyone can have everything they want. But the fact that intelligently designed landscapes can successfully accommodate different uses and values is and must be a cornerstone of modern resource management.

Today, landscape design of one form or another is increasingly discussed, modelled and employed in forested landscapes (though the D word is not necessarily used). For example, one of the certification requirements of the Forest Stewardship Council addresses the connectivity of landscapes. Habitat suitability models now look not only at stands but at landscape patterns, providing a crucial ingredient for designing managed landscapes that accommodate a wide range of biodiversity. Foresters in fire-prone areas are taking a landscape approach to balancing fuels reduction with habitat protection. Local communities are looking at maps and thinking about flows.

Ecological landscape design is a crucial component of ecosystem management, and like ecosystem management it has supporters and detractors. Some of the misgivings may stem from the two stages of any new idea (Stage 1: it's BS; Stage 2, we knew it all along), but most is probably rooted in a well-earned scepticism about the old command and control approach to resource management. The idea that we can "manage" ecosystems or "design" landscapes is just the latest version of the hubris that has led to trouble in the past. I have two responses to that. Firstly, there are precious few forested landscapes outside of the tropics and the far north that have not already been altered through management, whether it be past agriculture, logging or fire exclusion. The notion that left alone these will return to some desirable prior state is plain and simple wishful thinking. The legacies of past uses and misuses have laid a pattern on our forests, and the question is whether we let those historic patterns play out, or use our ecological and design knowledge to try and influence the outcome. We can ask the same question for the forests that will be managed in the future. Will our designs be guided by principles or will they be like throwing a handful of gravel against a wall?

My second response is that correctly done, neither ecosystem management nor landscape design is command and control (as Ian McHarg made clear in the title of his classic text on landscape architecture, *Design with Nature*). I like to think of ecosystem management and landscape design as akin to sailing. No sailor is under the delusion that the winds and currents can be controlled; rather a sailor studies the natural forces and uses them to follow a course. So it is with ecologically sensitive land management: intelligent and informed design in partnership with Nature.

Landscape design will be a critical tool to help us through the environmentally and socially challenging years ahead: shaping landscapes that allow migration during climate change, optimally distributing our management efforts to reduce fire hazard while protecting key habitats, restoring resistance and resilience to highly altered landscapes, protecting diversity within intensively managed landscapes and seeking pathways of reconciliation among diverse viewpoints and values. This eminently practical and informative book by two pioneers of modern forest landscape design could not come at a more opportune time.

David A Perry
Oregon State University

Notes

1   Turner *et al.* (2003) give a succinct history.
2   Forman, R.T.T. (1983) "An ecology of the landscape." *Bioscience*, **33**, 535.
3   Frazer, J.G. (1898) *The Golden Bough*. Reprinted by Avenel Books. New York. 1981.

Introduction

Designing Sustainable Forest Landscapes describes a process for deliberate management of the form, composition and/or pattern of forests in ways that take account of multiple goals and objectives. Forest landscapes are not normally "designed", but they are clearly altered as a consequence of human use and management. While natural, or unmanaged, forests may seem self-regulating, they are constantly changing and adapting, either to very gradual processes such as growth and succession, or more dynamic ones such as fire and windstorms. Natural and managed forests form spatial mosaics over large areas. It is the nature, form and distribution of these mosaics over space and time that can and should be consciously designed.

Northern temperate zone forests vary a great deal in terms of both their ecological character and how they have been changed through human use. The range of types includes: undeveloped, remote, primary natural forest tracts found today mostly in Canada or Russia; areas set aside and protected as wilderness reserves or parks covering large areas in North America; semi-natural forests exploited for timber harvest throughout North America and Europe; second or third growth forests naturally re-colonizing abandoned farmland; and plantation forests intensively managed for timber production. Apart from the untouched, inaccessible, remote forests, all the rest have been and will continue to be modified by human activity. Even forest reserves such as national parks and wilderness areas are affected by fire suppression, invasive species, recreational use and by management of neighbouring areas. Surrounding land uses often influence wildlife population dynamics within parks, such as in the case of the Greater Yellowstone Ecosystem in the USA.

While it is common to think of forests as being "managed", we rarely think of them as being "designed". This is in the same sense that we do not think of orderly rural, agricultural landscapes as being designed, even though they are intensively managed. Design is something more associated with a park, garden, residential development or college campus, typical jobs for a landscape architect. Landscape design frequently involves modifying topography, removing or planting vegetation, adding or removing water, setting out paths and so on. In doing this, a designer is engaged in a form of applied ecology, and is envisaging how to alter the patterns of a local ecosystem to facilitate its intended use. Likewise, when the mosaic of a forest is altered by changing the pattern of stands of trees across the landscape, there is also either a conscious or unconscious design element, though foresters who mainly work at the small-scale stand level may fail to see the implications at the larger scale of a watershed or landscape. Moreover, management for single objectives, such as timber production, generally produces landscapes that are very different from natural forests in both pattern and structure. Where forests are managed to achieve many objectives, as is normal at the present time, stand-level silviculture alone is usually found to be inadequate, since multiple functions must frequently be resolved at larger spatial scales.

*Figure 0.1
In the Yukon, in
Canada, there
are extensive
tracts of remote,
unmanaged and
unexploited
forest which
give a sense of
what natural
landscapes are
like. They have a
special quality of
wildness that is
difficult to
experience in
managed forests.*

In 1988 Chris Maser, in his book *The Redesigned Forest*, set out what he saw as the effects of industrial practice on the forests of the Pacific Northwest of the USA. He described natural mosaics becoming simplified and reduced to a plantation state, with consequent losses of ecological function due to the pursuit of more efficient timber production. He catalogued key differences between natural and plantation forests, particularly that of complexity versus simplification. In recent years, further research, spurred largely by controversy over old growth forest logging in the USA and Canada, along with similar issues raised elsewhere, has initiated worldwide changes to forest policies and practices. One result has been an increased interest in sustainable forest management, characterised by independent third-party certification of forest products. It is now more generally the case that commercial forests must be managed for a wider range of values than they had been. This change has presented forest managers with many new challenges. Firstly, there is the increased complexity of integrating non-market environmental values with constraints imposed by economic targets. For example, foresters may be asked to continue historic levels of cutting at high levels of efficiency, while at the same time increasing the protection of streams, wildlife and aesthetic values. Secondly, the outputs of multi-purpose forest plans cannot always be measured by easily quantifiable means. Timber volume production is easily quantifiable, but aesthetics, wildlife habitat quality and eco-logical integrity are more difficult to assess. Thirdly, forest stand pattern and composition modified through decades of management aimed at maximising timber outputs may not easily be adapted to a broader range of objectives (especially if the forest is of non-native plantation origin). Finally, the range of stakeholders involved in forest management has increased, so that many differing perceptions and values placed on forests by a wide variety of people now have to be accom-modated.

Why design forests?

We believe that all managed forests and most conservation reserves and nature parks can benefit from deliberate design. A key reason is to ensure that the larger landscape mosaic pattern is shaped in ways consistent with dynamic natural processes, and that the forest remains resilient enough to cope with inevitable yet unforeseen changes that will happen in the future. Through deliberate design as an integral part of forest planning, outputs of products and environmental services can be continued over long periods of time as part of a sustainable framework.

Whether a forest is of natural or plantation origin, managed to maximise timber or multiple values, and whether it has long been in management or has never been managed before, some shaping of pattern and structure through the influence of human activity is likely to occur. Visualising this potential pattern in order to guide management is the essence of sustainable forest landscape design.

At first, for some people, the very idea of "designing" forest landscapes may seem preposterous, especially in the case of relatively pristine forests. Certainly humans have altered and manipulated forests for many thousands of years, but these actions have not been consciously designed, have they? The idea of design implies a level of cultural control over nature that makes some people uneasy. The authors understand this concern. When we advocate the conscious design of forest landscapes as a means towards achieving economic, ecological and social sustainability, our intention is to produce a design that is deeply embedded in the processes that create and sustain natural forests, including landform, soils, climate, hydrology and ecology. We see design in a sense as a necessary evil. It would be much better if human culture was less at odds with natural forests.

However, the impact of modern human activities on forests has reached such an extent that a commitment to design may be necessary before true sustainability can be achieved.

Here and there, a few cultures survive in a pre-technological relationship with forests, mostly in remote parts of the tropics. Traditional ecological knowledge is still retained by some indigenous people in the USA or Canada, but more often than not this has been overwhelmed by technology and modern culture. Indigenous people typically used fire as a means of manipulating forests for various purposes, for example to attract game animals for hunting, to encourage grass and herbaceous growth, to produce woody shoots for basket making, arrows or other artefacts, to make travel easier and to reduce wildfire hazard. It would be an overstatement, however, to say that indigenous people *designed* forest landscapes, at least in terms of the way we use the term. We do not know if indigenous people designed in a conceptual sense, envisaging how the future landscape could look and then taking actions to make that vision reality. It is more likely that they worked with what was in front of them, observed how natural fire altered forest structure in ways that enhanced material gathering and travel, and gradually harnessed fire towards their own ends.

Forest design did not emerge from traditional approaches to forestry. While some northern and Central European cultures have retained a fairly intimate relationship with forests, particularly in Scandinavia, Germany and the Baltic States, forest management as it has developed in these countries does not involve design at the landscape scale. Active management of forests emerged as a result of deforestation that took place from the eighteenth to the early nineteenth centuries due to the pressures of population growth and industrialisation. These acts led in extreme cases to desertification (for example, the inland sand dunes in the North German Plain) and substantial erosion in the Alps. One of the first tasks was the conversion of native broadleaved forests to conifers in order to increase timber production. The present 33% forest cover of Germany has been maintained for at least two centuries, kept in place by stringent laws in combination with the institutionalisation and social recognition of the forestry profession. Forest sustainability has thus long been a key element in German and other Central European cultures, though forest landscape design has not yet caught on there, perhaps because of the moderate scale of forest management and the strong sense of silvicultural tradition.

Silviculture is not forest design. It is often defined as "the art and science of growing trees". Foresters have developed effective means for measuring and altering the growth rates and composition of forests. Cutting, planting, thinning, fertilising and pruning are all common silvicultural practices in commercial forests. Using these techniques, for example, German foresters converted the woodlands of their country from broadleaf dominated to conifer dominated in a few hundred years and increased the volume and economic utility of the timber they produced. Forest health problems, owing to loss of soil structure and air pollution, have more recently led towards the re-establishment of hardwoods in many German forests. However, this type of intensive silvicultural practice differs from forest design in several important respects. Firstly, it is practised at the site or stand level, rather than the landscape scale. Whereas a silviculturist works from a few to perhaps several hundred hectares at a time, forest designers usually work at scales from several hundreds to tens of thousands of hectares. Secondly, in most cases silviculture is aimed at achieving the rather narrow purpose of maximising wood production, though this is changing with the recent emergence of "restoration silviculture" in the Pacific Northwest of the USA. Thirdly, and perhaps most importantly, silviculture is practised from the "inside out" rather than the "outside in". Silviculturists work from within the forest looking up at the trees, while the forest designer works from above, or from outside the forest, in order to gain a wider perspective of patterns. Thus, while the forest is changed through

*Figure 0.2
Forest
management
traditionally
linked to
agriculture
in Norway.*

3

the application of silviculture on a stand-by-stand basis according to predefined objectives, there is normally no overall design vision or concept for the landscape as a whole.

Ecosystem management is not forest landscape design. Conservation biologists have developed new approaches to "forest reserve" design. In this case, an area of forest is deliberately planned to be preserved for wildlife, watershed protection or some other non-timber production purpose. The design of forest reserves is usually intended to identify areas that will be free from active management, so in a sense only the outside edge or boundary of the forest is designed. This has proved problematic in some ecosystems, where fire is often an important element in forest ecology, and invasive species can establish themselves in unmanaged forests. In most cases it is difficult to establish reserves large enough and remote enough to allow for a completely natural fire regime to take place unhindered. Thus, even many forest reserves can benefit from forest landscape design, as we will demonstrate later on.

Forest reserve design has a long history, dating from the establishment of hunting reserves in ancient Persia, a practice that continued through medieval Europe. It gained its modern momentum in the late nineteenth and early twentieth centuries in the USA when the first national parks were established in the West. The case for forest reserves was based on the conservation of "wild nature", primarily to provide spiritual and aesthetic benefits for people. It has only been within the past few decades that the idea of reserves to accomplish broader goals of the conservation of biological diversity has become established. These ecological goals have led to the development of various theoretical strategies for linking reserves together through corridors, for example, and for arranging managed stands around the perimeter of a reserve in ways that allow wildlife movement in and out. Some of these theoretical designs have been applied in a highly literal, mechanistic fashion, with no thought for the underlying character of the landscape involved. In a sense, they are the obverse of intensive plantation forests, in that both have narrow yet opposing objectives.

Forest landscape design should not be confused with strategic-level forest planning, though it often is. Forest planning, as practised in the USA, Canada, Australia, New Zealand and Europe, frequently focuses on allocating areas of the forest for different primary objectives or uses. Thus, one area might be zoned for recreation, another for wildlife protection, yet another for intensive timber production and so on. The attention is thus on how the area will be used, rather than on how it will look and function. These latter issues are only given secondary consideration, if considered at all. Forest planning is a necessary preliminary step that should precede forest land-scape design. It is often a political, legal, administrative or economic, rather than a creative process, but can set the ground rules within which a design is carried out.

Forest design therefore fills a gap between the strategic, often aspatial or simple zoning planning methods that ascribe objectives to different areas and the detailed stand-level application of silvicultural management methods. In planning theory, activities can take place at different levels – the strategic, the tactical and the operational. Most forest planning represents the strategic level and most silviculture the operational level. Design therefore takes the role of tactical planning, enabling the strategic objectives to be distributed in space and time and the operational activities of silviculture to be worked out in an integrated, not isolated, fashion.

What is forest landscape design?

So what exactly is meant by the term "forest landscape design"? Our definition is: *the deliberate alteration of the pattern and structure of the forest mosaic over fairly large areas and over long periods of time in order to meet multi-purpose objectives.* This can be applied, at one extreme, to intensively managed timber production plantation forests, where conservation goals may be fairly limited, or at the other extreme to forest reserves with no timber production at all, with a wide range of managed forests in between. The focus is on the manipulation of forest pattern and structure together with any necessary constructed infrastructure, such as roads, and the processes that shape or affect the landscape over time, both natural and human controlled.

The term landscape can be defined in at least two general ways. Firstly, there is landscape as the visual scene perceived by people, both externally, in terms of the views from an outside vantage point, and internally, in terms of the sensory experiences from within. People often judge things by how they look, and the visual effects of many forest practices have been the cause of much public concern, often forcing foresters to change their practices or move them over the hill (where they are out of sight). Much of the early development of forest landscape design in Britain was driven by visual concerns (see Chapter 3).

Secondly, there is an ecological definition of landscape: "a scale of ecological functioning and planning, smaller in area than a region but bigger than a stand or patch." The emerging field of landscape ecology is concerned with larger-scale patterns and processes. Concerns about wildlife conservation in forests have led to the development of forest planning and design based on an improved understanding of patterns and processes at the landscape scale. The developments in forest landscape design that have taken place since the early 1990s have been driven primarily by ecological concerns (see Chapter 2).

Mosaic patterns in forested landscapes are composed of a range of three-dimensional structures, including different species, ages and layers of trees, shrubs, herbaceous and other vegetation types. These result from the interaction of climate, landform and cultural practices. In the forest design process, we assume that climate and landform are unchangeable over the time-scales used in forest planning, though global climate change may increasingly need to be incorporated. Cultural practices are more flexible and adaptable, though not infinitely so. We attempt to understand what the range of possible structures is for any given area, and work towards sketching these, onto maps representing the landscape, in appropriate places, sizes, shapes and proportions that meet the needs of the owner/managers and the wider community, as well as those of the flora and fauna. We also try to understand the dynamics of the landscape as these structures change over time, due to growth, natural disturbances or human activity.

Forest design can incorporate a shifting mosaic, where regeneration forestry is practised, or reflect relatively stable conditions, where forests are managed lightly and selectively. The design must be tailored to each specific landscape, ecosystem, the local community and the needs of the immediate forest owners. It must also be within the capability of managers to implement.

The practice of forest design is therefore concerned with larger scales, with looking at the mosaic of stands rather than individual stands and with the landscape structure and composition as well as with the outputs from forest management. In some places the design has more of an ecological emphasis, while in others the visual appearance is more important. It is a means of ensuring that economic, social and environmental sustainability are achieved, by developing plans at appropriate scales and over suitable time frames. Design can be highly creative, producing a new landscape that does not currently exist, or it can be conservative, restoring a damaged landscape.

The role of forest design

Although foresters tend to be long-range thinkers in the sense that they typically envisage the life of a forest over a span of decades to perhaps hundreds of years, in our experience the actual practice of forestry has tended to react to specific social and site conditions in a piecemeal fashion over the short term. This practice has been driven both by society's need to obtain economic products from the forest, as well as by changing political pressures. We acknowledge the utility of this approach, which has seen some highly successful efforts at increasing wood production in many parts of the world. It has also been the case that some practices have both arisen and subsequently been prevented or restrained as a result of short-term economics and political pressure. Examples of this include limits to the size of clearcuts in British Columbia, and the banning of commercial forestry altogether across large parts of the US National Forest system.

Unfortunately, the long-term nature of forestry may mean that reaction to activities and the effects of changed practices often lag behind action on the ground. The understandable response of environmentalists and concerned citizens to questionable forest practices has been to develop a distrust of forest managers, to support the widespread establishment of protected areas, and failing this, to devise ever more complex and onerous regulations. This results in an increasing demand for forest products being extracted from a shrinking land base, and puts pressure on foresters to intensify production by shortening rotations or by growing genetically selected super trees, resulting in an ever-widening gap between managed and natural forests. Many people have thus come to see forest resource use and nature conservation as being essentially incompatible goals. We believe that the design approach put forward in this book offers a route away from this simplistic view of forests, and that it can provide an alternative that is more of a middle way.

The campaign to establish old growth forest reserves during the 1990s in the Pacific Northwest of the USA, for example, set urban environmentalists against rural forest workers and polarised politics. Each side found it necessary to exaggerate its claims about on the one hand the damage being done by clearcut logging (by unfairly calling it "deforestation"), and on the other the damage to the economy by ceasing timber production (the Pacific Northwest economy has boomed in the 1990s and early part of this century, despite a 90% decrease in logging on national forests). Thus, loggers and forest industrialists automatically opposed any wilderness or set-aside proposal, even those in remote areas on fragile, steep slopes with minimal timber production potential and maximum biodiversity, while environmentalists felt that they must continuously raise the stakes, from demanding less cut on public lands, to zero cut, to less cut on private lands and so forth. There has to be a better way of resolving these kinds of conflicts over resource use. Collaborative planning within a design process is a tool to help a wide range of stakeholders to see how the future could be, and where it has been properly applied has been successful in overcoming the kinds of problems described above. This creative approach to problem solving is what sets forest design apart from planning and zoning.

While forest design cannot resolve many wilderness versus plantation conflicts, which are as often as not about deeply held worldviews and issues, it does support a viable third option that has room for both, plus many shades of grey in between. The conscious design of forest landscapes, done well, can help reconnect cultural practices with forest ecology. We advocate achieving this by harnessing a rational, scientifically based, technocratic approach to creativity and imagination, using both sides of the brain.

Forest design borrows from a 250-year-old creative process developed by landscape architects and applies it to forest management with the help of modern technology and concepts from emerging fields of landscape ecology, conservation biology and restoration ecology. It matches the underlying, long-term aspects of the land (usually expressed in landform and soil) with the evolving structure of vegetation. It incorporates legitimate needs for human access and interaction with the land.

Our essential premise is that the human community needs or wants different services from the forest (wood, beauty, game, biodiversity, recreation, clean water) and that in the process of getting what we need we consciously shape the forest mosaic in ways that facilitate the optimum production of all of these over the whole landscape simultaneously.

To accomplish this lofty goal, clear objectives for the forest mosaic must be set by the land managers. These objectives must be as broad as possible and as specific as needed. The objectives must be within the ecological capability of the land, they must be practically achievable and they must have widespread social support. In the absence of clear, appropriate and agreed objectives, any forest design is likely to fail.

History and present use of forest design

The forest design approach as described in this book has a relatively short history. There are two strands in its development that were woven together to produce an overall stronger thread through the 1990s. This process has room for further research and development, although in some respects it has already proved its value.

The Forestry Commission of Great Britain began conscious design of forest plantations in the mid-1960s through the work of the well-known landscape architect Dame Sylvia Crowe. Great Britain was originally clothed in broadleaved forests after the retreat of glaciers 12,000 years ago. The spread of human settlement and agriculture resulted in the gradual clearance of forest. By the early twentieth century only some 5% of the land area remained wooded, and these fragmented remnants were highly degraded. Britain suffered severely in the First World War, when timber imports required for use in coal mines to feed her armament needs were blocked by German U-boats. Shortly after the war, the country began establishing plantations of mostly fast-growing non-native conifers, to help build a strategic timber reserve.

By the 1960s, goals for these plantations shifted from war strategy to economic development. The public was becoming increasingly uncomfortable with rectangular plantations of dense conifers placed on bare moorland hills of Scotland, Wales and northern England. Dame Sylvia Crowe was appointed by the Forestry Commission as a consultant to help match forest planting and harvesting to the landform and cultural patterns of the British landscape. This resulted in a series of design guidelines and policies that evolved throughout the 1970s, 1980s and 1990s, up to the present day. Initial efforts were primarily driven by aesthetic considerations, reflecting the concerns of the time, although nature conservation and water quality issues were also incorporated. The Forestry Commission systematically began developing 50-year "Forest Design Plans" for each management area. These plans, drafted by teams of landscape architects and foresters, redesigned forest patterns, and to a lesser extent, structure (monoculture spruce plantations were and still are the dominant element in Britain's national forests). They did so in clear graphic ways, using three-dimensional maps and drawings, to illustrate how the forest would look over time, supplemented by short written reports.

Figure 0.3
This is an old photograph showing how some of the early plantation forests in Britain were laid out, with rectangular shapes, lines running parallel with or perpendicular to the contours, resulting in a highly artificial appearance unsympathetic to the natural landform or semi-natural vegetation patterns.

Meanwhile in the USA, the focus was on developing land use plans for each individual national forest. The emphasis was on inventory, ecological description and use zoning, with little or no attention to design. Aesthetics were emphasised in some areas (scenic road corridors), and neglected in others. Integration of land uses was poor to non-existent. This system has persisted with some revisions to the present day.

In the early 1990s, in the Pacific Northwest states of Washington and Oregon, a landscape ecology-driven management approach was initiated by Nancy Diaz (now Molina). As lead ecologist for the Mt Hood and Gifford Pinchot National Forests, Nancy had organised a series of training sessions, using the pioneering landscape ecology work of Richard Forman and Michael Godron as a foundation for understanding landscape patterns, structures and processes. However, there was no operational method of applying what were largely still theoretical concepts, so ecologists attempted to incorporate them into forest plans in an *ad hoc* fashion. Forest planning at the time was two-part. Firstly, there was the strategic forest plan itself, which was basically a zoning map of the forest developed over a laborious ten or more year process, subject to constant revision and misinterpretation. Its results were largely a political compromise that set out a strategy for the forest: a bit of wilderness over here, some tree plantations over there, a scenic corridor in this area and so on, with over 40 distinct zones in Mt Hood National Forest alone, that had little or nothing to do with natural ecological units.

Figure 0.4 "Leoland", in Mt Hood National Forest, Oregon, USA, was the pilot area for the application of landscape ecology to forest management. This photo shows some of the small clearcuts which were helping to fragment the landscape.

Secondly, there was operational "project planning", which consisted of year-by-year timber sales, road building, trail reconstruction, campground development and so forth. The selection and implementation of these projects was driven by available budgets and politics. Once a forest plan was completed, the pressure to implement short-term projects left few opportunities to apply the ecological concepts Nancy and other ecologists had begun advocating. Individual projects also took place at a scale too small for landscape ecological approaches.

What was missing was a tactical master plan at the scale of individual watersheds or catchments that could bridge the gap between the too general forest land use plan and the too narrow and focused projects. In 1991, Simon Bell, then Senior Landscape Architect with the British Forestry Commission, happened to pay a visit to Mt Hood National Forest. He showed Nancy and Dean Apostol (then a landscape architect at the forest) a few of the design documents from Great Britain. These looked interesting, but were not particularly linked to the ecology. The areas of forest the British were dealing with were small by American standards, and they had distinct outer edges, being set within agricultural and pastoral landscapes. British silvicultural practices were very simplified: short rotations, clearcutting followed by replanting of non-native conifers. American forests, still mostly natural and with far greater complexity and scale, would clearly require more complex plans.

However, the idea was intriguing. Could a way be found to borrow from the British master planning approach, bring in emerging concepts of landscape ecology and adapt this to complex American forests? Simon, Nancy and Dean decided to give it a go.

The initial result was a pilot project on Mt Hood National Forest, the results of which were published in a small book entitled: *Forest Landscape Analysis and Design*. It describes the basics of landscape ecology, the applications of this field to US national forests, a rudimentary "design process" and the case study. This book was hailed in the Forest Service and elsewhere as a conceptual breakthrough in applied landscape ecology and its simple, effective communication approach has stood the test of time, so that we reproduce parts of it in this book (see Chapter 5). Over the next few years Nancy and Dean were able to convince some forest managers of the utility of the process, and were able to apply it to most of the large watersheds on Mt Hood National Forest.

In the meantime the peripatetic Simon Bell harnessed landscape ecology to British forest design and managed to export the process to British Columbia, Eastern Canada and some other European countries. From Mt Hood the process spread northward to national forests in Washington State, and south to the Applegate Valley in Southern Oregon. Our collective experience has been that the process works, and that it is highly adaptable to various cultures, laws, customs and landscapes. However, it is not, by itself, a panacea: it has to be applied thoughtfully, with appropriate and realistic objectives, consistent with local and regional ecology. Designs must also be implemented well on the ground. Follow-up monitoring and adaptive management are critical components, since the ecological learning curve is still steep.

Forest design and sustainable forest management

Agricultural and industrial societies have historically been pitted against the natural world, relying on the conversion of natural capital to economic capital to generate and build wealth. John Perlin, in his seminal book *A Forest Journey*, sums it up this way: *"Ancient writers observed that forests always recede as civilizations develop and grow."* Forests have provided fuel and building materials to humans since time immemorial, and civilizations have been built on their backs, or stumps of them. This unfortunate process continues today as the virgin forests of the tropics and far north succumb to chain saws. When civilisations decline or lay stagnant, forests often regenerate and spread back into formerly cleared areas.

Many cultures have been concerned with the sustainability of their forests, or at least their ability to provide timber, for hundreds of years. France, Germany, much of Central Europe and Scandinavia have long had strong customs and regulations that limit forest clearing and require regeneration after harvest. But it is only recently that sustainability of a wide array of non-economic forest values has taken hold.

The final two decades of the twentieth century brought much needed changes to the discourse about forest management that can be traced to several sources: the rise in influence of the environmental movement, increased awareness of the visual, spiritual and economic value of forests, recognition of the important role forests play in provision of multiple "ecosystem services" and increasing concern within civil society about the impacts of development on forest and the broader environment. The World Commission on Environment and Development in 1987 wove together economic growth, environmental protection and the welfare of present and future generations into the single concept of sustainable development: *"development that meets the needs of the present without compromising the ability of future generations to meet theirs."* The 1992 UN Conference on Environment and Development – the Earth Summit – committed national governments to a raft of actions towards sustainable development, founded on a set of over-arching principles set out in the Rio Declaration on Environment and Development. This Declaration established the principles of inter-generational equity in meeting developmental and environmental needs, environmental protection as an integral part of the development process, the participation of all concerned citizens in the handling of environmental issues (a point we shall discuss in greater detail in Chapter 4) and the adoption of the precautionary approach where development poses threats of serious or irreversible damage.

The Earth Summit established a new agenda for forests that recognised the central role that forests can play in sustainable development. Forests have the capacity to meet a wide range of human needs including wood and wood products, food, fodder, medicine, fuel, shelter, employment and recreation. At the same time they provide us with a number of essential services: protecting fragile ecosystems, watersheds and freshwater resources; acting as storehouses of biodiversity and biological resources and as sources of genetic material for biotechnology products; and sequestering and storing carbon.

Agenda 21 set out a comprehensive plan of action covering every aspect of sustainable development including far-reaching programmes for forest conservation. The chapter on forestry, one of 39 separate but linked programmes and supported by the Forest Principles, contains action points on many different aspects of forestry, including institutional capacity, protection and sustainable management of forest ecosystems and woodlands, protected area systems for wildlife, culture and heritage, conservation of genetic resources, greening of degraded areas, participation by communities, non-governmental organisation and other stakeholders, incorporating non-market values into decision-making and promoting efficient, rational and sustainable utilisation of forests. This programme has been developed further by the Intergovernmental Panel on Forests and the Intergovernmental Forum on Forests, whose work has resulted in a large number of additional actions to be implemented alongside Agenda 21 and the Forest Principles.

There are connections between forests and many of the other programme areas of Agenda 21. The chapter on conservation of biological diversity calls for the conservation of ecosystems and natural habitats, the rehabilitation and restoration of damaged ecosystems, the recovery of threatened and endangered species, and encouragement of a greater understanding and appreciation of the value of biological diversity. The programme is supported by the Convention on Biological Diversity, which requires governments to develop national biodiversity strategies and action plans, and to integrate these into broader national plans for environment and development. In 2002, at their sixth meeting, the Conference of the Parties (COP) adopted a work programme on forest biological diversity aimed at a number of goals, including the application of an ecosystem approach to the management of all types of forests. The COP's description of the ecosystem approach connects directly to the concept of sustainable forest landscapes.

The ecosystem approach is supported by 12 complementary and interlinked principles. The first two principles introduce human values and participatory decision making into ecosystem management:

> Principle 1: *The objectives of management of land, water and living resources are a matter of societal choice.*
> Rationale: Different sectors of society view ecosystems in terms of their own economic, cultural and societal needs. Indigenous peoples and other local communities living on the land are important stakeholders and their rights and interests should be recognised. Both cultural and biological diversity are central components of the ecosystem approach, and management should take this into account. Societal choices should be expressed as clearly as possible. Ecosystems should be managed for their intrinsic values and for the tangible or intangible benefits for humans, in a fair and equitable way.

> Principle 2: *Management should be decentralised to the lowest appropriate level.*
> Rationale: Decentralised systems may lead to greater efficiency, effectiveness and equity. Management should involve all stakeholders and balance local interests with the wider public interest. The closer management is to the ecosystem, the greater the responsibility, ownership, accountability, participation and use of local knowledge.

These are powerful statements. Forest management clearly is moving away from being the sole preserve of the technocratic forester. The ecosystem approach requires that management objectives reflect wider societal values that depend on the social, environmental and economic setting. These can only be taken into account if local people are given the opportunity to articulate them. This need to reflect ecosystem functioning and to involve communities in forest planning and design are central concerns in this book, reflected by chapters devoted to each and to several case studies.

This growing recognition of the need to sustain all forest values is a welcome development, but the practical effects of international declarations has not yet had much effect. There is strong evidence that forest ecosystems are still declining faster than they are recovering in most parts of the Earth. Good data is difficult to come by, but there is no doubt that primary forests are still being logged at high rates, particularly in the tropics, where population growth is rapid. In the developed world, the total acreage of forests appears to be stable or increasing, although much of the gain is in monoculture plantations. The Summary Report of the World Commission on Forests and Sustainable Development (1999) estimates that 15 million hectares of forest are lost annually, mostly in the tropics due to conversions to agriculture. The world has 3.6 billion hectares of forest left out of an original 6 billion hectares. In developed countries, many forests are in poor condition due to inappropriate management, including silvicultural errors such as conversion of broadleaf forests to conifer. Half of Europe's forests were classified as damaged in 1997, although the trends seem to be improving in the decade since then, showing that things are improving in some areas.

Looking on the brighter side, consider the case of the Pacific Northwest of the USA. Up until the early 1990s, primary old growth conifer forests, among the most biologically productive on Earth, were being liquidated by the US Forest Service operating under deliberate government policies supported by both major political parties. Conservationists were by and large not very interested or effective in conserving these forests, having focused their attention on the much more scenic sub-alpine parklands and wild river canyons. Researchers began studying the northern spotted owl, a rather obscure bird which seemed curiously dependent on old growth forests. They gradually uncovered a vast web of interrelationships between owls, voles, flying squirrels, mycorrhizal fungi, salmon and a thousand aspects of these complex ecosystems that had been completely overlooked by managers focused on converting "decadent" natural forests to "thrifty" plantations. The owl was listed as a threatened species, joined later by the salmon, resulting in a drastic reduction in old growth harvest under the Northwest Forest Plan.

A new monitoring report by the US Forest Service indicates that mature and old growth forests in the Northwest are now on the increase, having expanded by 250,000 hectares over the past 10 years. Forest management is changing across much of North America in large part because of new awareness generated by the old growth controversy. Regulations are getting tighter, and new approaches that try to mimic natural disturbance cycles and patterns are catching on, particularly in the vast and still largely undeveloped boreal forest of Canada.

Figure 0.5 The Northern Spotted Owl.

New ideas about what sustainability means continue to be put forth and tested. For example, there is a new advocacy for broadening how economists and nations define "wealth" beyond the traditional view that it is limited to goods and services. Natural forests provide numerous "ecosystem services" that are finally being accounted for; clean water, temperature moderation, biodiversity and carbon storage to name a few. In fact, the role of forests in storing carbon is bound to take on increased importance as global climate change is addressed.

The most important development in improving the outlook for forest sustainability may lie not in further government policies, which are always prone to too much compromise in any case, but to the emergence of certification. The Forest Stewardship Council, an arm of the Rainforest Action Network originally established to monitor tropical forestry, now operates globally and is having a huge influence in the establishment of market-based incentives. They have established ten principles of forest stewardship, covering issues including compliance with laws, indigenous people's rights, worker's rights, environmental impact and maintenance of forests with high conservation value. The principles are flexible enough to be applied to local conditions, but the standards are high enough that certification is not easy to achieve.

As a market defence measure, the North American timber industry has established its own certification network, the Sustainable Forestry Initiative (SFI). Critics have pointed out that in effect SFI results in industrial companies certifying themselves, but in January 2007 the SFI has become at least nominally independent of its founders. It remains to be seen what effect this change will have on industrial practices, but the fact that the timber industry has found sustainable certification at all necessary is encouraging.

The design process set forth in this book does not pretend to be sufficient proof that forest management will henceforth be "sustainable". Clearly, there are many factors at work, including international programmes, changing regulations, adoption of ecological forestry approaches and certification. What we believe is that design can and should play an increasingly important role in shaping forests around sustainability principles and practices.

Cautionary points

There are a number of reasons why forest designs can fail. Firstly, we simply do not yet know enough about the complex ecology of forests to make perfect choices about forest pattern and structure. As ecologist Frank Egler points out, "Ecosystems are not only more complex than we know, they are more complex than we CAN know." And to quote the nineteenth-century German forester Heinrich Cotta:

> Thirty years ago I prided myself on knowing forestry science well. Had I not grown up with it and in addition had learned it at universities? Since then I have not lacked the opportunity for increasing my knowledge in many directions, but during this period I have come to see very clearly how little I know of the depths of the science, and to learn that this science has by no means reached that point which many believe to have been passed.

Lest we think our knowledge is much more advanced today, consider a more recent quote from James Karr, an eminent ecologist from the University of Washington: "Because the road to ecology as a discipline is littered with the carcasses of discarded theories, premature advocacy of theory is at best foolish and at worst dangerous."

Many wildlife species have not been studied in detail, and we continue to discover new information about those that have been. Nutrient cycling, the role of landscape disturbances, unknown symbiotic relationships, the recent discovery of new species inhabiting the canopies of old growth conifer forests . . . all these and more should give pause to any of us who are convinced that our new, beautifully drawn plan is completely right. We may get new information next week or in the next ten years that compels us to adjust plans. Thus, we recommend that the goal should be an 80% correct design as a first effort, based on the present state of knowledge. We advocate well-planned, continuous monitoring, appropriate revision and a need for flexibility as more is learned about ecology.

Secondly, the social context within which forest design is done can also shift, sometimes gradually and sometimes rather suddenly. Since we advocate developing designs *firmly from within existing laws and plans,* we acknowledge that as these change, designs must be re-considered. For example, a forest design may be based on regulations that call for 50-metre riparian buffers. If, the day after the design is completed, a citizen's initiative results in a new regulation doubling the width of buffers, we obviously have to revisit the design and make the necessary changes.

Thirdly, a forest design can fail when a client, owner or manager of a forest sets unrealistic objectives. This could take the form of absurdly high timber output targets, ignoring the desires of the surrounding community. Since the client pays the bills, the designer must design to meet their needs, but if these needs are wrong-headed, the design will fail or otherwise prove to be unsustainable.

Fourthly, the design could fail to be implemented properly. A silviculturist could get too aggressive in stand management, removing too many trees in too short a time. A fire manager could misjudge the wind or relative humidity so that a prescribed fire burns too hot and damages an area inadvertently. A tractor operator could be careless with where he or she drives, resulting in soil compaction or erosion leading to stream pollution.

Fifthly, "nature bats last." This means that even if a design is well thought out and executed, there may be an unexpected and unwelcome surprise in the form of a historic drought cycle, a newly introduced pest species or rapid and severe climate change. These and other unpredictable events may disrupt the design to a degree where it has to be reconsidered.

Lastly, and most difficult for the authors to admit to, the designer could make mistakes. These could happen from being in too much of a hurry, from failing to read a map properly or from inexperience with the local ecosystem. It is important to spread this risk by working with a team of people, especially with ones who know the local area in some detail (see Chapter 5).

Who this book is for

This book aims to fill the gap of available knowledge about forest landscape design. It is designed as a handbook or manual, not a scientific text. Readers should therefore have some knowledge or background interest in natural resource planning and management. This book is thus primarily aimed at foresters, managers, planners, ecologists, landscape architects and any other sciences or professions involved in forest management, both practising and studying. Non-professionals involved in forest management as stakeholders or members of community groups may also find it useful, as will students in many fields.

Forest design is not a separate field of study and we hope it never will be. It occurs at the intersection of a number of fields, including forestry, ecology, landscape architecture, land planning and a number of associated resource sciences. It is interdisciplinary and participatory, rather than exclusive to any one discipline. This process is international in scope, and has been developed and applied in many countries. However, there may be limitations to its application in some regions, particularly tropical forests, where there is little tradition of silviculture and poor inventories of natural forest ecology. Thus we concentrate on temperate forests, while hoping that practitioners from tropical or sub-tropical areas may also find this book useful, at least conceptually.

The authors have not chosen to adopt any particular political position with regard to the future of forests and forestry, and we recommend against using this book to advance any position other than the advocacy of sensible planning and design. We have worked on projects for a wide range of clientele, including industry, private landowners, national, state or provincial agencies and conservation groups. We believe in respecting the major principles of good sustainable practices and on basing design in the ecology and social setting of the particular area in question.

This book is not pro-wilderness, nor should it be used to make a case for extending commercial forestry at the expense of ever dwindling wild forests. We believe that in every part of the world local communities have to come to terms with their own relationship with their forests, deciding how much to manage and how much to leave to non-human processes. We try to steer clear of any specific ideological or ethical positions on the rights of humans versus the rights of Nature. Humans and forests have coexisted for many millennia, and must somehow continue to do so. People will continue to take products from forests, live in and around them, need access to them for recreation and gain spiritual renewal from them. It is probably inevitable that commercial and social or environmental uses of forests will always be to some degree in tension, but we believe that our approach can minimise potential conflicts and resolve at least some of them.

Our intention is to provide both the theoretical background and practical application of forest landscape design. Part I presents major issues and principles, including sustainable forestry, forest landscape planning, landscape ecology, aesthetic principles, and social and community participation. These subjects have been covered more thoroughly in other books and papers, many of which are referenced at the end of each chapter. Our objective is to highlight the main issues and provide reference points. Part II presents the practical aspects of how to go about forest design, including the design process, managing the design, visualisation techniques, evaluation and implementation. Part III applies forest design to three different scenarios: ecosystem restoration, managed forests of natural origin and plantation forests. Each of these is supported by examples and case studies from around the world.

Simon Bell and Dean Apostol drew upon the expertise of several close colleagues when putting the book together. Many people are thanked in the acknowledgements but we need to make a special mention of several individuals who gave time and materials of direct use in the text.

Mike Garforth, an international consultant, an expert on sustainable forest policies and formerly with the British Forestry Commission, was responsible for working with a wide group of stakeholders to develop the UK Woodland Assurance Standard, the UK version of a certification standard. He supplied us with material on sustainability and information to help us embed forest landscape design within the sustainable forest management paradigm.

We drew heavily from the previous work and writings of the landscape ecologist Nancy Diaz (now Molina). She led the development of the applied landscape ecological analysis, first published in the book *Forest Landscape Analysis and Design* mentioned above. Formerly with the US Forest Service, she recently retired from the Bureau of Land Management (BLM) in Oregon. She contributed a detailed review of Chapter 3.

Max Hislop was until recently a researcher in the British Forestry Commission Research Agency, working on aspects of social forestry, especially public participation approaches. He has been collaborating with other researchers in the USA and Finland, bringing together recommendations for best practice. He helped us extensively with the content of Chapter 4.

Marcia Sinclair is a specialist in public involvement and communication who used to work for the US Forest Service, and presently works with the Willamette Partnership in Salem, Oregon. She has wide experience dealing with contentious situations in the field of forest planning and natural resources. Marcia contributed a detailed review of and ideas for Chapter 4.

Any errors, omissions, dubious concepts or questionable advice rests on the shoulders of Simon Bell and Dean Apostol. We do not claim to have all the answers for Designing Sustainable Forest Landscapes, but it is our sincere hope that this book will spark greater interest in and growth and development of the practice. We hope also that it will provide an important tool for all those involved in forest planning and management, and that it will prove to be a good read as well.

Part I
Key concepts for forest design

Chapter 1
Planning and forest design

Introduction

"Planning", according to Dr Robert Ribe of the University of Oregon, "is a natural social activity that grows out of the need for cultures to have some measure of control over their future." Thus, what is often thought of as an invention of the enlightenment of politicians and technocrats is in reality something that has been around for a much longer time.

Planning has come to be an essential tool in modern forest management. This is in large part due to the long time frame over which a forest grows and matures, as well as the desire by owners and managers to be able to predict how much and when they can harvest timber and manage investments. It is also due both to the increasingly public nature of forest management, particularly in western democracies, and the traditions of central planning of all aspects of the economy in the former Soviet Union and other Eastern European countries. Planning provides a structured process by which multiple conflicts can be addressed, while also (hopefully) providing a clear vision of how the forest will be managed over many years.

This chapter reviews a range of historical approaches to forest planning. It then demonstrates how forest landscape design can be used as a tool to incorporate and synthesise the best from them while also going further to address modern challenges. Alternative approaches to forest planning developed due to historical circumstances, although none has completely ceased to be used or to have an influence into the present day.

Traditional forest planning

Traditional cultures were (and remain) much more sophisticated in their interaction with the environment than we in modern, advanced cultures have appreciated until recently. In the Americas, for example, native people were not simply hunter-gatherers, but in many areas also practised agro-forestry for thousands of years. Methods varied, but included swidden (slash and burn) agriculture, maintenance of clearings and glades through repeated prescribed burning, and con-scientious game management. The character of North American forests, often open and dominated over wide areas by pine and oak species at the time of European contact, was largely a consequence of deliberate under-burning that facilitated and integrated food gathering, travel and hunting. Burning shaped and maintained woodland conditions that maximised their utility to human cultures. Can we consider that burning was deliberately planned in the sense that we understand the term?

Native Americans did not make maps of forests, nor did they divide them into rectilinear compartments. But they apparently had a keen sense of what to burn and when, activities that required forward thinking. Huckleberry fields in the Pacific Northwest may have been burned on a 50-year cycle, indicating a multi-generational sense of planning and execution. Ponderosa pine forests throughout the West were burned on a 7–20-year cycle. Oak savannas may have been burned every 3–5 years to help stimulate acorn production. Native peoples knew that fire travelled better up slopes, and used the receding snow line in the California Sierras as a seasonal firebreak.

Planning for fire use was probably largely intuitive, but was also based on generations of trial and error, with serious consequences if the wrong choice was made. Either too much or too little fire would damage fragile resources upon which the community depended. In the absence of a written language, oral traditions were followed to hand down forest management techniques from generation to generation. Typically this was from grandparent to grandchild, perhaps to ensure it was transmitted as far ahead as possible. By nature this approach was conservative. Since the economy was largely subsistence and transport of large objects out of the forest was not feasible, resource exploitation rarely exceeded the sustainability of the land. However, oral plans can be just as effective as the written plans we depend upon today, in the same way that oral contracts can have similar legal standing to written ones. Essential planning elements, including clear objectives, timing, location and application technique were all used. The maintenance of pine and oak for food and fuel, of grasses and herbaceous plants for food and medicine, and open ground for hunting and travel were essential resources of the day that depended upon planning and execution.

However, in addition to burning to clear areas for resource production, hunting and gathering, timber was and remains an important resource for native cultures. In the Pacific Northwest of the USA, native people were skilled woodworkers, and in particular harvested part or all of western red cedar trees, known as the "tree of life". From the cedar bark they made clothing, baskets and diapers. They split planks from standing trees for building lodges. They harvested whole trees to shape timbers and large, seaworthy canoes. They also harvested many other trees, including Sitka spruce, Pacific yew, grand fir, cottonwood and willows. Red alder was prized for woodworking and to smoke salmon for winter storage or trade.

Medieval European people followed a similar oral tradition approach to forest planning and management. Large tracts of European forests were gradually cleared to make way for farming over a 5,000-year period. Swidden agriculture was commonly practised in many places, where patches of forest were cleared by burning and farmed for a few years, until the soil fertility was exhausted. The patch was then allowed to regenerate over a variable period of time (depending on fertility and climate), while another was cleared and farmed. In some areas permanent clearings were made, and soil fertility carefully managed to prevent its exhaustion.

Figure 1.1 Ponderosa pine forests in the American West, such as this example in Sitgreaves National Forest in Arizona, USA, were managed by native peoples using controlled fire. This is an example of non-scientific forest management, which satisfied the needs of the time.

Gradually, a pattern of land use developed that included a village or hamlet located near fresh water, surrounded by fenced gardens, grainfields on the best soils and perhaps some pasture or hay meadow. A "forest common" lay just beyond the fields, and was used as a source of wood and grazed by livestock, and could serve as an escape area during periods of unrest. Beyond the commons was often a wild forest area reserved for hunting by the local royalty or aristocracy. In the later middle ages, the first foresters were law enforcement officers who kept the peasants out of the parts of the woods reserved for hunting to make sure that prized game was not illegally taken.

The working boundaries of medieval European forests were probably more precise than was the case with forests "managed" by Native Americans. Dating all the way back to Roman times, there was consistent use of a three-part zoning or division of the forest. The first zone was called the *silva caedua*, where coppicing for firewood and small round wood used for many purposes was practised. Cutting might be on anywhere from an annual to an eight- or twelve-year cycle depending on the productivity of the forest and the types of species present. The coppice woodland was divided into sections (cants) of roughly equal area, at least as many as there were years in the cutting cycle, ensuring a constant supply of small wood.

The second zone was called the *silva glandifera* or *silva saginacia*. This area was managed as wooded pasture, with domestic animals allowed to graze under tall trees. Acorns were particularly important forage, with pigs being the most important animals in many areas.

The third zone was a protected area where regular harvest of materials was prohibited. It was set aside and managed by higher authorities and landlords, and had its own special set of forest laws. Its primary purpose was a hunting ground for the royalty and aristocracy, but it also served as a buffer zone between settled areas and estates. It was also a conservation zone dedicated to the protection of favoured wildlife, most particularly deer. Sometimes it was referred to as *forestis silva*, and was thought of as separate from the tamed, managed landscape.

This three-part pattern of forest division continued throughout the middle ages. Coppice woods provided firewood, tools, stakes, basket-making materials and many other uses. Grazed forests provided tall trees for use in construction of important buildings. While the large trees could be taken for specific uses (e.g. construction of cathedrals), they could not be traded or sold. Other uses of the forest were allowed, including bee taking, and gathering of leaves, bark or other products for forage or soil amendments.

Medieval forest managers planned far ahead, and some mature trees were always kept to provide seed for the next generation. In some cases individual trees were retained for many generations, became "old growth" or "veteran trees" and were revered by the local community. Harvested trees were "rummaged", or selected here and there over a broad area. Some areas were managed to grow very large timber, on up to 300-year cycles for oak. Coppice-with-standards woodlands were true multiple-use forests that included firewood gathering, hunting, grazing and timber cutting.

According to Bechmann, the medieval forest management system was stable and sustainable for 1,000 years or more in many locations. One reason was that the population during the first few centuries after the fall of the Roman Empire remained low due to political instability, wars, plagues, famines and social unrest. In some areas lands once cultivated reverted back to forest and were then cleared again as the population gradually re-established itself. Since trade between communities was restricted due to lack of safety in travel, local communities were very self-reliant.

Relatively peaceful times in the eleventh–thirteenth centuries led to a significant increase in the population, which in turn led to pressure to clear more forest to make way for expanding agricultural crops. Forests began to

Figure 1.2
Pigs are still allowed to forage for beech mast and acorns in the New Forest in England, following the ancient practice of pannage. This is part of the historic development of rights held by commoners in the royal hunting forest, which still retains its own laws and officers.

*Figure 1.3
An ancient
coppice forest.*

become over-exploited, gradually leading to changes in the way they were administered. Decrees against forest clearing for agriculture were issued as early as the twelfth century in the Rhine region, and in the thirteenth century in France.

Silvicultural forest planning

Bechman notes that the first recorded instance where a forest was divided into definite compartments to help management was in 1359, in the City Forest of Erfurt. This idea soon caught on, and in fourteenth-century France under Charles V, "*coupes d'assiettes*" were designated. These were areas where forest agents managed selection cutting, a practice that replaced rummaging. This period marks the time when forest management by specialists began to displace the local community in decision making about what and where to harvest trees. "Coupe" is the word used to describe an area to be felled, from a French forestry term, derived from *couper*, the verb to cut. Coupes were specified to be 5–6 ha in size. The trees were to be cut flush to the ground, assuring there would be no waste, and they had to be cut level (in typical French fashion, a litre of wine had be able to rest on the cut stump!). A new term, "*coupe de reglée*", was used to distinguish cutting within specified sectors as opposed to traditional random selection.

Gradually forests and farms, once viewed as intertwined and integrated, came to be seen as very separate land uses. Many forests became reserved for the exclusive use of the lords and their agents, who developed into the profession of forester. Forests came to be seen as economic assets, with timber the primary economic product. This led to the development of "scientific forestry", where careful record keeping combined with experimental methods to improve yields were tested and improved. Clearcutting, which we often think of as a twentieth-century innovation that reflects industrial forestry, was actually introduced into forests in Germany in 1454.

The systematic recording of forest compartments led to the laying out of rectangular coupes to help calibrate harvest and predict growth with some mathematical precision. German foresters perfected this system, which became known as "*Schlagwaldwirtschaft*". Essentially, this is growing and cutting of trees section by section or area. The goal was to ensure the sustainability of trees over the long term at a rate of perpetual maximum utilisation. This was the first true scientifically based forestry, with field experiments to determine the physiological characteristics of tree species and optimal conditions for growth. The information from these experiments was transferred to the field, and silvicultural planning was born.

*Figure 1.4
An old photograph from
Russia in around 1905
shows extensive strip fellings
in the forest. This is a
particular form of
silvicultural system still
practised in many places. It
could be argued that this
repeated pattern is not very
visually attractive. Source:
Library of Congress.*

As the father of modern German silviculture, Heinrich Cotta pointed out in 1816 that silviculture developed in response to a scarcity of wood:

> The emergence of scientific forest management displaced not only agro-forestry, but also the hunting use of the forest, where forests that had been reserved as deer habitat were liberated from this exclusive use. Artificially high populations of deer in reserved forests had resulted in a lack of regeneration, which was unacceptable when wood became the resource of greater economic importance to the landlords.

There are three aspects of the *Schlagwaldwirtshaft* system that have influenced forest planning right up until the present moment. The first is the professionalisation and centralisation of forest management. Forests were taken out of the hands of local people and placed into the hands of trained foresters at the service of the state. This social separation set in motion a dynamic that still persists in the US Forest Service, the British Forestry Commission, the French Office Nationale des Forêts, the British Columbia Ministry of Forests, the forest administrations of the German *Länder* as well as most other state administrations with which the authors are familiar. Its legacy poses challenges for improving public involvement in forest planning and management, as will be discussed in Chapter 4.

Secondly, forests were defined as distinct from other land uses, particularly agriculture. Up until this time, the line between forests and farmland was blurred. Now forest boundaries were given more or less permanent status. This demarcation has continued today in the form of land use zoning in the USA, which usually distinguishes forest from farm lands. In some parts of Europe, particularly Scandinavia where farmers also usually own forests, this is not so much the case.

Thirdly, and perhaps most significant to the subject of this book, the division of forests into discreet management compartments, or "*coupes*", clearly persists and continues to have great utility. It has become part of the core psychology of forest planning, and likewise has permeated conservationist and even preservationist thinking – what is a wilderness if not a large bounded compartment or coupe where preservation is the goal?

While scientific forestry and silviculture improved productivity in managed forest stands, foresters could not keep up with the growth of an industrial economy and the forest resource continued to decline in many areas. Iron smelting, house construction for a rapidly expanding population, shipbuilding, the breakup of large estates (which often sold off their timber) and wars led to ever increasing forest clearing. In Great Britain, only 12% of the land area was still forested by the eighteenth century. This proportion dropped to a low point of 5% during the First World War.

Regulations to protect remnant forests increased as the forests themselves decreased. The first School of Forests and Mines was set up at Banska Sriebnica, in what is now Slovakia, in the eighteenth century by the Empress Maria Teresa of Austria-Hungary in order to ensure a regular supply of timber for the many mines in the region. Germany and France led the way with forest conservation laws, and saw their forested land area stabilised. America set aside vast areas in the west as public forest reserves in the late nineteenth century, establishing the US Forest Service to manage them in 1907. Great Britain established its own Forestry Commission after the First World War, and began the process of establishing forest plantations shortly thereafter.

Figure 1.5 Seeding felling has just taken place in this French forest in the Loire valley. The light admitted into the forest will allow regeneration of oak to take place.

Throughout most of this modern period of scientific forestry, however, forest planning had quite a narrow focus. Decisions were driven by site-scale silviculture, and focused almost entirely on the issues of harvest and regeneration. The working plan was the paper tool of the silvicultural forester, in which an area was divided into compartments, and a sequence of harvest and regeneration organised. The traditional working plan usually went directly from data about timber volume and growth to site prescription. There was little true analysis, and almost no testing of alternatives. The focus was on a single dominant use, or at most expanded to only a few additional issues (timber harvest, regeneration, some grazing and fire protection).

Figure 1.6 Arapaho National Forest in Colorado, USA. This is one of the many national forests established on public land.

This system has not only been used in the West. After the revolution of October 1917 in Russia, all forested land was appropriated by the state from former private estates and monasteries. Forest management based on German models had been introduced in the eighteenth and nineteenth centuries and was applied to the estates of progressive landowners in parts of European Russia. A state forest service had initially been established by Peter the Great in 1699 to help provide timber for the navy. This became the foundation for a new administrative bureaucracy during the Soviet era. Land was organised into "forest ranges" or *lyesnichyestva*, with a system of rangers and supervising foresters. Forests were zoned according to a tripartite system, which is still in use in modern Russia. Some 25% of the land is designated as reserves, including urban forest, nature preserves, recreation forests and strategic reserves. No regeneration cutting is allowed in these areas, but foresters can remove dead trees and conduct some thinning. Most of the forest land area (around 70%) is designated for commercial exploitation for timber. The remainder comprises a type of intermediate category. Within managed forests, compartments (coupes) are created, usually using a rectilinear pattern, with sub-compartments inventoried and detailed prescriptions prepared.

A further feature of silvicultural planning, throughout Europe and beyond, has been the development and often long-term implementation of classical silvicultural systems. Some are French, others German in origin and can be classified as even-aged or uneven-aged, depending on the expected structure and character of stands. "Classical" silvicultural systems are designed to ensure continuous regeneration of the forest and to produce timber at the same time, with emphasis on the former. The focus of these systems has been on trees in stands, although the organisation of stands by area and of yield regulation across a forest management unit has led to ideological notions of the forest as a whole as, for example, the "normal forest", expected to produce a sustained yield of timber over time. This idea of sustained yield has had a very powerful influence on foresters and has affected the development of concepts such as sustainable forest management. Increasingly, the classical silvicultural paradigm is beginning to give way to a new one based on ecological processes (see Chapters 2 and 9).

Land-use planning and forestry

Modern land-use control had its origin in England with the development of building codes in London. Prior to this time land use was controlled by custom rather than by law, except in a few areas of royal forest. Feudal land use was based on tenure rather than ownership, and common rights negated exclusive control by any one individual. The system was one of mutual interdependence, similar in many ways to aboriginal use "rights", where tribes or bands have agreements to share gathering of resources in certain areas.

The medieval and aboriginal land use systems arose from necessity rather than fiat. They were self-perpetuating, and quite stable, but lacked the vitality to be adaptable to modern times. As medieval Europe began to urbanise, municipal corporations were formed to deal with fire, sanitation and other issues that arise as a result of people living in close proximity. In response to the Great Fire of London in September 1666, the first modern building law was developed. These called for stone or brick exterior façades, street widths retained in proportion to their importance, retention of open space along the Thames in order to have access for firefighting and banishment of public nuisance activities from the city centre. The new rules also included height and bulk standards for buildings, and a number of other provisions.

As ownership of land became more concentrated by individuals, land rights evolved as "green sticks" and "red sticks". Green represents the things one has a right to do with their property, and the red represent prohibitions. Much of land use planning is in essence a negotiation between what one can and cannot do with one's property.

As mentioned earlier, the distinction of forests as a land use separate from agriculture has its origin in Germany and France in the fifteenth–sixteenth centuries. It was not until the late nineteenth and early twentieth centuries that forests were again subdivided into separate land-use categories (similar to the experience of the middle ages), emphasising different forest uses.

The setting aside of wilderness areas and national parks in the USA established a category of forest land use that broke away from the European view of the forest as a place to be managed primarily for timber. The late-nineteenth- and early-twentieth-century history of the environmental movement in America is one of a dichotomy between those who worked to preserve forests from economic exploitation (generally known as preservationists) and those who worked to keep forests in public ownership with the intent to manage them for economic use (conservationists). These two movements, usually represented by John Muir and Gifford Pinchot, respectively, at first worked together to retain forest lands in the public domain.

Figure 1.7
A Russian forest management unit.

From the time of the American Revolution up until the establishment of Yellowstone National Park in 1872, the federal land policy was to seize or acquire western land from whoever had it (native peoples, France, Mexico, Great Britain, Russia) and then transfer it as private property into the hands of settlers. There was no policy to retain any of these lands in the public domain. However, a series of events gradually raised public consciousness and eventually built support for the retention in Federal ownership of those lands that could not reasonably be developed or farmed.

The first factor to affect public opinion was the brutal and callous treatment of land by timber companies. Forestry did not exist as a profession in the USA until the late nineteenth century. There was no tradition of peasant forestry, as has been described in Europe. Once the native peoples were killed or driven off the land, timber companies had a free hand. What they practised was exploitive logging, with no provision for erosion control, reforestation or slash disposal. This led to a series of catastrophic fires that followed badly managed logging, particularly in the upper Midwestern states. In 1864, George Perkins Marsh published *Man and Nature, Or Physical Geography as Modified by Human Action*. This classic helped change the way Americans thought about land use, particularly the role forests played in protecting soil, which in turn protected rivers and harbours from siltation.

Gradually a movement to protect American forests took shape. A forestry division was created within the Department of Agriculture in 1876, but was given no budget or authority. A small number of dedicated (and usually wealthy) amateur naturalists took up the cause. A leader among

these was Charles Sprague Sargent, director of the Arnold Arboretum at Harvard University. Sargent was commissioned to survey American forests for the Federal Government. This study led to an endorsement for keeping federally owned timber in public hands, at least until completion of a more comprehensive study. A little noticed but critical amendment to a general land law was passed in March 1891 that authorised the President to create "forest reserves" by withdrawing them from the public domain. This led to the setting aside of millions of acres of land that eventually became national forests, parks and wilderness areas.

There was no consensus on how to manage these lands. Muir, Sargent and others advocated "absolute protection" of public forests. They wanted to see these areas patrolled and protected against poachers and sheep grazers by the army (as had been the case in Yellowstone). It was Gifford Pinchot who suggested that supervision of the reserves be done by a "forest service, a commission of scientifically trained men". Pinchot, Sargent, Muir and a few other notables were formed as a government-sponsored commission and charged with touring the newly established reserves, providing recommendations on how they should be managed. The final report called for strict protection, including a ban on sheep grazing. However, at the insistence of Pinchot, the door was left open for the eventual introduction of regulated lumbering and mining.

Politicians in western states gradually woke up to the implications of what was happening and began resisting the establishment of forest reserves. In 1894, Congress went against the Presidential establishment of reserves by passing the Forest Management Act, which suspended the President's right to create reserves, and explicitly opened existing ones to grazing and mining. Pinchot was appointed as a special forest agent by the Department of Agriculture, and he put in place a policy to promote regulated logging and grazing.

Sargent and Muir were embittered by this turn of events, and the split between preservationists and conservationists was initiated. This split has only deepened and widened over the past 100 years.

Land use planning in public forests was at first focused on deciding which areas ought to be set aside from exploitation and which should be managed. The areas to be managed were then subject to more detailed silvicultural planning, usually in the form of working plans as advocated by Bernard Fernow and Gifford Pinchot.

Multiple use planning

For decades, the US Forest Service managed the vast federal forests primarily as caretakers. There was little demand for public timber until most of the private forests had been logged. This point came just after the Second World War, coinciding with the post-war suburban housing boom and the need to rebuild Europe. The Forest Service was eager to demonstrate its ability to provide the public with timber. This was at a time when the concept of "tree farming" had just taken hold in the USA. Plantation forestry was already well established in Europe, but in the USA and Canada there were still significant amounts of virgin forest available for cutting, leading to a lack of investment in plantations.

Forest Service managers firmly believed that they could have it all. They could log, replant and at the same time protect watersheds, wildlife and recreation opportunities. The US Congress believed it too, and gave the agency large budget increases to build roads to transport logs to the mills. The cut from public lands doubled during the first half of the 1950s. "Multiple use" was the guiding principle. It was first developed as a concept in the 1933 Copeland Report titled *A National Plan for American Forestry*. This report was more a suggested set of policies than a plan. S.B. Show, the Forest Service head of operations in California, referred specifically to the question of multiple use in his section regarding future forest land ownership. He felt that timber production, watershed management and recreation could all be served at the same time. He did, however, acknowledge that some situations called for one use to be designated as dominant over others.

This idea that a forest could and should be managed for a number of competing purposes was of course not new. As described earlier, traditional forest practices in medieval Europe and by indigenous people in North America demonstrated the desirability of multiple use of forest areas. However, the Forest Service's embrace of the old concept was a departure from "modern forestry", as developed in Germany and France, which had elevated timber management above all else.

After much debate and development, the Forest Service convinced Congress to pass the Multiple Use Sustained Yield Act in 1960. This law provided a legal mandate to balance timber, recreation, wildlife, watershed, mining and grazing uses on national forest land. The law left it up to the Forest Service to work out how to do this.

*Figure 1.9
This example shows an area of forest in the Pacific Northwest after it has been extensively cut over and replanted.*

However, the tension between competing uses only grew worse. The expansion of industrial forestry on public lands, with dense road networks and large-scale clearcuts, came at the obvious expense of recreation and scenery. Lawsuits over clearcutting on the Monongahela and Bitterroot National Forests in the 1970s threatened to derail the entire federal timber programme. A legislative compromise was reached in 1976, with the National Forest Management Act (NFMA). This law explicitly told the Forest Service to zone its lands for the use that was most appropriate. It also called for increased and improved public involvement.

NFMA led to the development of complicated zoning maps for each national forest. In essence, the attempt was (and still is) to resolve competing uses by creating overlays and associated standards, very similar to what one finds in local land use planning. The Mt Hood National Forest in Oregon, for example, identifies over 40 separate zones, including such designations as timber emphasis, developed winter sports, scenic viewsheds, a big-game winter range and late successional reserves. Each zone has a set of standards that specify whether logging can occur, and if so at what rotation, the size of clearcuts, the allowable density of roads and so forth.

The level of zoning typical on US national forest today demonstrates the limits of what can be accomplished through zoning-based plans. At its worst, it represents an absurd division of the forest to a point where it cannot even be understood, let alone competently managed. The managers are tied in knots, and neither they nor the public who owns these forests can effectively visualise what the forest will look or be like.

Much of this problem has its roots in the over-specialisation of natural resource management that has become an unfortunate side effect of modernism. In the late nineteenth century, it was possible for one individual to be an environmental "renaissance man". All the technical knowledge in the fields of forestry, biology, geology and other natural sciences could fit into the study of a single wealthy naturalist. Thus George Perkins Marsh could write *Man and Nature*, a very influential book that linked the condition of forests to the protection of soil and watersheds. Then natural science professions evolved, developed their own university departments and degree programmes, their own peer-reviewed journals and research priorities. Gradually, information became so dense and complex in each field that the various disciplines have found it nearly impossible to understand or communicate with each other. James Karr, of the University of Washington, describes this as each discipline digging a hole straight downwards. They believe that by going deeper and deeper they are learning ever more, but in fact are losing contact with everything else that is going on. "Hyper-zoning" of forest lands is a direct reflection of this over specialisation. Forest managers

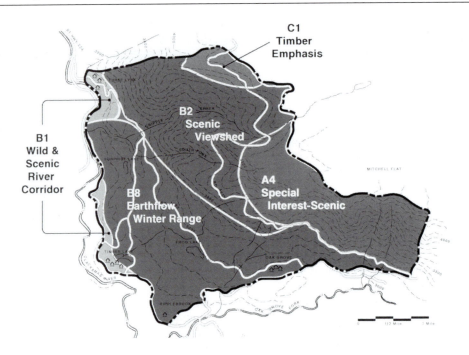

Figure 1.10
This is an example of a zoning map for Mt Hood National Forest in Oregon. The area is that of the pilot study for the application of landscape ecology. It shows how each zone has a single major management objective.

are attempting to overcome this problem through interdisciplinary teams. But the results have so far not been encouraging.

Harvest planning

In places where virgin forest was available in large tracts and an industry based on logging could be developed, a great deal of unregulated exploitation took place. However, many countries have seen the need to regulate harvest levels so as to reduce the rate at which timber reserves become exhausted, to ensure that regeneration takes place and to help develop sustainable industry by guaranteeing long-term supplies of timber to mills. In some countries, such as parts of Canada, the land is retained in government ownership and tracts are licensed to timber companies. The provincial forest service regulates harvest levels but the planning is carried out by company foresters.

Harvest planning is carried out based on the volume of timber available expressed as an annual allowable cut (AAC). The level of this AAC is derived from a calculation based on the total increment of the growing stock in the area, with or without allowances for protected areas, non-merchantable timber or inoperable sites. In theory it is possible to cut a volume equivalent to the annual increment without ever depleting the forest capital, hence use of the term "sustained yield". Such calculations require an accurate inventory of the area and information on site quality, growth rates and so on, rarely available in primary forests. Once the AAC is agreed, it is the task of the forester to plan where and how the cut is to be achieved. Typically, until quite recently, this has been a combination of road engineering and harvest operational planning, set within regulatory constraints such as "adjacency rules", where one coupe, cutblock or harvest unit (terms vary) cannot be logged adjacent to another one where the regeneration has yet to reach a certain age or height.

Harvest planning is strongly driven by cost considerations – to keep the cost per unit volume of timber harvested as low as possible. Since roads are expensive to build and maintain, their layout must be considered and the total length of roads kept to a minimum, although not at the expense of too much costly extraction over the ground, which may affect the total cut. Foresters normally use aerial photographs to select primary stands or sub-compartments and then, together with road engineers, consider options to obtain access for harvest. Logging equipment costs, haulage distances, costs of road construction, culverts and bridges must all be considered, and mathematical tools such as linear programming have been used to calculate the optimal solutions given information on these variables. Increasingly, geographical information systems (GIS) can help by identifying difficult terrain, or by more easily calculating distance, all related to the database of stands and their timber volumes that can now be linked directly to the maps held on the GIS system.

Frequently, harvest planning has been carried out only 1–2 years in advance of logging, to allow road construction to take place before the timber is needed at the mill. No long-term planning is undertaken in many places. However, in order to optimise the harvest across a landscape, it makes sense to look well ahead and consider the total road network, the global effect of rules and regulations and the spatial pattern of blocks in terms of their timings and the effect of this on cost. Thus a form of planning called "total chance planning" or TCP has evolved in Canada. The name refers to an old term for an area as a "timber chance", that is the potential to harvest timber surveyed by an inventory forester.

TCP considers a valley or other land unit and develops a complete pattern of harvest units, their timing, road construction and all the territorial constraints. Typically, using a GIS, various constraints will be identified, such as inoperable areas, riparian zones, areas prone to landslide, water sources, stand types of no economic interest and so on. The timber in these constraint areas will be "netted out" of the cut and zones where these constraints apply marked on the map. Then a complete road plan will be prepared for the operable area, including primary and secondary haul roads, tracks, landings and so on. This will be checked on the ground to some degree. After this the operable forest will be divided into blocks or units each capable of being logged and extracted to one or more landings by a certain type of logging equipment. Once the TCP has been approved in outline, the optimum timing of this, using computer techniques, can be carried out, involving the adjacency rules, predicted green-up times and road construction schedules. Then the first blocks are selected, laid out on the ground, the roads constructed and the timber harvested.

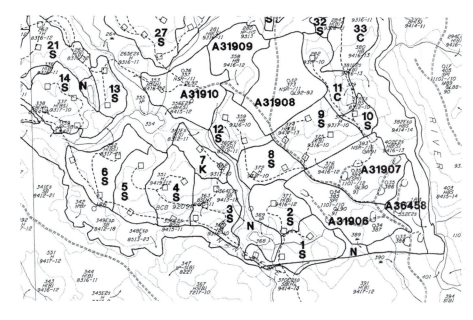

Figure 1.11

A total chance plan for a forest in British Columbia, Canada. The plan is developed over the forest cover map in order to relate the harvest units to the underlying stands. Areas to be left unlogged are delineated and then access roads, tracks and landings are developed. With an analysis of terrain and economic yarding distances the operable area is divided into harvestable units that are combined to form the cutblocks. Source: Ministry of Forests.

27

TCPs have the advantage of being long term and landscape scale. They also generate valuable information and analysis. Their main drawback is that they are based solely around harvest optimisation and do not take account of ecology or aesthetics beyond the legally required protective zones. However, as illustrated in Chapter 5, the basic development of an engineering plan can be an important element of a forest landscape design, and the idea of dividing the whole operable area into blocks with phased timing is at the heart of most forest design. Thus, it is often possible for forest harvest planners to understand and accept the idea of forest design since it contains ideas similar to those in a TCP, expanded to incorporate many more issues and resource values.

Ecological forest planning

The 1980s witnessed the development of three new fields that represent attempts to integrate forestry better with ecology: landscape ecology, conservation biology and ecological forestry. These three fields will be considered in detail in Chapter 2. In this section we will focus on how these fields, particularly conservation biology, have initiated a new type of forest planning.

Conservation biology is an activist-based field. It was born of the modern crisis in biological diversity surrounding the loss of species and ecosystems. Much of the impetus came from *Island Biogeography* by McArthur and Wilson. Their breakthrough research noted that smaller and more isolated islands had higher extinction rates than larger ones closer to mainlands. This eventually led to similar studies and conclusions in terrestrial habitats, which have increasingly become "natural islands" within a sea of developed or altered landscapes. Conservation biology draws from a number of related fields, including population ecology, community ecology, wildlife ecology, biogeography, forestry and wildlife management. The central challenge is the design and management of reserve systems, often including large areas of forest.

Landscape ecology provides much of the analytical support for conservation biology, so in conservation circles the two are often considered to be a team. The goal of conservation biologists and their activist allies is to design and implement a broad system of reserves and connecting corridors adequate to protect all native species. Their sphere of influence includes all of North America, as well as parts of Europe and Australia. A number of key planning principles have emerged:

- Large reserves are better than small ones.
- One large reserve is better than multiple small ones that may in sum be equal in area to the large one.
- Multiple reserves should be placed close together rather than far apart.
- Reserves should be clustered rather than linear.
- If possible reserves should be linked by corridors.
- The shape of a reserve should be blocky rather than linear in order to reduce edge impacts.

These principles represent an idealised picture of reserve planning. In fact, conservationists have to be satisfied with establishing reserves wherever and whenever they can, rather than rejecting candidate areas because they are too linear or not connected to a corridor. In addition, much conservation work is focused on river systems, which are by nature linear and seldom clustered.

While planning is a critical element of conservation biology, the goals are narrowly focused on biodiversity conservation. Thus there is little integration with socio-economic issues. Generally, reserve systems are established at the expense of economic development. Most conservation biologists are less concerned with reforming forest management practices than they are with removing forest management from large expanses of land. In their defence, they point out (accurately) that systems of large reserves are in all likelihood the only method for conserving biological diversity in a comprehensive way, that economic forestry has consistently failed to take adequate measures to support ecosystems and that human culture has appropriated far more land area for its purposes than it has an ethical right to.

The planning and design of reserve systems should not be confused with previous efforts to protect national parks and "wilderness". In the latter case, the traditional goals have been to protect landscapes of high scenic or historic value. Protection of ecosystems and biodiversity has rarely been the goal of wilderness area designation. This is easily demonstrated by looking at a map of wilderness and national parks in the USA and Canada. They are almost invariably located in high mountains or deep canyons, places of sublime landscape scenery. Areas of the highest biological richness, such as estuaries, low elevation river valleys and prairies have rarely been protected, since these compete with high economic value land uses.

Ecologically based plans have developed significantly over the last decade on both sides of the Atlantic. In Canada, many projects, including those where the authors have been involved, have taken landscape ecological principles and developed an analytical process to understand the forest patterns and dynamics (see Chapters 2 and 5). The identification, firstly, of a network of potential areas within the landscape unit is followed by the development of management units, coupes or cutblocks based on natural shapes, sizes and types of natural disturbance within the operable forest or matrix. Silvicultural systems developed from the natural disturbance regimes are then fitted into these units (see Chapter 9). This approach, using a combination of fixed landscape features (a forest ecosystem network) with the more dynamic elements has also been reflected in Scandinavia, especially Sweden, although with a slightly different emphasis.

Following research, especially with influence from the Pacific Northwest of the USA, and by foresters and ecologists from forestry companies, several case study areas were set up on company and non-company lands in Sweden. These have had a significant impact, so that ecological planning is now a very popular tool applied on scales of 5,000–25,000 hectares. There are two approaches in use. One is a natural landscape approach, where management aims to maintain or establish natural patterns of stand types and successional stages, based on mimicking fire patterns. At a stand level, structure and composition should reflect that found in natural successional stages. It is also necessary to protect key habitats. One example of this natural model is called the ASIO model, where stands are classified into four groups. "A" stands for "Absent" because due to natural circumstances, such as wet conditions, they almost never burn. "O" is at the other extreme, meaning "Often" burns, so that regular clearcutting tends to reflect natural disturbance. "S" and "I" are intermediate – Seldom and Infrequent. This approach suits the extensive, more natural forests of northern Sweden.

In southern Sweden, the forests have been heavily altered and cultural heritage, aesthetics and recreational values are considered to be as important as ecology. The approach developed here is the "supportive features model". Forest owners identify the supportive features of their forest, such as streams, old farmsteads, sites of particular biological or aesthetic value and so on. These are then preserved or developed, using appropriate silviculture, at the stand level. Since there are so many small owners in a landscape in southern Sweden, landscape-level planning is difficult unless carried out in a participatory way (see Chapter 4).

Sustainability at the forest landscape level

Over the last 10–15 years, the concept of sustainable forestry has developed, supported less by government regulations (though these exist) than by the market, operating through the system of certification. This is having an impact on forest planning and it seems to lead naturally towards a state where landscape-scale forest planning and design is a major tool for ensuring sustainability at the correct level of resolution. One of the main ways that sustainability is measured is by applying criteria and indictors which have been developed, for example, by the Pan-European or Montreal Processes (in Europe and North America, respectively). Some regional planning processes have recognised the substantial conceptual gap between criteria and indicators designed for national-level monitoring and reporting and the practice of sustainable forestry at the landscape level. The Pan-European Process, for example, has developed a set of Forest Level Operational Guidelines that give more practical meaning to its six criteria. For each criterion, the Guidelines prescribe management processes and outcomes, the underlying assumption being that if the prescriptions are implemented in a generic European forest, the result will be sustainable forest management. The Guidelines come with a cautionary note that they should be adapted to the specific local, economic, ecological, social and cultural conditions and to the forest management and administration systems that are in place. The note adds that all interested parties should be encouraged to participate in the process of adapting the Guidelines for local use.

Many countries have developed codes of practice, some on the initiative of government, some on the initiative of industry. They come in different forms: comprehensive single documents such as the Forest Practices Code of British Columbia. Others are sets of separate, linked documents such as the UK Forestry Standard's Forest Practice Notes and related forest management Guidelines and Forest Practice Guides. Forest certification schemes enable forest managers to make independently verified claims that they are meeting a specified standard.

The Forest Stewardship Council (FSC) operates a global accreditation scheme for forest management and chain of custody certification, endorses national forest certification standards that have been developed in accordance with its rules of process and content, and licenses the use of its logo to companies whose purchasing, production and sales processes have been certified against the FSC's chain of custody standard.

The Pan-European Certification Council (PEFC) administers a system of mutual recognition between European national forest certification and chain of custody schemes. National schemes that conform to the PEFC rules for accreditation, forest certification standards development process, standards content and chain of custody certification are authorised to license the use of the PEFC logo.

Some national entities set forest certification and chain of custody standards, provide accreditation services and license the use of their name or a label (e.g. the schemes operated by the Indonesian Ecolabelling Institute and the Malaysian Timber Certification Council).

How do all of the criteria and indicators, guidelines, codes of practice and certification standards address the two central tenets of sustainable forestry:

1   Sustaining forests' capacity to provide multiple goods and services;
2   Meeting the needs of present and future generations?

Sustaining the forest's capacity is the traditional realm of the forest manager: maintain the health and vitality of forests; maintain forest soils in a condition favourable to tree health; maintain ecosystem dynamics and nutrient cycling; prevent harm to the tree crop; avoid damaging impacts on the forest ecosystem and the wider environment. Criteria and indicators, codes of practice, guidelines and certification standards are packed with best practices that responsible managers have been employing for decades. The assumption has been: get all this right and foresters can ensure a continuous flow of timber and non-timber forest products and all the non-consumptive benefits that forests provide simply by being there, for example, carbon storage, clean and constant water, a pleasant place to walk.

If it is as simple as this, why have forest managers been under an almost constant challenge? We suggest that this has most often been because forest managers have taken decisions about the goods they will produce and the importance attached to other goods and services with too little thought to the value that other people attach to these goods and services.

Forests are affected by external impacts that are beyond the control of forest managers. We are not thinking here of dramatic events such as fires, hurricanes, avalanches or volcanic eruptions, but gradual change in the environments in which forests function. The first significant occurrence that entered public consciousness was the impact of air pollution in Europe leading to forest degradation and, in extreme cases, the death of forest stands. The situation has since improved as a result of national measures to reduce toxic emissions. Climate change will have uncertain impacts but in some regions of the world they will be significant between the time a stand is regenerated and felled. In Britain, for example, a recent study made the prediction that autumn frosts will be more damaging, temperature extremes will kill off some species with the result that the mix of trees that make up our forests may well change and deer and grey squirrel numbers are likely to rise, posing an even greater threat to trees unless control measures are stepped up. Temperature rise will also encourage the development of insect pests. Increased winter rainfall may raise the water table sufficiently to kill off roots, making trees more prone to summer droughts and diseases caused by fungi.

Whether the inhabitants of those regions affected by climate change like it or not, some tree species will become less well suited within a specific forest landscape and others will be better suited. Some plant and animal species will decline and may become locally extinct, while others prosper. The value and mix of goods and services that forests in a specific region will be capable of providing in the future will change. Forest managers need to start thinking now about how climate change will affect the mix they will be producing in 50 years time. Thus, calls for more sophisticated forest planning at the correct scale.

The forest landscape design process is aimed at a scale of a few thousand to perhaps 100,000 hectares in size. In a very real sense, this is too small a scale to achieve true sustainability, whether we are talking about biodiversity or economics. Yet the issue of sustainability, while often addressed as broad policies at the national or international levels, in fact plays itself out at the scale of forest design. This is true for several important reasons. Firstly, the landscape level lies between the site or stand scale and the sub-regional scale. The former is clearly too small to provide habitat, protect water quality or even provide much economic benefit. The latter is too large for design to be effective, but is where larger land use decisions are made. For example, the question of deciding on areas to set aside as reserves is best made at the regional and sub-regional scales.

Viewed as part of a larger landscape, the area to be designed may be a critical link for some wide-ranging wildlife species. Or it may lie in the path of a proposed cross-country highway.

Figure 1.12
This view of northern Sweden shows a forest landscape of the sort of scale at which forest design might be carried out. It is not a suitable scale for measuring sustainability but it is the scale at which sustainable management is planned because the regional scale is too large and the stand is too small.

It may be part of a bio-geoclimatic type that experiences stand replacement fires on a scale far larger than its own area, or it may be part of a timber supply network crucial to sustaining a small town that is 20 kilometres distant.

In later chapters of this book, we will illustrate various approaches and tools for measuring sustainability at the landscape scale. For now, let us simply state that it is implicit that sustainability be a clear goal of any forest design, and that this must be nested within larger scales and policies to help ensure that wise decisions are being made.

Forest design and forest planning

The authors view forest design as an evolutionary step consistent with our historical sketch of forest planning. It encompasses most of the types of forest planning illustrated in this chapter: traditional, silvicultural, total chance harvest, reserve design and ecological. In addition, forest design relies upon a basic element of land use planning, the three-part zone. It is also ideally suited to satisfy sustainable forest management needs.

Forest design incorporates important elements of "traditional" forest planning. By developing designs through a process that builds on local needs and knowledge (Chapter 4), traditional use, access and management techniques regain or retain their place in the woods. These can include grazing of domestic animals in ways that maintain a desired forest understorey, planned harvest of "special products" such as greenery, mushrooms and berries, and most importantly (in western North America at any rate) the restoration of open woodland through prescribed fire and selective cutting. While the forest design process does not presuppose any of these actions, it does make room for them by relying on local community involvement.

Forest design is closely linked to silviculture, and in fact has been called silviculture at the landscape scale. Silviculture is the primary tool for shaping the pattern and structure of the forest landscape in commercially managed areas. Increasingly, silvicultural systems and prescriptions are moving away from classical examples towards those based on natural processes, such as "variable retention" as advocated by Jerry Franklin, Dean Berg and others in the Pacific Northwest. Essentially, these are prescriptions that mimic small- or large-scale disturbances, with careful attention to leaving "legacies" (downed wood, snags, live green trees).

One way of incorporating harvest planning into forest design is through the development of the "total chance plan" described earlier, which looks at a whole area in terms of roads, harvesting access and operability. This also helps to ensure that any designs are completely practical and operationally possible. Since, in the majority of cases, forest designs are being prepared because managers who wish to change the forest through harvest or silviculture (except in ecosystem restoration or reserve management), road access and harvesting systems are fairly universal.

Where timber harvest is not a primary goal of the design, minimising or even eliminating roads is often the better option. Much of the forest restoration work in the Pacific Northwest and Rocky Mountains is based on removal of existing roads. Other solutions include temporary or seasonal roads, which can even be created over ice in the far north, leaving little or no trace once they melt away.

Ecological planning is now incorporated into most forest designs at the landscape scale. This usually includes an integrated network of protected areas, incorporating riparian vegetation, steep slopes, fragile habitats and key corridors such as low elevation mountain passes. The forest design planning process also represents an important opportunity to augment or complement larger reserve networks established at regional scales.

A basic three-part forest zoning system comprising intensive forestry, extensive forestry and wild land or wilderness forestry areas strongly complements forest design. Conversely, forest design can be used to craft a zoning plan where one does not yet exist. However, zoning may be more flexible and reflect different emphases in terms of objectives, design concepts or silvicultural practices. A forest design strategy map, which may be presented as a series of zones with different visions, is also an essential element of a forest design plan.

The most developed version of forest design has arisen as a fusion of planning and design. In Britain, the land owned and managed by the Forestry Commission is subject to what are known as Forest Design Plans (FDPs). These plans incorporate all the elements of forest planning, together with the design approach described in this book. These plans are relatively long term (30–50 years) and implemented over a series of phases which usually coincide with harvest planning periods, thus enabling plantations to be restructured and significantly improved (see Chapter 10).

Conclusions

We have described design as stemming from a forest planning tradition that involves a conscious shaping of the forest in a way that facilitates flows of the goods and services that the human community wants and needs. We have tried to show that design and planning involve choices, and that the choices we make as designers and managers are a reflection of the value we attach to attributes that forests have now or could have in the future, to the goods that forests are able to produce and to the services they can provide. The attributes, goods and services that healthy forests are capable of sustaining will change over time, because the environment in which they grow will change. Individuals, communities and nations view forests from different value frames, and we should not assume that future generations will share any of them in the exact way we do today. All this might make the designer's task appear impossible. But there are some basic principles that, if followed, will enable us to design forest landscapes that are sustainable. These principles are:

- The design should maintain the forest's capacity to provide multiple goods and services. In other words, at a landscape level a forest design should not result in an overly specialised or simplified ecosystem. This is a particular challenge in the design of plantation forests, which by definition are highly specialised.
- The attributes of the forest and the mix of goods and services that it provides should be decided by negotiation between the forest owner or public manager and local community interests, and should at least take account of wider interests in forest issues.
- The design should build in resilience to external impacts that are beyond the control of forest managers so that future generations inherit a healthy resource which they can choose to use in whatever way makes the most sense to them. This implies that forest managers making long-term decisions need to be aware of and factor in issues such as climate change, as well as episodic large-scale disturbance patterns and time frames.

As illustrated throughout this chapter, planning in some shape or form has been an integral aspect of forest management for hundreds, if not thousands of years. The long time it takes for trees to grow and mature makes planning a necessity, not a luxury. Traditional forest planning was done on the ground rather than on paper or computer, but it was planning nonetheless. People have long managed forests with an eye to conservation and future structure. The development of intensive, economic forestry brought with it paper plans and mathematical precision aimed at ensuring long-term timber supplies in the face of growing population and wealth. Since the turn of the twentieth century, there has been a struggle between those who want to protect natural forests from exploitation and those who want to utilise them. The twenty-first century finds us in a new era, where a more holistic form of planning needs to be adopted. We believe that the forest design process meets this need.

The following chapters concentrate on setting out the theory and background behind the main concepts needed for the practical application of forest landscape design. While planning must be incorporated, forest design takes aspects of forest planning described above and adapts them, adding these further elements, to create a system that not only plans for the outputs from the forest, but also the future pattern and structure of the forest landscape.

Chapter 2

Landscape ecology, conservation biology and ecological forestry

Introduction

The emerging fields of *landscape ecology, conservation biology* and *ecological forestry* provide key underpinnings for designing sustainable forest landscapes. In this chapter we introduce and summarise some key concepts and a practical method of analysing forest landscapes from an ecological point of view. This approach was pioneered by Dean Apostol and Nancy Diaz of the US Forest Service in the early 1990s and developed and applied more widely since then. The chapter presents the theoretical basis of this applied approach, then describes the analysis process before illustrating it with a case study. Subsequent chapters also describe aspects of the process either in the way it is integrated into the forest design process as a whole or as part of the case studies described, especially in Chapters 8 and 9.

Landscape ecology is a subset of the older field of ecology, and is concerned with understanding the interactions of organisms and their environment across kilometre-wide mosaics. It is an inter-disciplinary field that includes contributions from biogeography, anthropology and landscape architecture as well as ecology. Landscape ecology has been applied across a wide range of land use issues, including agriculture, urban areas, wild lands and forests. Conservation biology is a multi-disciplinary science that has been developed to deal with the crisis of declining biological diversity. It is part of the larger field of applied ecology, which includes forestry and game management, and draws upon many of the same sources as landscape ecology. Conservation biology is not a neutral science, but rather is self-described as mission orientated. Its practitioners are driven by the ethic that biodiversity is a good thing, and that they have a duty to conserve it.

Island biogeography is a theory relating to extinction and repopulation, first advanced in the early 1960s by Edward O. Wilson. It brought forth a number of concepts that have strongly influenced both landscape ecology and conservation biology, particularly the design of conservation reserves, including size, spatial arrangement and connectivity between them.

The term "ecological forestry" is used here to describe new approaches to commercial forest management that seek to emulate natural disturbances in terms of landscape pattern, harvest rotation schedules and biological legacies. It has its roots in the Pacific Northwest of the USA and its key proponent has been Jerry Franklin of the University of Washington. As with conservation biology, ecological forestry can be described as mission orientated, but in this case the mission includes the economic use of forests in ways that foster conservation.

The importance of these new fields to forest design lies in the way they help to synthesise complex information regarding multiple variables in ways that are fairly easy to understand by practitioners and the public. They provide the key conceptual framework for generating, evaluating and selecting a design strategy for a given area of forest, and they also establish the bases for measuring ecological sustainability.

The field of landscape ecology can be traced back to the work of pioneering natural history scholars and authors, such as George Perkins Marsh, who linked concepts of geography and ecology. The advent of aerial photography in the 1920s allowed geographers to obtain a clearer view of landscape patterns, and the term "first landscape ecology" came into use. Over the years many concepts have become incorporated into the field, particularly that of island biogeography. The recognition and study of "land mosaics" developed in earnest during the 1980s through the work of Richard Forman, Michel Godron and others. This work investigates landscapes as pieces of a puzzle linked to human activities, whether as a result of conscious planning and design or exploitation of resources.

Ecologists have long been concerned with identifying patterns, and connecting these to processes that generate or maintain them. Ecology has evolved from a discipline based largely on empirical observation to one with a strong multi-branched theoretical content, and landscape ecology can be viewed as one of these branches. Other branches include population, community, conservation and restoration ecology. Restoration ecology is the process of assisting the recovery of an ecosystem that has been degraded, damaged or destroyed. This is also a fairly new field of work that is rapidly coming of age.

While landscape ecology is a key foundation stone for forest design, it must be noted that there remain important gaps between many of the theoretical concepts and the hard experiential information required to prove these. There is a lack of practical experience applying recommendations in the field, so theories are ahead of practice. Because managers are often looking to ecology to help navigate the difficult challenges of sustainable forestry, there is a tendency to put new theories into practice quickly, without waiting for applied research to determine if they are correct or not. The authors recognise this danger, and while we advocate that forest design should take place using applied landscape ecological principles, we also urge caution by the reader and practitioner not to confuse theory with proved practice.

The main theoretical basis underpinning the landscape ecology framework used in forest design is provided by Forman and Godron. Their work laid the foundations of the subject and they present it in impressive depth. Taking this theoretical foundation and applying it as a practical tool for forest ecosystem design and management has proved to be a challenge, but we believe that this has been done with sufficient evidence of success to advocate its wider use.

New concepts and empirical studies are continually emerging, and we encourage the reader to explore the literature on this subject in greater detail. New tools that advance its application include computer-based analysis and modelling to help consider different ecological factors, subjects touched on in later chapters (Chapters 6 and 7). It should also be stressed that landscape ecology must be integrated with a wider knowledge of forest ecology. Work by researchers such as David Perry, Hamish Kimmins, Jerry Franklin and P.J. Burton provide important textbook references on the specific ecology of forests as well as ecological management strategies.

Key concepts from landscape ecology, conservation biology and ecological forestry

In this section we introduce and explain the major concepts that are used in the practical application of landscape ecology as a design tool. These concepts are drawn mainly from landscape ecology as a neutral discipline describing the functioning of a given landscape, with additional concepts from the more mission-orientated fields of conservation biology and ecological forestry.

In landscape ecology the main conceptual distinctions are drawn between the physical structure of the landscape and the functions that take place within it. Firstly, what do we mean by the term landscape, how does it differ from other scales of resolution and why is it important to use this scale in planning and design?

Landscapes

Landscapes are defined by Forman and Godron as heterogeneous land areas composed of clusters of interacting ecosystems that are repeated in a similar form throughout. A second definition is: *A kilometres-wide mosaic over which particular local ecosystems and land uses recur.* For most forest areas this often translates into an area defined by a significant landform and/or drained by a major stream system, within which the climatic regime, geomorphological processes and natural vegetation types are fairly uniform. A landscape is larger than a stand (an area with a relatively homogeneous vegetation in terms of species composition and structure), and smaller than a region (defined by climatic, geological and/or vegetation zone boundaries). Thus a landscape can be populated by many stands of different composition and structure, and it may be one of many similar landscapes within a region. In terms of relative size, stands may be found in the range of one to tens of hectares, landscapes tens to thousands of hectares and regions thousands to hundreds of thousands of hectares.

From an analytical and design standpoint, landscapes can vary greatly in size. The authors have worked on projects of as little as a few hundred hectares in some European countries, to hundreds of thousands of hectares in the USA and Canada. Yet all have met the criteria of being clusters of interacting ecosystems, and all included mosaics of varying structure. It is probably less important to define precisely what a landscape is or how large it should be than to understand the relationships between the processes that occur in the landscape and the structures necessary to sustain them.

Landscape structures

Landscape structures are the physical, tangible elements of ecosystems. They can be living (biotic) or non-living (abiotic), mobile or fixed. Trees and forest stands, bodies of water, bogs and wetlands, rock and scree are all examples of landscape structures. Later in this chapter, structures will be described in more detail and certain characteristic aspects of them explained.

Ecological functions

Ecological functions are the activities, roles or processes performed by structures. Functions can be classified in a variety of ways, with five main types generally recognised:

1   *Capture* (*input*): resources such as organisms, materials and energy are brought into the system from outside, e.g., rainfall or snow caught in a drainage system, photosynthesis by the forest trees capturing energy, animals migrating into an area.

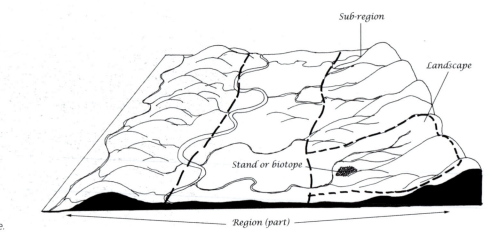

*Figure 2.1
This diagram shows the ecological definition of a landscape, set within different scales of resolution – the region, sub-region, landscape and stand or biotope.*

Sub-region

Landscape

Stand or biotope

Region (part)

2   *Production*: resources are "manufac-
tured" within the system, e.g., animals
producing young, trees regenerating,
growing and producing wood, complex
soil development.

3   *Cycling*: resources are transported within
the system, e.g., nutrients released by
rotting vegetation being taken up again
by growing plants, rain water taken up
by trees and transpired back into the
atmosphere, animals moving from place
to place within the system.

4   *Storage*: resources are conserved for a
time within the system, e.g., carbon (as
wood in trees), water in lakes or wetlands, sediments in lakes or wetlands.

5   *Output*: resources leave the system,
e.g., outmigration by animals, water flowing out of an area or timber being harvested and
transported away.

Structures are normally involved in more than one function, and a function may involve more
than one structure. For example, a stand of trees may capture sunlight, produce seed, cycle nutrients
and water and become an output as timber. Water storage may involve rivers, lakes, ponds,
wetlands, soil and vegetation.

Interactions
Interactions are a key aspect of the way we understand ecological systems. There are three main
kinds of interactions to consider.

1   *Interdependencies exist among functions*: e.g., capture and cycling must take place in order
for production to be sustained.

2   *Structure and functions are co-dependent, and act to change each other*: As Forman and
Godron state, "Past functioning has produced today's structure; today's structure produces
today's functioning; today's functioning will produce future structure."

3   *Transfer between scales*: Ecological systems should be viewed as nested scales that interact
with larger and smaller systems. No ecosystem is completely isolated, so in order fully to
understand an ecosystem it is also important to understand its connections to other systems
at larger and smaller scales. These interactions may operate between spatially separate systems
at the same scale, for example several eco-regions or several watersheds, or they may operate
between different scales, such as a watershed within an eco-region, or both simultaneously.

Ecosystem resistance and resilience
Ecosystem resistance and resilience are interrelated aspects of stability. Resistance refers to the
ability of an ecosystem to absorb and contain disturbance in ways that prevent large-scale alteration.
For example, an open Ponderosa pine stand keeps fires at ground level, and avoids damaging
crown fires. Resilience is a description of the ability of an ecosystem to return to a reference state,
productivity level, or composition and structure following a significant disturbance. Resilience is
especially important in the temperate and boreal forests that are the primary concern of this book,
because such forests are prone to periodic stand replacement natural disturbances.

Ecosystems vary in the degree to which they can absorb change and still retain their variety of
functions. Some forests, such as temperate broadleaved forests in Europe or the temperate
rainforests of the Pacific Northwest, are highly diverse in terms of tree and other plant species

*Figure 2.2
This view shows
an area large
enough to be
considered
suitable for
analysis at the
landscape scale.
It contains
several landform
units and a
significant
drainage system.
It can also be
taken in from
one viewpoint,
reflecting the
link between
the aesthetic
and ecological
concepts of
the term
"landscape".
Alaska Highway,
Fort Nelson
Forest District,
British Columbia,
Canada.*

composition and in structure. They are not naturally prone to cataclysmic disturbance, except on a relatively infrequent basis (a few hundred to one thousand years return interval). Post disturbance, they tend to be very resilient except where changes wrought by poor management reduce the ability of the system to rebound. Examples include excessive soil compaction or erosion, road systems that disrupt natural drainage patterns and logging that triggers dense hardwood regrowth that suppresses conifer regeneration. In the boreal forest zone, for example, in Canada or Siberia, the forest ecosystem tends to be simpler in composition and structure and is visited more frequently by catastrophic, large-scale natural disturbance. Resilience may be dependent on the maintenance of key refugias that are retained through multiple cycles of disturbance, such as topographically sheltered old growth patches or wetlands.

Figure 2.3
Two examples of resilient and less resilient forest. a) shows a coastal forest of British Columbia, Canada, where catastrophic disturbance, while it occurs, is very infrequent and the forest is resilient to many effects; b) shows a boreal forest area in northern British Columbia, Canada, where fires recur quite frequently. The resilience of the ecosystem lies in the areas that escape the fire.

Since ecosystem functions are dependent on the structures that preform them, it follows that changes which eliminate certain structural elements or features can lead to a loss in function. Thus, management to sustain ecological resilience involves identifying and protecting structural elements, functions and relationships, with the objective of maintaining the function of the whole.

Ecological forestry usually attempts to retain key structural elements during and after forest harvest. These are referred to as biological legacies, and can be considered at multiple scales. Some of the ecological planning approaches described in Chapter 1 endeavour to maintain resilience during forest management by protecting key structures or by emulating the processes that maintain compositional and structural diversity. One of the major sustainability issues resulting from the changes brought to forests by managers over the last 100 years is the simplification of diverse structures, which has reduced the resilience of ecosystems.

Describing and understanding landscape structures

An important first step in attempting a practical application of landscape ecological analysis is to describe in some detail the characteristics of the existing landscape structure. A well-established classification system can be used and it is best to prepare a set of maps with accompanying descriptions as part of this (see illustrations in this chapter and also the process described in Chapter 5). The three main types of structure are defined as patch, matrix and corridor.

Landform
Landform is an often overlooked landscape structural category, at least in hilly and mountainous areas, where it is often the driver of the landscape pattern. Ridgelines and valleys exert strong influence on vegetation patterns, water flows and species movement. Valley floor riparian areas are at the receiving end of energy, nutrients and material that gradually move downhill. Ridgelines are natural movement corridors for many species, including raptors and large mammals. Landforms also establish the conditions that allow various patch types to occupy a landscape. For example, south-facing slopes often have drier, more open structured patch types than north-facing ones. Steep ridges have shallow soils, and thus may have rocky openings or grassland patches within a forested matrix. Valley bottoms or terraces may be the only places where bogs or wet prairies can form. In hilly or mountainous areas an analysis of the landform will demonstrate how dependent the pattern of the landscape mosaic is on terrain. Thus, it can be particularly important to consider the analysis of the landscape structure within the context of the landform rather than as an independent variable.

Figure 2.4 Rocky Mountain trench area dominated by strong landform.

Patches
Patches are areas of vegetation that are relatively homogeneous internally (in terms of composition, three-dimensional structure, successional stage, etc.). Forman describes patches as being relatively homogeneous, and non-linear in shape. A particular patch should be readily distinguished from others in its vicinity. In a forest landscape, clearcuts, young, mature or old growth forest, grassy areas, wetlands or rock outcrops are common patch types, along with rare or unique stands, such as an aspen grove in pine woodland. In fragmented landscapes the mature forest may exist as a series of disconnected patches. Some more detailed characteristics of patches are presented later in this chapter.

A matrix
A matrix is normally defined as the most dominant patch in a landscape, either by total area or degree

*Figure 2.5
In this diagram
patches are
defined, both
natural in
character and of
human origin.*

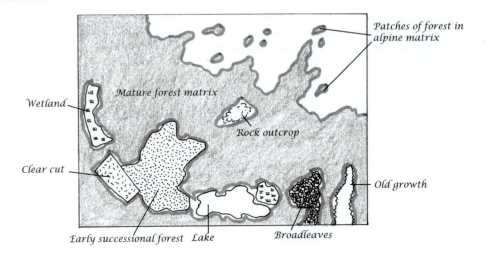

*Figure 2.6
This landscape in
Norway contains
many patches,
such as fields,
a farmstead,
patches of forest
within the
farmland, a
felled area (in
the foreground)
and, in the
distance, rock
patches.*

of connectivity. A very important ecological feature of the matrix is that it is thought to exert strong control over landscape dynamics (movement of material, energy and organisms).

In most natural forests, mature or old growth age classes/successional stages usually comprise the matrix. This is increasingly less true as forests become fragmented by clearcutting, and in some cases it may not be possible to discern a matrix at all because the degree of connectivity has been drastically reduced. In extensively managed forest landscapes, the matrix may have shifted from mature to younger, early successional forest, or even non-native plantations.

It is worth noting that what constitutes the matrix depends on the scale of resolution. A study area may be wholly forested, so that the forest clearly performs matrix functions. However, at a larger scale this same forest may be a patch or island surrounded by prairie or open grassland. In some complex mosaic areas, the matrix may be difficult to identify, and there may be competing matrix elements that divide a landscape into several sub-areas, with each element connected to extensive areas beyond the study boundary (see discussion of the Latvian example later in this chapter).

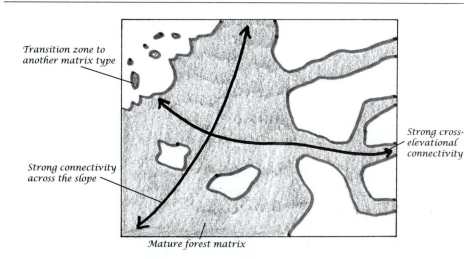

Figure 2.7
This diagram shows how a landscape matrix can be defined. It demonstrates the concept of connectivity in relation to the area covered by it.

Transition zone to another matrix type

Strong cross-elevational connectivity

Strong connectivity across the slope

Mature forest matrix

Figure 2.8
An example of a forest with a strong matrix, both extensive and highly connected. North-eastern British Columbia, Canada.

Corridors

Corridors are relatively narrow, linear landscape elements that provide connectivity among similar patches through a dissimilar matrix or aggregation of patches. The patches connected by corridors are often called nodes, meaning that they are larger or wider habitats linked within a system. Roads and streams are corridors that may connect early successional patches (along the roadside verge), or aquatic habitats, respectively. A protected riparian zone may be a corridor connecting patches of relict mature forest in an otherwise heavily logged landscape. Different kinds of corridors facilitate flows (the cycling function) of different materials or organisms across the landscape. This functionality is what distinguishes a corridor from a linear-shaped patch that may provide no flow facilitation. The effectiveness of a corridor to provide connectivity often depends on how wide it is (how much is actually edge) and how frequently breaks or discontinuities are encountered.

Providing habitat corridors as a conservation tool was first advocated by Edward O. Wilson (*Island Biogeography*) in the 1960s, and has now become an article of faith among many landscape planners and conservation biologists, especially those working in urban and agricultural areas. The concept of wildlife corridors has been expanded to regional and even continental scales. However, it remains unclear how to design or retain a functional corridor successfully, particularly given the limitations imposed on conservation by economics and other social issues. In most cases, we lack detailed information on species movement patterns and edge impacts. Obtaining good information is difficult, since it would require landscape-scale experiments over long periods of time. Are narrow habitat corridors sometimes death traps, or population sinks for species that have nowhere to go to escape from predators? Are they avenues for invasive species? Are they even needed in forested landscapes, assuming the matrix is in decent condition?

Even with these unknowns, corridors are a powerful conceptual landscape planning tool that make strong intuitive sense. As such, they are an important component in most forest design projects, as will be demonstrated later on.

Stepping stones

Stepping stones are usually thought of as clustered patches that act as corridors in terms of allowing wildlife to move between habitats that would otherwise be unreachable. They may be particularly

*Figure 2.9
This diagram shows the types of corridor and pathway found in a forest landscape.*

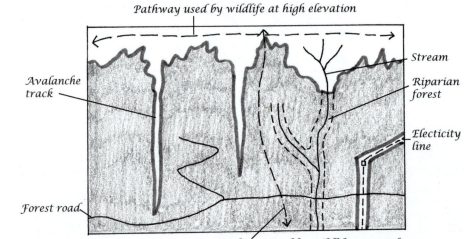

*Figure 2.10
This landscape is dissected by a number of corridors, such as forest roads, rivers and streams and a powerline. Some of these, such as the river, are natural and perform important ecological functions. Others, such as the powerline, may interfere with ecological functioning by acting as a barrier to movement. Vancouver Island, British Columbia, Canada.*

important for birds and butterflies in urban and agricultural landscapes, but can also be important in forests.

Landscape Continuum Model

Not all landscapes can be defined into clearly recognisable patches, matrices and corridors. The Landscape Continuum Model was developed to help a better understanding of landscape function where the boundaries between separate patch types are not clear, or where they cannot easily be distinguished from a background matrix. It recognises that many landscapes are more complex than the patch-matrix-corridor model implies and it emphasises that the matrix itself may be the most important habitat within a landscape. This concept calls for distinguishing "variegated" landscapes, where patches are of low contrast, from "fragmented" ones, where they are of high contrast. It is a very important feature of ecological forestry, which seeks to blur boundaries and establish a more seamless landscape that may not need connecting corridors.

Landscape structures derived from conservation biology models

With its goal of ensuring the conservation of biodiversity, conservation biology has developed other ways of considering landscape structure, or rather of placing values on certain elements or combinations of elements. Part of the analysis of landscape structure may be aimed at identifying these areas as a precondition for management.

Core areas

Core areas, or reserves, are the central backbone of conservation biology practice. Essentially, these are strictly protected zones, or zones managed with biodiversity conservation as the primary goal. Natural disturbances are allowed to occur, or may be mimicked by managers. Core reserves can be fairly small or quite large, depending on context. Generally, they are established at centres of the greatest biodiversity.

Biological hotspots

Biological hotspots are areas that have special importance for either organisms or ecosystem processes. They may or may not be recognised as distinct patch types, but for some reason play a key role in sustaining local biodiversity. For example, there may be an old growth patch that is in a sheltered area near a wetland or creek, and thus it has much greater productivity than an old forest patch of similar structure elsewhere.

Refugias

Refugias are areas that occupy positions that protect them from large-scale disturbance, thus they have an important role as a repository of biological legacies. These may include areas on the lee side of hills or mountains sheltered from prevailing winds, or where soils are particularly deep and moist.

Landscape mosaics

The term "landscape mosaic" describes the pattern of patches, matrices, corridors and stepping stones associated with the underlying landform, and is of key interest, since it is the spatial arrangement of these elements that largely determines ecological functions. A few of the ways to describe these elements are presented below. When applying the steps of the landscape ecological analysis as part of the integrated forest design process (Chapter 5), it is important to include a full description of each identified element. The following section explores the various aspects that can be used to describe landscape structures.

Composition

Composition is the physical and biological expression of a landscape element. It includes the types of vegetation present, in terms of the different species, age/size class (in the case of trees) and the

three-dimensional structure. Composition largely determines how a landscape element interacts with various landscape flows. For example, a Douglas fir forest offers different food sources from a Ponderosa pine forest, and thus will facilitate the flow of different species of mammals and birds. A wetland of willows and herbaceous plants provides different feeding opportunities for elk than does a conifer forest. Fish use a stream segment shaded by forest differently from one open to the sun.

Origin

Origin refers to the means by which a landscape structure was created. It could be natural (fire, blow-down, beaver dam, avalanche,

Figure 2.11
This aerial view of south western Norway shows a mosaic landscape formed of many different patches and corridors.

natural succession from young to mature forest) or caused by human activity (logging, planting, building a dam, constructing a road). Origin is important for understanding landscape dynamics from the perspective of the rate of change. How likely is it that more structures originating from the same cause will be created and at what rate?

Early successional forests can be thought of as ephemeral systems in terms of composition and structure that are rapidly changing towards something different. By contrast, the old growth forest stage may persist for hundreds of years with little change. Some wetlands may dry out and be taken over by forest while others may last indefinitely. Landscapes naturally dominated by ephemeral elements function differently from those dominated by persistent ones. For example, forests dominated by a mosaic of early successional patches may be home to a range of fauna comprising species capable of using many elements and suited to changing landscapes, while those dominated by old growth may favour fauna with much narrower habitat requirements, and may be unable to cope with rapid change.

The landscape mosaic pattern itself can be classified in terms of its stability, which is a reflection of the combined stability of different landscape elements and their relative positions. For example, the outcome of a forest fire where susceptible stands are split up by patches of less-flammable forest will be different from one where susceptible patches are concentrated in only one area.

(a)

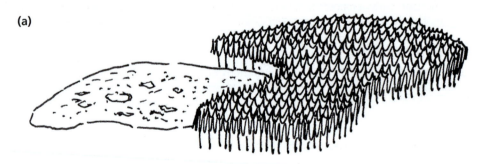

Figure 2.12
This series of sketches demonstrate some of the ways that landscape elements can be defined:
a) patch contrast: a stand of trees with strong vertical structure compared with a wetland of low vegetation.

(b)

Figure 2.12 (continued)
b) patch edge variation: a hard edge against a cutblock versus an ecotone against a naturally open area.

Contrast

Contrast refers to the degree to which adjacent landscape elements differ from each other in terms of species composition and physical attributes. For example, there is a high amount of contrast between a sedge/willow wetland and late successional Douglas fir forest, not only in plant species but also in structure, height, soil wetness/dryness, origin and stability. A late successional Douglas fir forest and an old growth Douglas fir forest display much less contrast – the species may be very similar, the tree height almost the same, the main difference being dead wood content and openness of the canopy (greater in the old growth). Contrast exerts an influence over landscape dynamics to the extent that it controls landscape flows. The two contrasting examples described above may mean that animals moving through the forest are blocked by the wetland, or that certain birds may rest in the forest yet hunt for food in the wetland. The two less contrasting forest types may present no apparent differences for animals moving through the landscape, yet the old growth may provide more feeding and nesting opportunities for insectivorous birds than the late successional forest, due to the higher proportion of dead and decaying wood. Low-contrast landscapes may be best understood as all matrices, with subtle variations.

Edge

Edge refers to the character of the interface between landscape elements of different composition and structure, for example between a clearcut and closed canopy forest. Edges have environmental conditions (temperature, light, atmosphere, humidity, wind) that are different from either of the adjacent elements. In abrupt edges to high-contrast elements this may be less marked than where there is a more gradual transition, although edge conditions affect each adjacent element to a degree, such as shade cast on the clearcut or light penetrating into the edge of the closed canopy forest. Very often, the plant composition of the edge is a mixture of those belonging to both components.

(c)

Figure 2.12 (continued)
c) patch shape: the difference between regular and irregular shapes, including the amount of interior habitat possible in proportion to the edge zone.

Some animals may use edges differently from the interiors of each adjacent element and some species use edges as their habitat. This can mean that edge species benefit when a landscape is fragmented and contains more edges than usual, while those species needing interior habitats suffer.

Patch shape
Patch shape is important because of the effect it has on the proportions and amounts of edge and interior habitat. Regular-shaped patches, which are closer in shape and proportion to circles or squares, have the highest ratio of interior conditions to edge, whereas those with highly convoluted shapes have proportionately more edge. Patch shape also tends to reflect the origin of the landscape element. Regular, and especially geometric, shapes are almost always of human origin, whereas irregular shapes are more likely to have been created through natural processes. However, shape is related to size – small patches of forest are limited in terms of shape variability, whereas larger ones can be highly convoluted and irregular. The shape of naturally created patches

(d)

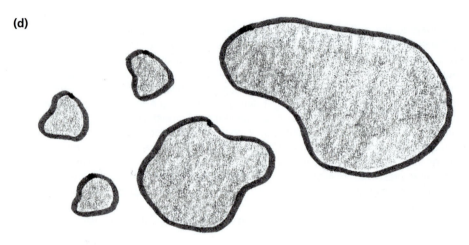

Figure 2.12 (continued)
d) patch size: small, medium or large patches.

frequently reflects the influences of landform and microsite or microclimate conditions and tends to be simpler where landform is simple and more complex when it is more broken and contains a greater complexity of microsites. Patch shape can be measured by its fractal dimension, which is the degree to which the shape differs from a simple circle or other geometric figure.

Patch size
Patch size refers to the scale of individual patches. Size is important for the habitat requirements of some species. The greater the size the greater the proportion of interior habitat will be available for a given degree of complexity, assuming that the edge zone width is constant. Some patch sizes are controlled by landform, and it is difficult for large patches of a single forest type to occur in a complex topography full of wet hollows or steep peaks. Other patch sizes may result from types of disturbances. Some man-made patches have often been of a standard size due to regulations on permitted maximum sizes of clearcuts.

Connectivity
Connectivity refers to the spatial contiguity within a landscape. It is a measure of how easy or difficult it is for organisms and energy to move about without encountering insurmountable

(e)

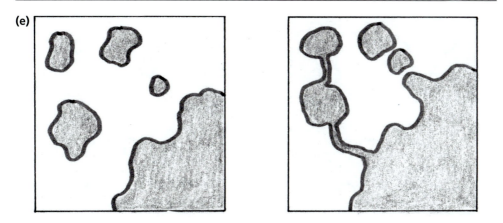

*Figure 2.12
(continued)
e) connectivity:
low versus high.*

barriers. Connectivity occurs both within a matrix and via corridors. Some species may prefer to use the matrix while others might be confined to particular corridors. The degree of connectivity is a functional definition which changes from one species or flow to the next. Some may need completely connected corridors while others, such as birds, may fly from one patch to another, using them as stepping stones, as long as the patches are not too far apart.

(f)

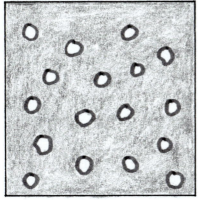 

*Figure 2.12
(continued)
f) landscape
grain: small
patches give a
fine grain while
large patches
give a coarse
grain.*

Landscape grain

Grain is the average size of elements, which provide a texture to the landscape. Fine-grained landscapes consist of numerous small patches, while coarse-grained ones have fewer, larger patches. Grain may change as a result of species composition, landform, microsite variation or shelter. Often human alterations create a landscape of more regular, uniform and fine grain than natural examples, which tend to have a wider range of patch sizes.

Porosity

Porosity is the density of a particular patch type within a matrix. A landscape with many small patches of a similar type distributed through a matrix is highly porous and has a lot of edge, although the matrix may still display a high degree of connectivity. If porosity becomes too high, resulting in too much edge and too little interior habitat, then the landscape is fragmented.

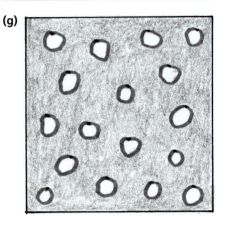

Figure 2.12 (continued) g) porosity: many patches give a high porosity while few patches result in a lower porosity.

(g)

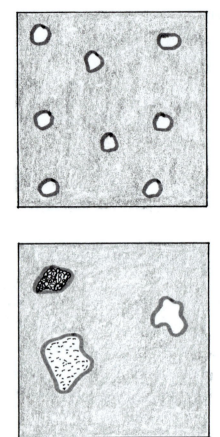

Figure 2.12 (continued) h) patchiness: many different patches and small matrix compared with fewer patches in number and character.

(h)

Patchiness

Patchiness is the density of all patch types compared with the area of matrix. It is a similar measure to porosity but takes all patch types into account. Some landscapes are naturally very patchy, such as areas in Newfoundland that are full of small ponds, bogs, rock patches and scrubby forest. Other landscapes are not naturally patchy yet may have become so through human activities such as logging, farming or suburban sprawl.

Heterogeneity

Heterogeneity refers to the variation in aggregations of elements. If there are many different elements per unit area, the landscape can be said to be micro-heterogeneous. If there are different combinations of elements in one area compared with another, the landscape is macro-heterogeneous. It is useful to be able to determine what the average degree of heterogeneity would be for a particular landscape under natural conditions. As will be seen later, some landscapes are currently highly heterogeneous at the micro-scale, which would not naturally be so, and vice versa. Evidence of past natural landscape structure may not lead to the correct conclusion of current natural structure, yet ecosystem management and restoration requires a knowledge of this to be developed. A study of the landform, soils and microclimate will help gauge the correct range of variation but this might have been overridden by disturbance events (see below).

(i)

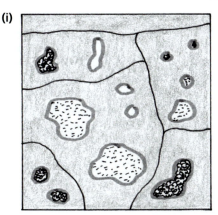 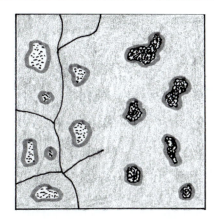

Figure 2.12 (continued) i) heterogeneity: micro-heterogeneity versus macro-heterogeneity.

Landscape functions and flows

Landscape functions are best understood by relating them to flows and their interaction with different elements. Landscape flows are phenomena that move across and interact with landscapes, moving on, under or above the ground. They often respond to multiple elements (a pond, stand of trees, or road, for example). Flows include both organic and inorganic phenomena such as mammals, seeds, birds, water, air, soil, etc. As each flow passes through or interacts with an element, one or more ecological functions (capture, production, cycling, storage or output) occur. For example, an elk (the landscape flow) might use a wetland for food, and a mature forest stand for thermal cover (production and storage). Water might flow to a pond, where some of it is used by aquatic plants (production), some evaporates back into the atmosphere (cycling) and the rest recharges the ground water (storage). A landscape mosaic of multiple elements therefore provides an animal such as a deer (the flow) with food, cover and breeding sites (production, storage, output).

It is relatively easy to identify, map and describe landscape elements at a scale of resolution useful for analysis. It is more difficult to select the appropriate flows for diagnostic purposes. In part this is because there may be hundreds if not thousands of flows in any given forest. The authors have found it most useful to choose flows that are most representative of the general ecosystem type, that use a significant part, if not all, of the landscape and that are in some way indicative of ecological health. The following categories can be used to help to decide which flows are best to use for analysis.

Keystone species

Keystone species play prominent roles in forest ecosystems, affecting the habitat of other species by their presence or absence. For example, carnivores may affect the populations of their prey, which in turn may alter grazing pressure. Recent research in Yellowstone Park, since reintroduction of the timber wolf, indicates that some riparian woodlands are now recovering, since elk spend less time grazing in risky places. At the opposite scale, the presence of mycorrhizal fungi facilitates more rapid tree growth by extending the effective reach of root systems. Keystone species are those that affect the linkages in food webs so strongly, and their role is so specialised, that their removal may result in a radical reorganisation of the food web. This theory was first advanced in 1966 based on research in inshore marine ecosystems. Many believe it has been over used and too broadly applied, yet it is still a useful concept, if applied to all species that play a critical ecological role.

Umbrella species

Umbrella species are those with large area requirements, and often represent the top of the food chain. Many conservation biologists believe that if natural resources are managed to ensure the range and distribution of these species, then species at the lower trophic levels will also benefit. Typical examples of umbrella species include large carnivores such as lynx, grizzly bear and timber wolf.

Rare, threatened and endangered species

Rare, threatened and endangered species are represented in the biodiversity legislation of most countries, and their potential presence in a particular area of forest is usually public knowledge.

Indicator species

Indicator species are those that if present, tend to mean that a set of other species will also be present. They are generally used as a surrogate for other species too difficult to track or measure. The Northern Spotted Owl was selected as an indicator for old growth dependent species in the Pacific Northwest, because it was easily tracked and dependent on older forest structure.

Flagship species

Flagship species, or "charismatic megafauna", often capture the public's imagination. Flagship species are also often rare, threatened and/or endangered. Typical examples in temperate forests include timber wolf, grizzly bear and woodland caribou.

Socially or economically important species

Socially or economically important species are usually game animals, and may be important for subsistence hunting, the tourist economy or simply as representative of what local people feel should be in a healthy forest.

Wildlife guilds

Wildlife guilds are groups of species that have similar habitat and dispersal characteristics. If carefully selected, they provide a wider perspective on ecosystem function than single species.

A practical method of analysing the interactions between flows and different elements is to use a table or spreadsheet, the structural elements being the columns and the flows the rows. The Latvian example provided later in this chapter illustrates this approach.

Landscape dynamics

The terms defined so far relate to the existing landscape. It is important to note that all landscapes change over time and an understanding of the rates, scales, intensities and types of dynamics is vital. Two main types of landscape or ecosystem dynamics are particularly important: disturbance and succession.

Disturbances

Disturbances are events that result in significant change to vegetation pattern and structure, often over a short space of time. Disturbances can be natural or human-caused events that interact with and initiate changes to landscape structure. No ecosystem or landscape is forever unchanging. In fact, change is crucially important to maintaining ecosystem functioning and resilience. Without plants growing, dying and decaying there could be no production, storage, cycling or outputs. It is therefore necessary to understand the dynamics of the natural and human-caused processes at work in any given landscape. This is particularly important for the purpose of developing management systems for forests that enable the production of outputs such as timber while maintaining the ecosystem functioning. Disturbances can be described in terms of their type, intensity, frequency, duration and effect.

Natural disturbances include fire, flood, wind, avalanches, landslides, insects and pathogens, all of which affect forest landscapes. Some affect very large areas while others tend to be more localised. Windstorms may do their work in only a few hours while root rot can take years. Most disturbances follow somewhat predictable cycles of occurrence, with the most extensive and severe usually having a low frequency of occurrence. Disturbances mutually interact with the landscape mosaic in that the current pattern and structure in part determines the location and spread of disturbance, while the disturbance sets the stage for future landscape structural development. This is a chicken-and-egg process, where cycles of growth and disturbance go far back in time, and in most cases it is impossible to identify the initial cause. In fact, disturbance cycles take place within larger climatic cycles, such as warming and drying, or increased wetness and cooling, so the cycles themselves are subject to change over time. Global climate change as a consequence of carbon increase may have an effect on such cycles, but in ways that will vary from place to place.

Human-induced disturbances can be as or more important than natural disturbances in terms of impacting landscape pattern and structure. These may include gradual forest change brought on by over-grazing by domestic livestock, fire suppression, introduced pests, rapid change through the introduction of large-scale clearcuts, and forest fires. In most cases, human-caused disturbances produce very different effects on the landscape compared with natural disturbances. For example, a natural fire may affect a large area but it will leave behind pockets of unburned trees, fringes of partly burned edges and even occasional thick-barked single trees that survive, together with ash, carbonised wood, standing dead but not burned trees and so on. A clearcut of the same area may leave few or none of these biological legacies. Thus there may be a lack of long-term nutrients, and little habitat for organisms that use charred wood, perching posts and tree cavities, as well as insufficient seed sources for natural regeneration.

Succession

Succession is the process by which plant communities grow and gradually change in composition and structure over time. Classic succession models refer to open fields or barren areas becoming a pioneer forest of sun-loving plants, then mature forest and finally a "climax" community of shade-tolerant plants that is internally stable until the next large disturbance comes along. In reality, most temperate forest communities never lock into a stable climax. Ecologists usually develop model successional pathways for each ecotype to chart the expected rates of development, changes to plant communities and structural characteristics. A particular area might follow various trajectories, depending on a number of factors, including intensity and scale of disturbance, distance from seed sources, presence of legacies and the vagaries of the weather. Computer-based modelling tools enable managers to project successional development so as to see what patterns are likely to develop over time. By combining models for both disturbance and succession it is possible to determine the "average" pattern of the landscape over time, in terms

(a)

(b)

(c)

Figure 2.13

Three examples of different types of disturbance: a) an example of the aftermath of a natural disturbance event – a forest fire - showing the structure left behind in Labrador, Canada; b) the results of a catastrophic wind storm in the High Tatra mountains in Slovakia with a huge amount of debris (Source: Peter Fleischer, High Tatras National Park); c) by contrast, a clearcut area, an example of man-made disturbance, is devoid of any of this structure. Aldewood Forest, Suffolk, England.

of the proportion of different patch types and matrices across different areas or sub-zones. This can be very useful when it comes to deciding the ecological "desired future condition" (see below).

Cultural disturbance and land use may result in accelerated or arrested succession. In many European countries with a long history of forest management in a landscape where the forest only covers a small percentage of its original extent, it is the managed or semi-natural landscape elements that have vital importance for ecosystem functioning. Areas of heather moorland (*Calluna* spp.), managed by burning and prevented from recolonising by natural forest, represent an important semi-natural habitat in Scotland for many species.

Issues to be considered in the study of disturbance and succession

Once the dynamic processes have been understood there are several important questions to be raised, especially when considering the differences between natural and human-caused processes. One of these is to consider how the current state of dynamics fits into a longer time-scale, the degree to which fragmentation is an issue and the role of legacies in handing genetic material from one generation to another.

Historical variability analysis

Historical variability analysis is the study of how the disturbance regime of a forest has affected the landscape mosaic over long periods of time, usually before intensive disturbance initiated by modern human cultures. It is therefore easier to carry out in the USA, Canada, Chile, Siberia, New Zealand or Australia than in most European countries. Understanding historical variability requires much more detective work as compared with describing and analysing the current pattern and structure. However, clues in the present landscape can provide glimpses of past disturbances that can be used to build up at least a partial understanding. Tools for understanding historic disturbance patterns include old photographs, early survey records, pioneer journals or ancient literature, oral tradition (if indigenous people are still in the area), fire scar dating and pollen analysis. Where it is difficult to find data on the specific forest being analysed, one can usually extrapolate from data collected on nearby areas of similar landform and vegetation.

Fragmentation

Fragmentation is the breaking up of a large habitat into smaller elements, to the point where the remnants lose their ecological integrity. It is a landscape characteristic most associated with dispersed clearcutting on national forests in the USA. It begins when relatively small clearcuts are

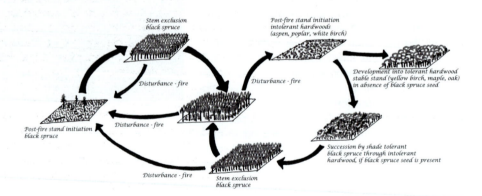

Figure 2.14
This diagram shows the different stages of natural succession, with potential variations depending on the circumstances, following natural disturbance. The example is taken from black spruce forest type in Nova Scotia, Canada (see the case study in Chapter 9).

*Figure 2.15
A fragmented
forest in Maine,
USA, where
alternate squares
have been
removed in an
attempt to
provide white-
tailed deer
habitat but
which has
drastically
reduced
connectivity
across the
landscape.*

placed within an unbroken matrix of mature or old growth forest, creating openings and edges. As more clearcuts are added, the remnant forest becomes all "functional edge", meaning it no longer has interior habitat conditions. Models have shown that logging 50% of an area in this way completely eliminates interior habitat.

Permeation
Permeation happens when roads, powerlines or other linear features penetrate into an otherwise undisturbed matrix, providing avenues for pests, invasive species, poachers and so forth. It is a particularly important issue in the Amazon rainforest of South America.

Biological legacies
Biological legacies are traces and features that remain in a landscape following a disturbance. They include organisms, organic structures (such as downed logs and standing dead trees), soil, and patterns that persist and link the old landscape structure with the emerging one. Legacies are a key concept within ecological forestry, and have been much studied in the wake of the 1980 eruption of Mt St Helens in Washington State and the 1988 wildfires of Yellowstone Park. Researchers in both areas have found that biological legacies were very important in colonising surrounding areas from deep within the disturbed zones.

Interior habitats
Interior habitats are those buffered from "edge effects". In essence, an interior habitat is far enough away from an adjacent, contrasting patch that it is not influenced by it. Some mammals and birds nest deep within forest interiors in order to be safe from competitors, predators or brood parasites (e.g., the brown-headed cowbird in Pacific Northwestern forests). Generally, edge effects (light, wind, temperature) penetrate up to 300 feet (100 m approx.) into a forest edge, depending on aspect, slope and vegetation density.

Ecological linkages

The study of a given landscape cannot proceed as if it were an island, unless it actually is one. But even in the case of islands, it is important to consider how the area under investigation is linked to others in the immediate vicinity. Ecological linkages are the points or areas where one landscape meets another, and energy and materials are exchanged. In particular, managers will want to know how and where the most important flows interact with lands outside the study area itself. Since landscape processes operate at multiple scales, different landscape flows will require varying land areas, which may be larger than, smaller than or coincide with the study area.

The way to determine what linkages exist is to ask the question "Which key flows cross the study area's borders, and how do these flows occur?" Some flows, such as water and aquatic species, will link to surrounding landscapes in predictable places, such as the mouths of streams. Determining the territory or key routes of large mammals may be less precise, and require some guesswork if direct field knowledge is unavailable. Migratory birds may be seasonally linked to landscapes on other continents. It is probably unnecessary to map their complete sequence, but it may be important to know that they use the study area as a stepping stone on the way north or south. Open ridgelines are often key connectors for raptors, while deep gaps in these ridges may provide avenues for forest-dependent mammals to get to the next watershed safely.

In terms of the relationship of linkages to ecological processes, it may be important to know if the study area is a key population source, a sink or part of a network of similar habitats across a region. The quantity and quality of wildlife flows over time within the study area might be a result of factors at work elsewhere, and have little to do with the changing landscape structure of the study area itself. Equally, if flows originate and move out or migrate from the study area it may be important to consider the effect a reduction in production of these flows might have, even if there appears to be no problem within the study area itself.

A second aspect of linkages concerns the landscape structure. The wider landscape sets the context for the study area. Are there any structures in the study area that are critically important in the wider context, as an intact matrix area for example, or does the pattern of structures create problems, perhaps where the study area is more fragmented that the surroundings? Does it contain a portion of a critical migration route for a particular species or does it act as a node in a wider network?

Desired future condition

The purpose of all the analysis is to lead to an understanding of where the landscape is at present and where it should be directed by management in the future so as to ensure the maintenance of ecological functioning or to restore damaged or lost functions as far as possible. Thus, while the landscape has an existing condition, this may not be what is desirable for the future. The term "desired future condition" describes what we want a landscape to be like at some future date. It can reflect an idealised, sustainable system or a modified one. There can be more than one option for a desired future condition, but generally the aim is to envisage a future pattern and landscape structure that is more desirable for sustaining ecosystem functioning. This can be described in terms of the amounts and locations of the landscape structures that should be in the landscape over given periods of time, perhaps with special core areas for specific conservation management also identified. A map depicting zones related to topography and drainage, elevation or soil types accompanied by text may be the most useful outcome for the incorporation of this analysis into the integrated design process.

Case study: Taurene in Latvia

In order to demonstrate the practical application of an ecological analysis, the following case study is presented. Latvia is one of the Baltic States, lying between Estonia, Belarus and Lithuania. It is a country with some 40% of its land covered in forest and most of the rest in agriculture. This case study was a pilot project to evaluate the application of the landscape ecological method described in this chapter to a sample area of mixed land use in the Latvian countryside.

The landscape of the Latvian countryside is presently undergoing significant changes. In the period since Latvia regained its independence from the former Soviet Union in 1991, land ownership has been radically restructured, as many state holdings and collective farms have been restored to their former owners, or their descendants. In agriculturally marginal areas much of the land has ceased to be farmed and has either become abandoned or is being actively converted into forest. Predictions as to exactly how much land will eventually fall surplus vary, due to the lack of national land use or agricultural policies. Current estimates are that at the most, 1.5 million out of the 2.5 million hectares of agricultural land will be required for continuing agricultural production. This leaves 1 million as potentially surplus to requirements.

The forests themselves are being exploited for timber production much more intensively as part of the means of rebuilding the Latvian economy. The result of this increased rate of harvest is that the structure of forests is likely to change significantly with more intense management and a shift towards a greater proportion of younger age classes. Fragmentation of forest areas may reduce the extent of continuous mature forest and increase the amount of forest edge, thus favouring wildlife species that can cope with early successional forest and edges at the expense of those needing deeper forest and no edges.

These changes are having profound effects on the appearance of the countryside, on its traditional character and on its wildlife.

Scale of resolution and selection of boundaries for the planning area

Latvia lies in the climatic zone known as the nemoral zone, a transition between the northern boreal zone and the central European broadleaf zone. This defines its main ecological character, although there are variations across the country according to latitude and distance from the Baltic Sea and the Gulf of Riga.

Figure 2.16
The location of the case study area of Taurene in Latvia.

Planning generally takes place at the level of the region (rajon) and district (pagast) and it is at these levels that the scale of resolution of the project was defined for practical purposes. The rajon can be used for strategic planning but vegetation types could only be defined in broad categories. The pagast is where most physical and landscape planning takes place and was chosen for the analysis scale.

A pagast, that of Taurene in Vidzeme, a central province with upland character of rolling hills, was chosen for the study. The reason for this was that much development of physical and economic planning had already been completed and it made sense to see how the landscape ecological analysis would fit into the established planning methods. Taurene is an interesting example to study because it represents many of the issues facing Latvian landscape planning. The area lies in central Latvia, to the east of Riga. The landscape is quite hilly with around 52% forest cover, which is higher than the average for Latvia, and there is around 25% abandoned farmland. Agriculture used to include mixed farming of crops and cattle. Some of the land was improved by drainage and the rest was unimproved pasture and meadow. The upper part of a major river, the Gauja, together with a chain of lakes, runs through the area. The forests are mainly spruce (*Picea abies*) and birch (*Betula pendula*) with some pine (*Pinus sylvestris*) (spruce 44%, birch 31%, pine 15% in the study area).

The study followed the steps described earlier in the chapter very closely. The results of each of these steps are described below:

1. Landscape structures

The classes of structure defined for Taurene are:

Matrix
At first glance Taurene appears to show a complex mosaic of forest and open land. The boundaries of the two main land cover types of forest and agriculture are highly irregular in shape and it is difficult to say where the mainly agricultural area stops and the predominantly forested part begins.

However, taking the degree of connectivity as the main criterion, closer examination shows that there are three distinct areas, two of forest matrix and one of farmland. The line shown dividing the study area into the three zones is not a hard one, but indicates the general location of the division. In general terms the matrix elements can be described as follows:

Forest matrix This is mainly composed of mixed forest aged between pole stage and maturity, dominated by spruce with varying proportions of birch, a little pine and areas of broadleaved woodland. The area includes forest that has been there for many centuries, but a significant proportion has developed over the last 100 or so years. Thus, large areas are only reaching late successional or mature conditions at the present time. The connectivity of the forest has also increased significantly as the area has expanded, potentially enabling a wider distribution of forest wildlife, while some agricultural areas have become islands surrounded by forest and increasingly influenced by it, for example in terms of microclimate, drainage, tree seeds and wildlife use. The forest generally occupies the land least suited to agriculture due to soil, drainage or topography.

Agricultural matrix This is currently mainly grassland with some arable, much of the area abandoned or only marginally used for agriculture. Some of the area was ameliorated by drainage during the Soviet era. The land is very open with few features subdividing it, although there are patches of forest, some quite substantial, and smaller features. The agricultural land lies mainly on the better soils, although some sandy areas are prone to water erosion. Some places close to the forest edge and abandoned for a few years already show signs of becoming colonised by trees.

The major ecological significance of the division between the two matrix types is that there are different ecological processes at work in each, different groups of wildlife are associated with each and there are different landscape dynamics at work. Thus they can be considered as two different but interconnected ecosystems. There are also many edge zones where each type influences the other.

Within each matrix type there is a range of different patches that are different enough from their surroundings to affect ecosystem functioning. The patches that are identified for the study area are divided into those found in the forest zone and those associated with the agricultural zone.

Forest zone patches

In this zone there are two sub-sets of patches – forest and non-forest types:

Forest patches: The most significant forest patches are:

- *Felled areas* – These are regular in shape with well-defined edges and are either completely clear of trees or retain a small proportion. They retain some of the shelter of the surrounding forest unless they are next to farmland.
- *Replanted/regenerated areas* – Herbaceous vegetation and young trees characterise this patch type, presenting browse for a number of herbivores.
- *Thicket-stage stands* – where light competition kills the undergrowth and trees are densely packed together.
- *Very old forest* – Includes dead wood, dead snags, a multilayered canopy and plant species restricted to ancient forest where light levels and site disturbance are low. This is to be found within the largest areas of matrix and along the lake edge.

Non-forest patches may include natural features that persist in the landscape, or those resulting from human activity and reliant on management for their continuing existence. The main types of patches are:

- *Lakes, forming part of the river system of the area* – These are sub-glacial lakes, quite deep, averaging 3–4 metres but up to 15 metres deep, and containing cold, clean water. They are mainly surrounded by old forest or are edged by wetland. Some lakes lie in relatively deep valleys and retain some shelter from their surroundings. Also, where rivers have been straightened there are some small ponds and remnants of old watercourses.

- *Bogs* – There is one example of an acid sphagnum raised bog in a basin. Vegetation includes a scatter of stunted pine trees and cotton grass (*Eriophorum* spp.). The water has a pH 3.5. This is a unique biotype for the Vidzeme upland area.
- *Swamps* – These lie in shallow basins in base rich clay morainic soils where ground water collects. They contain black peat and are vegetated with grasses, sedges (*Carex* spp.) and reeds. They are base rich, pH around 7 and are usually fringed with birch and black alder. They are extensive, up to 50 hectares in size.
- *Wet forest* – These are swamps that have become colonised by trees, some mainly by pine, and others by spruce and birch. The tree cover is gradually drying the wet soils and in some places these areas have been drained to improve tree growth.
- *Forest meadows* – Clearings in the forest that have been present for a long time. Some are completely surrounded by forest, others linked to farmland. They comprise a mixture of grasses and have been used over the years for hay production. Thus their open character is dependent on continued management. They frequently have an edge of birch and shrubs next to the forest.
- *Farmland* – As the forest matrix has expanded over the years some large patches of farmland of highly irregular shape have become isolated. Some of these areas have been managed as extensive permanent meadows, while other areas have been used for arable production. Many of the areas were intensively managed with drainage and fertiliser application although this has reduced recently.
- *Wet meadows* – These occur in association with parts of the river valleys and have traditionally been used for hay production. As permanent grassland they have high biodiversity value.
- *Castle mounds* – These are two grass- or shrub-covered man-made mounds dating back to the medieval period and now surrounded by forest which has encroached on them. They are unmanaged.

Agricultural zone patches
There are fewer patch types in this area, where the matrix is perhaps the most homogeneous. Because of the management history, most patches are either significantly affected by human activity or are a direct result of it. Compared with the forest patches these often have a long duration and remain firmly fixed in the same place. While they may be largely man-made, they may possess significant biodiversity value. The main types of patches are:

- *Patches of forest on old farmland, on steep slopes and in wet hollows* – These may date back a long time or may be younger, the result of land rationalisation when the collective farm system was set up. They represent the places where agriculture was not worth carrying out. They may be quite isolated or close to other areas of forest so that their biodiversity value is likely to vary.
- *Farmsteads with orchards and clumps of trees, gardens, etc.* – These are often quite old, perhaps several hundred years in terms of their use and development by the generations of occupants. The trees may be very old and of considerable biodiversity value if they contain hollow trunks. The old buildings may also provide homes for bats.
- *Village of Taurene – buildings, park, surfaced area, etc.* – The total area of the village is significant and is quite diverse. There are old trees and buildings that may provide habitat in the same way that the old farmsteads do. There may also be some pollution and disturbance by industry.
- *Old collective farmsteads – buildings, yards, etc.* – These are mainly abandoned and partly overgrown areas of dereliction, which may provide some habitat. Residual pollution from the

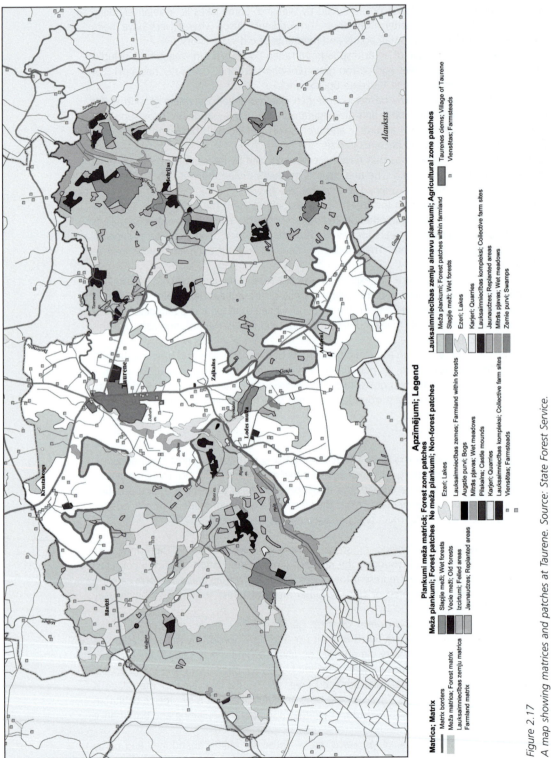

Matrica; Matrix

— Matrix borders
Meža matrica; Forest matrix
Lauksaimniecības zemju matrica
Farmland matrix

Plankumi meža matricā; Forest zone patches
Meža plankumi; Forest patches   Ne meža plankumi; Non-forest patches

Slapjie meži; Wet forests   Ezeri; Lakes
Vecie meži; Old forests   Lauksaimniecības zemes; Farmland within forests
Izcirtumi; Felled areas   Augstie purvi; Bogs
Jaunaudzes; Replanted areas   Mitrās pļavas; Wet meadows
                                       Pilskalns; Castle mounds
                                       Karjeri; Quarries
                                       Lauksaimniecības kompleksi; Collective farm sites
                                       Viensētas; Farmsteads

Apzīmējumi; Legend

Lauksaimniecības zemju ainavu plankumi; Agricultural zone patches

Meža plankumi; Forest patches within farmland   Taurenes ciems; Village of Taurene
Slapjie meži; Wet forests   Viensētas; Farmsteads
Ezeri; Lakes
Karjeri; Quarries
Lauksaimniecības kompleksi; Collective farm sites
Jaunaudzes; Replanted areas
Mitrās pļavas; Wet meadows
Zemie purvi; Swamps

Figure 2.17

A map showing matrices and patches at Taurene. Source: State Forest Service.

cattle and pig sheds is probably a lot lower than it used to be when slurry, rich in nitrates, once polluted some of the watercourses.

- *Quarries (sand and gravel)* – These are used by industry. They are constantly presenting freshly exposed sand and gravel to the wind and rain and may cause pollution if runoff can find its way into nearby streams.
- *Swampy areas* – These are areas of developing semi-natural vegetation where field drainage systems installed during the Soviet era are becoming blocked and ceasing to function. They represent a restoration of wetlands that were there previously and, in time, will add to the biodiversity potential of the area.
- *Shrubby areas* – These are developing on steep slopes and along now-blocked ditches as a result of less intensive management. They may represent early successional forest or remain as a shrubby structure for a long time.

Corridors

Corridors in Taurene also differ between the two matrix zones. Across the pagast as a whole there is a widely connected river system running through and connecting both matrix types. As well as the water, there is a well-connected series of riparian forests on steep slopes and wetlands along flatter valley bottoms, including some wet meadows. This forms a complex linear habitat. There are also the small water bodies associated with the straightening of some of the rivers such as the Pisla.

Forest zone corridors

The density of corridors in the forest matrix is lower than in the agricultural area at present. The corridors are:

- *Roads* – There are some main roads that pass through the area and a number of gravel roads, some of which connect agricultural patches within the forest matrix. There are few forest roads, although their number and density might be expected to increase as forest cuttings accelerate. There are some extensive areas, especially the larger matrix sections, that are almost road-free.
- *Rivers and streams* – These have a mainly natural structure and the riparian forest is well connected to the rest of the forest and to many of the wetland patches, etc.

Agriculture zone corridors

The majority of the corridors by length and density occur in this zone:

- *Road corridors* – Most of the road system in Taurene is associated with and lies in the agricultural land because it connects all the scattered settlements. There is a limited amount of tarred roads, the rest are gravel. These roads have calcium carbonate spread onto them to keep the dust down in summer. This can result in a high pH in the soils of areas next to them from water running off and dust blowing off. These road corridors also affect the distribution of weeds and invasive plants, such as the Siberian hogweed. These roads cross the rivers in surprisingly few places.
- *Forest corridors* – As the forest area has increased, some of the larger isolated patches have become linked by physical connections of strips of forest or by smaller patches creating a potential stepping-stone effect.
- *Electricity lines* – These mainly follow the roads to the settlements, only occasionally passing through forested areas to create linear openings. The transmission poles are used by storks and sparrow hawks for nesting or perching.

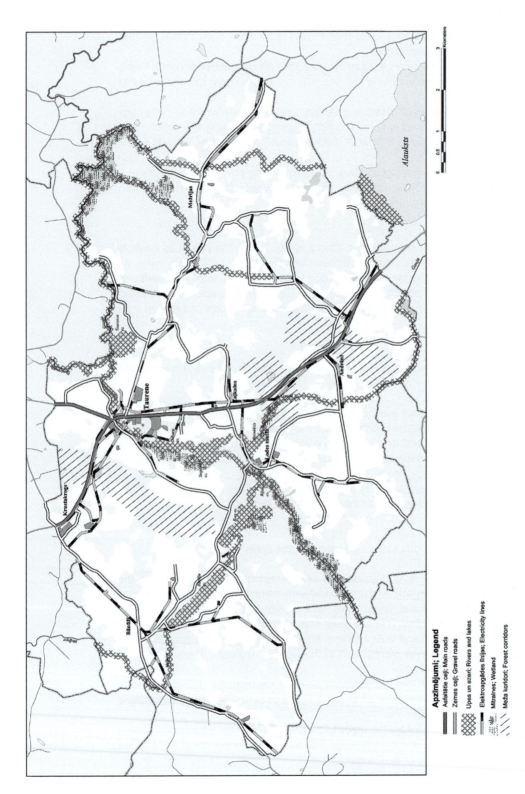

Apzīmējumi; Legend

Asfaltētie ceļi; Main roads

Zemes ceļi; Gravel roads

Upes un ezeri; Rivers and lakes

Elektroapgādes līnijas; Electricity lines

Mitraines; Wetland

Meža koridori; Forest corridors

Figure 2.18
A map showing corridors at Taurene. Source: State Forest Service.

2. Landscape flows

For the study area at Taurene a range of wildlife flows were selected to represent the main taxonomic groups that use the landscape, as well as water, pollution and people. The flows are as follows:

Abiotic flows
- Water is important given the hydrological system that is present. Water arrives as rain or snow, depending on the season, and is collected by the landform, penetrates the soil, is stored in the lakes and swamps and flows out through the river system as well as being lost through evapotranspiration. The water is clean and pure, more so now than in the past.
- Pollution arrives from a number of sources, as atmospheric pollution, from vehicles and generated within the study area. It may be less of a problem now than in the past due to the changes in industry and farming that have taken place in Latvia in general and the study area in particular.

Biotic flows
These are divided into mammals, birds, fish and plants. Whether the species is a keystone, umbrella, etc. is also noted.

Mammals

- Beaver and otter are both users of the riparian system and associated habitats. Beaver are active in affecting the structure of the streams and riparian forest while otter rely on fish from them. Both are examples of keystone species.
- Foxes live in the agricultural area or in the forest edge, preying on many small mammals and birds. The sandy soil is ideal for their earths.
- Elk supply part of the prey for wolf and are users of forest containing a range of successional stages as well as wetland habitats. Their browsing changes the forest structure. They are a keystone species.
- Wild boars are found in Taurene, though their numbers may have been significantly reduced in recent years due to the effects of swine fever. While relying on different successional stages of forest they also affect its structure. They are a keystone species.
- The racoon dog is an invasive predator that has recently moved into this part of Latvia, preferring to stick to the watercourses. It carries rabies and may become a serious pest, severely altering the relationship among many predator and prey species.
- Roe deer live in the forested areas, especially the mosaic landscapes where they can use open areas within reach of the forest edge.
- Grey hares live in the farmland areas, thriving in the mixed agricultural landscape of the open country where predators are fewer than around the forest edges. They provide a prey species for many animals. They are sensitive to changes in farming practices and may have benefited from the reduction in intensity of agriculture over the last few years.
- Pine marten prefer coniferous or mixed forest and hunt small birds and mammals.

Birds

- Common buzzards and northern goshawks live in the forest zone where they hunt for smaller mammals and also thrive on carrion. They need a mixed forest landscape to provide them with year-round supplies of food. They are an umbrella species.
- Sparrow hawks use the open farmland, young coniferous forest and forest edges where they hunt small mammals and birds. They often perch on the electricity poles. They are an umbrella species.
- Forest passerines are a guild of small birds that inhabit the forest. They include seed and insect feeders. The group includes robin, blackbird, song thrush, mistle thrush, warblers,

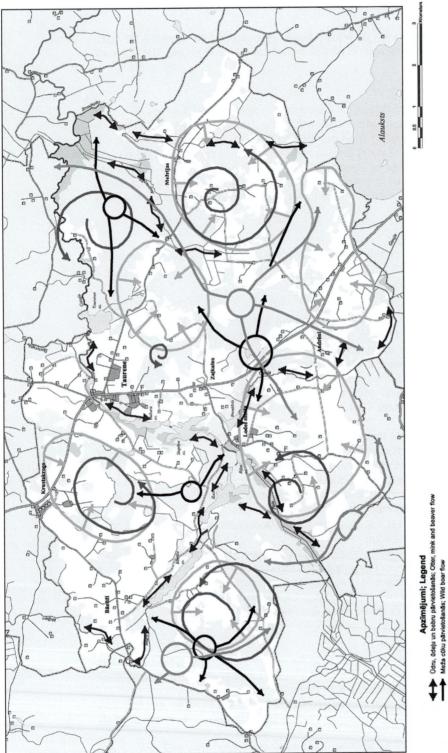

Odru, ūdeļu un bebru pārvietošanās; Otter, mink and beaver flow

Meža cūku pārvietošanās; Wild boar flow

Alņu pārvietošanās; Elk flow

Stirnu pārvietošanās; Roe deer flow

Pelēko zaķu pārvietošanās; Grey hare flow

Jenotsuņu pārvietošanās; Raccoon dog flow

Figure 2.19

A map presenting diagrammatic, rather than actual, flow patterns for mammals at Taurene, Latvia. Source: State Forest Service.

chiffchaff, goldcrest, tits, pied flycatcher, raven (not a passerine but added to the group), chaffinch, nutcracker, jay (not a passerine) and golden oriole. Taken as a whole they have a wide range of habitat needs and are best catered for by a mixed forest of all successional stages.

- Farmland passerines are a group or guild that mainly use the farmland and associated habitats. The group includes skylark, meadow pipit, yellow wagtail, pied wagtail, trash nightingale, warblers, corncrake and sparrow. Some of these are rare and endangered in other parts of Europe but remain common in Latvia.
- Game birds are mainly found in the forest and also need a mixed landscape of different successional stages and associated habitats such as wetlands. The group includes woodcock.
- The white stork is a typical and characteristic bird of the farmland zone, often nesting on electricity transmission poles. It eats small fish and amphibians from the wet areas and ditches that are currently increasing in area.
- Waterfowl are associated with all the bodies of water found in the area. The group includes great crested grebe, grey heron, swan, goose, mallard, teal, garganey, goosander and coot.

Fish

Fish include species that use only still water, only running water or both. The group includes tench and crucian carp that prefer still water, trout, chub, grayling and gudgeon that prefer flowing water and pike, pike-perch, perch, roach, rudd, ruffe, bream and burbot that are happy in both waters.

Plants

The sole plant that is included is the invasive Siberian hogweed, once tested for use as animal fodder and now spreading along the roadsides and on the point of arriving in the study area.

People

People use the landscape in a variety of ways: they live there, they work on the land or in the forest or they hunt and fish and take part in other pastimes. There is less use of many areas, especially the farmland, than there was in the recent past and there may be more intensive use of the forest in the near future than there was in the past.

3. Interaction between structures and flows

This was undertaken using the spreadsheet approach described earlier in the chapter (see Chapter 3, Table 3.1).

4. Linkages between the study area and the wider landscape

In the Taurene study area one of the main linkages is the headwaters of the Gauja river system which rises in the area and flows out from it, taking clean water captured by the landscape. There are also roads that pass through, the main one being the Cesis-Madona road. Many other roads are minor and do not have much traffic on them. A major linkage is also the matrix forest that provides the connectivity to allow wolves to move around the landscape. The agricultural matrix also connects beyond the study area boundary though not many flows will use this wider connection.

5. Current functioning of the landscape

From the analysis undertaken and on the basis of the scale of resolution, the choice of flows and the assumptions used, it is possible to come to a view on the current ecological state of the Taurene study area. The following conclusions can tentatively be made:

1   In general terms the ecosystem is in quite a healthy state. There appear to be few major problems and, in fact, things are probably better than they were some years ago. Due to the reduction in intensity of management of the farmland there is less pollution. For example, there is now a lower risk of nitrates entering the ground water, in the absence of pesticide use, and a reduction in the efficiency of field drainage systems. All this is likely to be helpful to the population of plants and insects upon which birds and their predators feed as well as other herbivores such as mammals. If farmland persists in its abandoned state and gradually converts to forest, the early successional stages that develop will contain high biodiversity potential for some time to come. Eventually some habitat loss will occur to affect population levels of open ground species and this needs to be evaluated.

2   The forest matrix includes some large contiguous blocks, some of which are part of more extensive areas beyond the study area boundary. These tend also to be the older stands, so that the considerable value of extensive old forest is well represented. However, there is little in the way of truly ancient forest, so that if it is desirable to increase the proportion in an "old growth" condition, some key areas will need to be set aside and managed as such. At present forest fellings cover only a small area, but this could easily increase quite quickly. Without any guiding principles to follow, such cutting could fragment the forest and lead to a loss of its quality. At the same time, however, there is also a possible shortfall in the amount of early successional forest, so that by using a combination of new afforestation/ natural regeneration and carefully planned forest cutting and replanting a wider range of successional stages could be introduced.

3   The structure of the landscape pattern is such that there is a high degree of connectivity. This means that wildlife can move freely around in both the farmland and the forest zones. Continuous corridors are valuable in some places, especially the river systems, together with riparian forest, swamps and wet meadows. In other places it is sufficient to maintain patches of forest that are close enough together to use as "stepping stones" to move from one forest area to another. The exact nature of the size and distances between these patches cannot yet be ascertained. Such an arrangement also maintains connectivity between the farmland intact.

4   With the exception of two invasive species, the racoon dog and Siberian hogweed, all the wildlife chosen as representative of ecosystem functioning appears to have their habitat requirements satisfied. However, in the absence of on-site studies to measure population it is not certain to what extent the population levels of these and other wildlife are maintained. Some, such as those using open farmland, could easily be expanding while others, primarily game species, could be depressed by hunting. Thus, while it is possible to conclude that habitat diversity is such to provide for a high species diversity, the actual position may not be as good. However, the landscape ecological process is about managing landscapes, not species, so this is a natural limitation of the planning process.

5   Two invasive species, racoon dog and Siberian hogweed, were examined. The racoon dog is an alien predator and may pose a serious threat to the population of some species and to the predator-prey balance. It is not clear what effect it currently has but it clearly needs to be watched. The hogweed is only coming into the study area but it also has the potential to alter the biodiversity value of certain biotopes and its progress too must be studied.

6   From the analysis of the interaction of flows and structures, there are several structural types that appear to be more important for a range of biotic and abiotic flows. Some of these naturally form an integrated network, others need to be within a matrix but otherwise do not form part of a network. These critical elements are as follows:

    •   Ancient forest in connection with the riparian network and the large forest matrix.
    •   Riparian network of forest, meadows, water bodies, swamps, rivers and streams.
    •   Matrix forest in late successional condition.
    •   Forest meadows.
    •   Wet forest.
    •   Bog.

This concludes the description of the existing landscape; the next stage is to explore the dynamic aspects of the landscape.

6. Disturbance and succession

In Taurene, because of the fact that it is a managed landscape, the disturbance types are divided into those that are natural and those caused by human activity. The natural ones are: fire, insects, fungi, mammals (elk, beaver, wild boar), floods, snow, soil erosion and cold air/frost. The human-caused disturbances are: fire, forest cutting, cultivation, replanting trees and forest management. Their effects are also separated between the forest zone and farmland zone.

Disturbance
The main conclusions from the analysis of disturbance are as follows:
Forest zone. The natural disturbances mainly act as endemic processes gradually changing the forest structure and composition. They are mostly small in scale and while they are intense locally, they only affect a small proportion of the forest at any one time. They can be described as gap-phase agents creating circumstances where the successional processes can advance as the canopy is opened, allowing understorey and advance regeneration to develop, thus making forest structure more complex. In this way the forest is indeed mainly a matrix with the patches being either produced by human activity or mainly non-forest cover types. All the different agents ensure that different parts of the structure are disturbed – the canopy trees, the understorey, the ground layer and the soil itself. They also contribute to the dead and decaying wood content of the ecosystem.

By contrast, the human activities tend to be more concentrated, with cutting being more intense in terms of what is removed and at a larger scale so that the microclimate and light levels are changed quite considerably. Management tends to result in simpler structures and less diverse species composition as well as removing much of the candidate dead wood material.

The implications are that the forest could be managed differently to favour the ecology of the landscape. Instead of clearcutting, a silvicultural practice based on small-scale group fellings would be much more soft on the forest and help maintain the compositional and structural complexity found naturally. Larger-scale compartments or coupes could be set out, within which a number of small groups could be felled. Several years would elapse before a repeat visit was made.
Farmland zone. Here, the disturbances are primarily those of grass cutting for hay and cultivation of arable land. This has the effect of preventing forest succession from taking place, and its cessation can be seen to allow forest succession to start very soon. Other disturbances include flooding, soil erosion and some mammalian effects.

Succession
In Taurene there are two pathways to consider and compare in the forest zone: natural regeneration following natural disturbance and planting after felling. In the farmland zone, natural colonisation of the abandoned land follows another successional route.
Succession after natural disturbance starts in the most open areas, the stages of development being as follows:

1   Stand initation occurs when, for example, a rotten tree falls and breaks a number of surrounding trees to create a gap where light enters. This allows seed to germinate and for a dense stand of pioneer species such as birch or alder to develop. Any spruce that has survived in the sub-canopy or as an understorey will be released into faster growth, so that a multilayered stand may develop. This stage may endure for 10–15 years. Perhaps some 10–15% of the canopy of the forest is affected in this way at any given time. Dead wood accumulation becomes an increasing feature of the forest.
2   Where a stand is dense, stem exclusion starts and may continue in the group of regenerating trees for the next 20 years, until only a few survive, including the spruce which has a greater

competitive advantage, so that the pioneer species may be fewer than at the time of initial colonisation. At the same time, new gaps are being created and new stands initiated.

3 The even-aged group continues to grow and becomes a part of the continuous canopy of the forest by some 40–60 years of age, casting a fairly dense shade beneath, while older trees on the margins continue to die and gradually give way to new groups.

4 The forest assumes a matrix structure where there is a more or less continuous canopy represented by cohorts of even-aged trees interspersed with gaps containing regeneration at various phases of development. The oldest cohorts contain dead and dying trees and present a lot of dead wood on the ground, while the younger canopy elements have a more open stand structure with less dead wood and sparse advanced regeneration. This structure has become old growth and has the stable yet dynamic character of a shifting mosaic. This is the structure that provides high habitat values for species such as the black stork.

Following *timber harvest in managed stands*, a very different successional process takes place:

1 The felled site is open to the sky and sufficient light is available to enable a flush of herbaceous and grass species to invade and create a field layer. Spruce trees are planted at a regular spacing. Dead wood is restricted to small diameter branches and tops, which soon rot and provide only temporary habitat value. Over a 10–15-year period the spruce grows until it reaches thicket or canopy closure stage, at which point almost all the ground layer of vegetation has been shaded out. Birch may have become a component, depending on the degree of management, such as cleaning, that the stand has received.

2 Stem exclusion proceeds for the next 10 to 20 years, although because of the initial density of planting this is not so ruthless at first. The light levels beneath the canopy remain low and no understorey can develop.

3 Between 25 and 70 years, depending on management, the stand may be thinned in such a way that timber volume is concentrated on fewer, larger trees and the light levels beneath the canopy remain low, continuing to exclude an understorey, unless an element of birch has been allowed to remain in the canopy.

4 At some time between 80 and 100 years the stand will probably be felled and replanted once more. At this stage, harvest takes place when the stand is at a mid-successional stage when compared with a naturally succeeding stand.

5 If the stand is not felled, then canopy break-up will eventually start and gaps open into which spruce seed will be able to germinate and develop, starting the process of understorey reinitiation. If birch has been a component, it will also thrive and probably assume a greater proportion of the canopy in the next phase of development. Eventually, over the next 100 years, a multi-age forest stand will develop that begins to contain dead and dying trees and dead wood, and starts to function more as a natural stand, providing greater habitat value.

The main stages of *succession from abandoned field to forest* are as follows:

1 Stand initiation starts with spruce, birch and white alder seed blowing onto the abandoned field, so that by 10 years a shrub layer has developed. Agricultural species have become shaded out and some herbaceous perennials are a component of the herb layer. The alder and birch grow quickly and dominate the stands at this stage.

2 By between 30 and 39 years, the stem exclusion stage is well under way. Depending on the site and balance of seed sources the canopy may consist of white alder and spruce as co-dominants in varying percentages, or be a white alder/birch canopy with a spruce understorey. Some stands, perhaps on the wetter sites, may also be pure white alder.

3 By around 40–49 years several variants may have developed, depending on site conditions. The pure alder stands will have developed a spruce understorey; the alder/spruce canopy will

be maintained; the alder/birch canopy will probably have become mainly birch, with either a dense or a sparse spruce understorey.

4   After another 10 years, by age 50–59, most stands will have a strong component of spruce in the canopy, apart from the birch stands with a sparse spruce understorey. It is likely that some management will have affected some stands, perhaps by removing the alder, so that the spruce becomes the main canopy element.

5   After some 70–79 years of development the stands will probably have become one of three variants: a more or less pure spruce canopy, with some understorey reinitiation of white alder, a mixed spruce/birch canopy with a sparse understorey, mainly a shade-tolerant ground layer and a birch canopy with some spruce still in the understorey or sub-canopy. There will be some accumulation of dead wood in the stands, both in terms of snags formed from trees shaded out and wood lying on the forest floor.

6   By between 80 and 100 years the stands mature. A spruce understorey appears beneath the stands dominated by a spruce canopy, while the other stands retain their canopy composition. Dead wood continues to accumulate.

7   After a variable number of years, up to 200 since colonisation, canopy break-up and greater degrees of disturbance start, so that the second generation of the stands start to develop within the gradually more open pockets. Dead wood, including large old trees and whole fallen trees, becomes a significant component of the stand structure.

The implications of this analysis for Taurene are that the managed forests are not allowed to follow all stages to old growth and that management practices such as thinning and felling both advance succession and revert it to the beginning earlier than natural rates. This prevents old growth from developing and maintains mid-successional stages, which are not necessarily the most ecologically valuable, in a higher proportion in the landscape. Management should consider these implications and the need for added old growth; future management directions are considered further in the next section.

7. Desired future condition

From the analysis it is clear that there are two main, significantly important elements that should comprise part of the desired future condition. These are the hydrological/riparian network and old growth forest.

1   *The hydrological network* comprises all the streams, rivers, lakes, swamps and the riparian forest and meadows associated with them. These are delineated as a continuous network on the map. However, they are not currently complete or protected and clearly include many owners. As well as the network itself there needs to be a buffer zone. This may be more difficult to achieve in the farmland zone than the forest zone, although with the low intensity of agriculture it may not be a problem.

2   *Old growth.* As timber harvest in Taurene increases, the maturing forests now entering the understorey reinitiation stage are likely to be harvested in patches and replanted for future timber production. This means that old growth is unlikely to become a feature of the landscape unless specifically included in management plans. The questions are how long will it take to develop, where should it be located and how much of it should there be?

Old growth normally develops in places least likely to suffer catastrophic disturbance, of which there are few in Taurene. It also develops on richer sites and in inaccessible places, given the long history of human use of the landscape. Small isolated areas are of little value and places where colonisation of stands by species that rely on undisturbed forest is unlikely can also be ruled out. Thus, the places where old growth should be located are in association with existing undisturbed sites, within large contiguous forest matrix areas and in places where management for old growth is unlikely to conflict with other land management objectives.

On the basis of the analysis of landscape function, the low proportion of old growth was noted. Since the major large-matrix forest areas are owned and managed by the state forest enterprise, these offer the potential to provide substantial reserve areas to develop into old growth for the species which require low levels of human disturbance. Road construction should be avoided and existing forest roads removed from these reserve areas.

The riparian forest also offers existing ancient forest on steep slopes above water bodies and expansion of these areas by management of the surrounding forest should be a priority. The areas for old growth management are recorded on the map.

3   *The forest matrix* should be managed by smaller-scale group selection systems, rather than clearcutting. This will emulate the natural structure developed by the endemic small-scale disturbance regime and ensure greater connectivity across the forest. This management is especially important around the old growth reserves and to maintain corridors through the matrix and across the farmland matrix by using the forest patches as stepping stones. It is less important in other areas where patch clearfelling can be retained as a system. Road construction and disturbance to the matrix, especially near or into the old growth reserves and their buffers, should be carefully planned and controlled.

4   *The farmland matrix* should be maintained in a large-scale open condition by restricting afforestation or colonisation to the edges of or extensions to the existing forest. Small, isolated patches of forest are of less value to wildlife and fragment the open farmland. Management by annual mowing to retain strategic open areas will be necessary if agriculture is not active. Forest meadows in the forest matrix area are also valuable and should be managed for their preservation.

Conclusions

This chapter has listed and explored some of the most important emerging concepts from landscape ecology, conservation biology and ecological forestry that are incorporated into the forest landscape design process, as demonstrated in the Taurene example. There are many additional concepts that we have failed to touch upon, but the ones listed are the basic building blocks for forest landscape design. They provide a descriptive system for identifying the key landscape components that the design will seek to organise, and a conceptually sound, though still rapidly evolving technical basis for evaluation of alternatives. In Parts II and III we will go into greater detail on the use of these concepts in formulating designs and evaluating alternatives, and use more case studies to demonstrate their application.

Figure 2.20
a) and b) Views of Taurene showing many of the features of the landscape, such as forest, farmsteads, fields (some worked, some abandoned), small patches of bushes, a road and powerline.

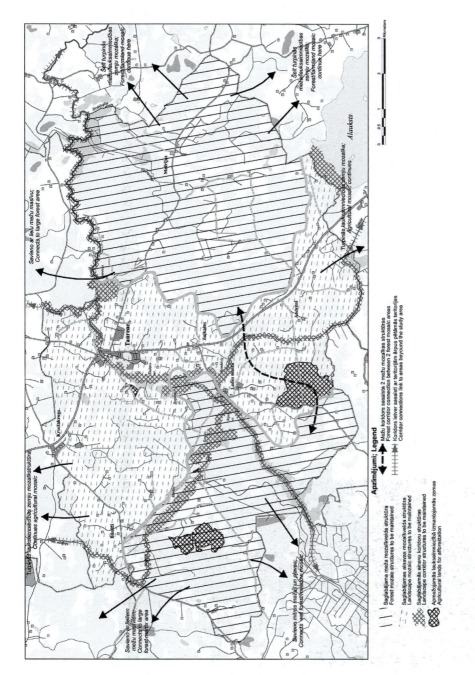

Figure 2.21

A map showing the desired future condition for Taurene after the completion of the analysis. Source: State Forest Service.

Chapter 3
Key principles of forest aesthetics

Introduction

As we have already seen, the term landscape can be used in a variety of ways: as a scale of planning, as an approach to forest ecology and as the perceptual realm of the senses. The earliest use of the term, originally a Dutch word, means a "prospect of scenery that can be taken in at a glance from one point of view." Thus landscape is both a physical and an experiential concept. The last chapter concentrated on its physical and ecological attributes. In this chapter we are going to examine some of the concepts and practical aspects of the experiential sense of landscape, review the different approaches adopted to try to manage landscape change and/or protect scenery and present a method for analysing landscape character and sensitivity that can be applied as part of the integrated design process to be presented in Chapter 5. One of the important interactions we have with our surroundings or environment is the aesthetic.

The nature of forest aesthetics

Aesthetics refers to things perceived by our senses and our reactions to such perceptions. Forests are particularly important elements in any landscape because of the way they tend to affect what and how much we perceive. The main reason for this is the height of trees, which are able to control views and enclose space. It is useful to consider two related aspects of forestry aesthetics. These are the traditional, scenic mode of viewing and the participatory, or multi-sensory mode.

The scenic mode represents the typical static viewpoint offering a prospect of scenery. The forest in this aesthetic mode acts as a component of the vista; it is an external view over the canopy of the forest. We perceive the landscape primarily visually. We comprehend patterns made up of shapes, colours, textures or lines and their different arrangements. In a natural forest, unaffected by deliberate human activity such as logging (but possibly affected indirectly through fire or disease control), these patterns arise from the interaction of ecological processes with climate and landform.

In landscapes changed by human activity the patterns remain influenced to greater or lesser degrees by natural processes, but also include a range of human-caused patterns. These include logging, clearing for agriculture, plantation forestry and sundry associated activities such as road construction. Some of these patterns may have a long history, such as those of small woodlands linked by hedges in England or France, and may never have been planned with a pleasing aesthetic result in mind, although many people find them very attractive. Other patterns result primarily from commercial forestry activities such as logging. There may be some relationship between logged areas and landscape, for example if the loggers have been only interested in a particular species or size of tree, or the pattern may be a product of the machinery employed to harvest the trees and bears no relationship with the landscape.

The aesthetic experience of landscape tends to be one of two varieties – a sense of beauty or of the sublime. The sense of beauty is a positive emotion induced when we find the patterns we see arranged in such a way that they fit together harmoniously and seem all of a piece with no jarring elements (how theoretical beauty relates to peoples' preferences is discussed below). The sublime is experienced when we are faced with limitless expanses of a scale and magnitude that dwarfs us, and when the forces of nature seem to overwhelm us. This is not necessarily a positive emotion but it can nevertheless be powerful.

A good deal of work has been carried out over the years attempting, quite successfully, to find out what kind of forest landscape people actually prefer. This preference research has often been based on the use of photographs of different scenes, the researchers asking members of the public to evaluate them by expressing which they like best or least, for example, and why that might be. It can be argued that most people tend to prefer certain kinds of scenery although the extent to which this is cross-cultural is not completely clear. Two of the main researchers in this field, the Americans Steven and Rachel Kaplan, have identified some key factors, which they believe account for much of the expressed preferences. These factors are not objectively measurable attributes of the scenes under evaluation, but are cognitive factors, that is, they are concerned with how we make sense of the scene and interpret it. These factors can be summarised as follows:

- *Coherence*, where the scene is comprehensible, the patterns are understood and the structure is clear.
- *Legibility*, where the landscape is "readable" and the relationships between different elements are clear.
- *Complexity*, where there is an abundance of variety, where the structure is not simple.
- *Mystery*, where parts of the scene are hidden from direct view, where it has to be explored, where not all of it is visible at once.

It should theoretically be possible to identify characteristics in scenes that contribute to expressed preferences by invoking the cognitive factors in a positive way. In fact there are some useful links between the cognitive factors and design principles, as will be explained later in the chapter.

There is no easily definable rule that equates good landscape as being natural or bad landscape as being caused by human activities, although some people might see it that way. This part of the argument gets easily confused with ethical debates on, for example, the means, scale and sustainability of timber harvest. Ecology and aesthetics are related, but they are not necessarily mutually dependent. An area of forest can look attractive, but have serious ecological problems. Conversely, it can look unattractive (immediately after a fire for example) but be in a good state of health. However, it is a challenge in its own right to manage change in landscapes while retaining, or in some cases enhancing, the aesthetic quality of the scenery (this will be further explained later in this chapter).

The second mode of aesthetic interaction is the participatory aesthetic. This makes use of all our senses (sight, hearing, smell, taste and the complex of *haptic* or *kinaesthetic* senses such as

*Figure 3.1
This view in Austria represents a prospect of scenery where we concentrate on the visual impression of the exterior of the forest canopy. There are various patterns visible in the forest as a result of clearance for farming, timber cutting and so on.*

*Figure 3.2
This scene shows
the internal
landscape
beneath the
forest canopy.
This example is a
primeval forest
of beech and
silver fir in
Slovenia, which
creates a space
amongst the tree
trunks.*

touch, temperature/moisture detection and balance). It is participatory because we are actually in the landscape as opposed to seeing it from a distant viewpoint. This is where the forest encloses us and can shut us off from external surroundings. Such close proximity to the forest, experienced perhaps from a trail, can be very rewarding. We can be assailed by many stimuli – the sight of the trees, shrubs, flowers, rocks, wildlife; the sound of the wind in the trees, of bubbling water, of birdsong; the smell of hot tree resin, flowers, fresh soil; the tactile surface of bark or rock, the coolness and dankness of deep shade and the warmth and dryness of bright sun. In a forest all these stimuli work together harmoniously (in comparison with many urban landscapes, perhaps) and induce the sense of beauty described above.

Depending on the topography, some places lend themselves more towards the scenic mode of aesthetic experience, others to the participatory mode. In mountainous or hilly countries, where unforested mountain tops, lakes or cleared valley bottoms enable large-scale distant views to be obtained, the scenic mode is often dominant. Forested hills or mountainsides may surround settlements; scenic routes used by tourists may exploit these areas and both the beauty and sense of the sublime make them prime candidates for national parks or other designations of protected status. It is also possible to experience the participatory aesthetic but somehow the open vistas and dramatic views are far more significant. The Canadian and American Rockies, the Swiss and Austrian Alps, the Southern Alps of New Zealand, the smaller ranges of the Appalachians, the Scottish Highlands and so on all yield examples of splendid forest scenery that are best experienced in this way.

Countries with lesser topography often offer a completely different experience and the relationship of the inhabitants to the forest is often more direct and integral to their everyday lives. Finland is dominated by forests and all the farms seem to occupy spaces carved out of it. The forest is an ever present, enveloping experience. The same applies to large parts of Sweden, Russia west of the Urals or the northern Canadian Shield.

Thus, the consideration of forest aesthetics must deal with both modes of experience. In the past, aesthetic concerns were raised by the public over large-scale logging in highly visible mountainous areas of the US National Forest System, or the extensive new conifer plantation re-afforestation on the open hillsides of Britain, New Zealand or Chile. Thus the major developments in forest aesthetics tended to concentrate on scenery management and designing the appearance of the forest in the wider landscape. It is much more recently that significant attention has been paid to the internal design of flatter areas where the participatory aesthetic tends to dominate.

Most early attempts at forestry scenic management involved trying to screen or hide forestry activities from view. This included retaining belts of trees along roadsides to screen logging from view or, after identifying the most scenically important travel corridors, ensuring that all logging was out of sight over the hill, in side valleys or in hollows. These strategies persist in many places but they can have unfortunate side effects, such as interrupting ecological functioning. Forest managers who rely on these techniques also eventually find there is nowhere left to hide logging, yet wood still has to be cut from somewhere to supply the mills. In addition, many members of the public are outraged at the notion that forest managers are "hiding" what they are doing behind vegetation or topographic screening.

More recent approaches have been much more proactive in planning and designing for forest landscape change. This was first developed in order to direct new afforestation in Britain in the 1960s, but has since been successfully applied to planning and design of logging in many countries. This approach accepts that landscapes are going to change (as the understanding of ecological sustainability demonstrates, all landscapes change anyway; preserving them in their present state is not a real option). The issue is how and at what pace they change. The key is to plan human-induced change, whether through logging, prescribed burning or other means, in ways that will be socially acceptable and ecologically responsible.

Approaches to visual aesthetics in forest landscapes

The issue of visual resource management in forestry largely came to prominence after the Second World War, as the increasing network of highways and mass car ownership enabled large numbers of people to explore the countryside or natural landscapes of North America, Europe and other developed countries. This period also coincided with greatly increased forestry activity such as afforestation programmes in Britain, Ireland and New Zealand and with increasing levels of timber harvest in the USA, Canada and Scandinavia, especially on public lands. By the mid-1960s, public concerns over the appearance of both newly planted forests and logging operations had increased, prompting agencies such as the British Forestry Commission and the US Forest Service to look for ways in which to safeguard the landscape. The models originally developed in Britain and the USA followed different routes, partly due to the scale of the forests and forest operations but also reflecting the type of forestry.

In the USA, logging took place (and still mainly takes place, at least in the National Forest System) in extensive natural forests, where the visual impact of sudden changes to the scenery, occurring over a large-scale landscape, can be very great. While the impact of an individual cutblock could have a negative visual effect, the cumulative impact over large areas was often considered to be greater still. This prompted the development of an approach to suit the scale of the landscape, the extent of the forest and the need to try to control the rate of landscape change and its degree of impact, an approach that is generally referred to as "Visual Resource Management" or VRM. This approach aimed to manage the level of impact of logging on the natural scenery, especially as seen from key viewpoints, and this led to a highly developed visual management system intended to prioritise areas within large tracts for different levels of scenic protection.

In Britain, the programme of afforestation led to significant landscape change but each new planting project was relatively self-contained and there was some degree of flexibility over the layout of such forests. Owing to the fact that conserving an existing landscape was not an option, and considering the freedom to create new landscapes offered to the foresters of the time, a design-led model was developed following the appointment of a landscape architect, Sylvia Crowe, to be a consultant to the Forestry Commission, which aimed to use creative design to produce new forests of good visual quality to fit into the landscape, especially as seen from significant public viewpoints. This is referred to as the "proactive design approach".

Although these approaches are generally quite different, they do have some similarities. Both models emphasised the value of the scenic, external mode of landscape experience (see above). Viewers observe the scene from a distance and the aesthetic quality is associated with the notion of

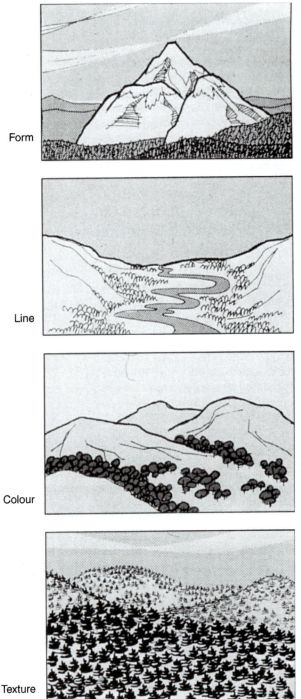

Form

Line

Colour

Texture

the picturesque, the landscape aesthetic model that emerged in the late eighteenth century and which for the first time celebrated the wild beauty and sublime qualities of untouched Nature. Viewers expect to see "natural" scenery and the presence of unnatural elements disturbs the quality (although in many countries, especially Britain, there is no actual natural scenery left). Design aspects were included in the American system, though not so highly developed and not so rigorously applied, while in Britain a degree of management based on relative sensitivity or landscape importance (based on visibility and numbers of likely observers) meant that some areas were designed more carefully than others.

Both Britain and the USA have hills and mountains that are highly visible from roads, settlements, hiking routes and mountain summits, which partly explains the importance given to the scenic external mode of viewing and appreciation. Forest policymakers and managers in other heavily forested countries with less dramatic topography, such as Sweden or Finland, where views of the landscape are confined more to forest interiors, did not feel a need to develop visual resource management programmes to the same degree until much more recently. Nor did countries with forested mountains and high scenic qualities, such as Alpine countries, where the forests are managed by continuous cover or selection types of silviculture system. In these areas, although the landscape is highly visible and valued for its scenic, picturesque qualities, dramatic changes to the landscape tend not to take place because there is no clearcutting, so that the perception of a never-changing natural landscape can be maintained.

The visual management approach

The US Forest Service developed a comprehensive system, starting in the 1960s and continuing through the 1970s. The first stage of development was to hire a researcher, R. Burton Litton, Jnr., to look at forest landscapes and to develop a visual resource inventory and evaluation programme. Following this a basic description, analysis and classification system was developed and implemented. The system was presented in a series of booklets, the first of which explained how to consider a landscape, and how to describe and analyse its various characteristics. It introduced some basic design principles, such as line, form, colour and texture and considered the effect of scale, light, time, viewpoint, etc. on how landscapes are perceived in order to introduce some rigour and rationale into the subject.

Figure 3.3
An example from the early US Forest Service visual landscape management system, presenting basic design principles for use in forestry. Source: US Forest Service.

The second part presented a method of landscape inventory and analysis so that a forest area could be divided into landscape units, emerging from an overlay analysis, each with its own combination of visual character, visibility and level of sensitivity. This enabled planners to ascribe a series of visual quality objectives (VQOs) to each landscape unit. For example, in a highly sensitive landscape – one that was very visible to a lot of people, with a high visual quality (often meaning an undisturbed canopy of climax mature forest) and no visible logging – the foreground area might be given the VQO of "Preservation" where no logging was allowed, and the middle ground area, a little further from the viewer, given a VQO of "Partial Retention", where some logging was permitted as long as it did not disrupt the character of the area and any visible logging was subordinate to the rest of the landscape. The base level of VQO for the least sensitive areas was always "Maximum Modification", which meant that logging could proceed free of visual constraints. Subsequent books in the series dealt with more detailed aspects such as the design and layout of logging, of roads, utilities and range issues to minimise visual impact.

The implementation of this system was often prioritised for the visual portion of the landscape as seen from roads and key viewpoints (given that Maximum Modification was the baseline VQO supposed to be applied everywhere). Viewshed analysis identified what was and was not visible from a set number of viewpoints and the landscape was divided into units, analysed and given visual quality objectives. Visual quality objectives, set by experts, therefore represented the desired requirements for the visual resource, to be incorporated into the wider forest planning process which included many other resources, such as timber, water, wildlife, recreation, range and so on. All these competing requirements had to be balanced and this led, in some circumstances, to the visual resource being overridden by other values.

During the 1990s the system, then some 25 or more years old, was in need of an overhaul and the result was the Scenery Management System. This incorporated much of what was good about the original system but placed more emphasis on landscape character, which recognised the uniqueness of every area and the contribution made to it of cultural as well as natural features.

The current system starts with an Ecological Unit Description, sometimes called a mapping unit description, which represents the common starting point for both the Scenery Management System and for Ecosystem Planning. This was introduced in order to integrate scenery and ecosystem management to some degree and to overcome perceptions that the two aspects could be in conflict with one another (we consider that the integrated design process we present in this book is better at achieving this fusion). From the Ecological Unit Description an objective (i.e. factual) description of biological and physical elements is extracted, and combined with attributes for landscape character to produce a landscape character description. The idea is to be able to group the combination of scenic attributes that make each landscape distinct, identifiable and unique. This description provides the baseline or reference for the next stages, of defining scenic attractiveness classes and degrees of scenic integrity.

Scenic attractiveness classes are used to determine degrees of relative scenic value of different areas within a particular landscape character zone. There are three possible classes: A – Distinctive, B – Typical and C – Indistinctive. The method of calculating relative scenic value is to describe the landscape elements that make up each character zone in terms of line, form, colour, texture and composition. Scenic integrity indicates the degree of visual disruption of landscape character. If

Figure 3.4
An example of a map showing Scenic Classes derived using the Scenic Management System. Source: US Forest Service.

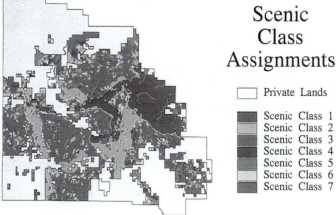

Scenic
Class
Assignments

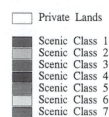

☐ Private Lands

Scenic Class 1
Scenic Class 2
Scenic Class 3
Scenic Class 4
Scenic Class 5
Scenic Class 6
Scenic Class 7

a landscape has very low disruption it has a very high degree of scenic integrity and vice versa. There are six classes of scenic integrity, from Very High to Unacceptably Low.

The next stage examines the visibility of the landscape and takes into account two factors – firstly, the relative importance to the public of various parts of the landscape and secondly, the relative sensitivity of the scene based on its distance from observers. Relative importance to the public may come from a variety of sources, including special perception and preference studies. Constituent analysis (a technique for evaluating the views of different people) is used to gauge a level of public concern about aesthetic qualities, assessed as high, medium or low. Distance zones of foreground, middle ground or background are used to classify relative sensitivity.

The scenic attractiveness classes and landscape visibility data are combined to create Scenic Classes, ranging from 1 to 7, which indicate the relative importance or value of discrete landscape areas. These scenic classes are used during forest planning.

It is also possible to prepare a landscape value map using overlays in a GIS, in order to present the information spatially for use in various planning procedures. During the development of a plan for a forest area, the descriptive aspects of landscape character are used to develop landscape character options that are deemed to be realistic within the overall multi-objective forest plan. Once the forest plan has been adopted, the landscape character description becomes a management goal and the scenic integrity levels become scenic integrity objectives. These are similar in nature to the previous visual quality objectives. The idea is that a given level of scenic integrity should not be reduced by forest activities, although it is also recognised that the degree of scenic integrity can change over time through natural landscape processes. In order to meet a specific integrity level and to carry out logging or road construction some design is needed. This is where the system is weakest: it provides very little guidance and few examples of how to achieve a satisfactory result.

The VRM approach was also adopted by the Ministry of Forests of British Columbia, Canada. British Columbia is mountainous, densely forested and also relies heavily on the timber industry for its economic well-being. In the early 1970s, managers recognised that as more and more logging became visible on prominent mountainsides and from significant tourist routes, whether roads or shipping lanes, some kind of visual resource management was necessary. A landscape forester was charged, in 1979, with developing this based on the US Forest Service system, with many adaptations.

The system as originally developed and applied consisted of five steps. Step 1 is landscape inventory, where three elements are identified. The first is the identification of the extent of the landscape visible from established viewpoints such as roads, settlements and recreation areas. The second element is the suite of landscape features present, both natural and man-made. The third element is landscape sensitivity, calculated from physical factors and viewer-related factors such as numbers of viewers, viewing distance, duration and perception.

Step 2 is landscape analysis, consisting of detailed mapping, the recommendation of VQOs and the final establishment of VQOs by the forest manager, taking into account the other resource factors that have to be balanced against aesthetics. Step 3 is design and layout of roads and cutblocks, step 4 is logging and silvicultural practices, and step 5 is follow-up.

*Figure 3.5
A section from the British Columbia Forest Landscape Handbook demonstrating the five steps of the original version. Source: Ministry of Forests.*

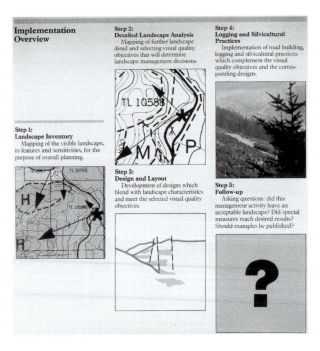

In 1997 a revised method was approved, moving to a more quantitative, numeric and prescriptive system of Visual Sensitivity Classes (similar to VQOs) within pre-identified Visual Sensitivity Units.

The information is incorporated into the forest database on the GIS at each forest district and used to inform planners working for both the Ministry of Forests and forest licensee companies. By the early 1990s, managers realised that the concept of visual management only achieved a limited level of landscape quality: it could direct how much harvest was possible in a given landscape unit but was not very successful at achieving good design. Thus a form of the proactive design approach was introduced to supplement the visual management system (see below and in a case study in Chapter 9).

Visual impact assessment is the logical development of the visual landscape management system in British Columbia. Where a timber harvest is being planned in a known scenic area with established VQOs, a visual impact assessment is required in order to prove that the VQOs will be met. It is considered to be an integral part of step 3 of the process described above. For this it is necessary to prepare a design for the proposed activity and to illustrate how it will look from established viewpoints using various simulation tools, and then to justify how it will meet the VQOs. This includes a qualitative assessment based on the adherence to design principles and also a quantitative assessment based on a concept of "percent alteration", where the proportion of the visual scene that will be altered by the proposal is calculated (in perspective, not plan). A table is used to assign allowable percentages in relation to different VQOs.

The US Forest Service system was also introduced into Australia. The states of Victoria and Tasmania adopted it and, especially in the case of Tasmania, developed their own variations. In Tasmania, the whole island has been divided into broad regional landscape character types. This was an integral part of the Visual Management System following US methodology, giving priority for visual management and generalised visual objectives to be achieved. It did not include design-based principles as in the UK but rather specified generic guidelines for various viewing situations and forestry-related operations. The system is controlled by staff in the Forestry Practices Board, who regulate the environmental quality of logging proposals.

The most recent version of the Tasmanian Visual Management System incorporates a new tool for rating the importance of any area of the landscape and setting visual objectives to guide management, especially of the location and management of plantation forests. This is called the Rural Landscape Priority rating and has accompanying Plantation Landscape Objectives, which relate to three scales of the landscape and viewing. The new visual objectives are "Integrated effect", "Co-dominant effect" and "Dominant effect", based on the acceptable degree of influence of the plantation forest in the landscape. In the most sensitive landscapes the assumption is that the integrated effect would be appropriate using strongest adherence to positive design. However, all three objectives require the application of principles taken from the proactive design approach (see below).

As an adjunct to the new visual objectives, local visual units are identified and corresponding landscape character attributes defined as a guide to forestry. The aim is generally to retain (and in some cases improve) the character diversity between different visual units by achieving specific designs for harvesting and establishment within each visual unit. This is being applied progressively to operations but could be most useful for strategic planning. Thus the management system evolved into a series of local design guidelines as well as incorporating the concept of VQOs based on a viewing rating.

The proactive design approach

In Britain, with its programme of large-scale afforestation and, more recently, felling and replanting of its plantation forests, an approach has been developed based on designing forests to fit into the landscape. This originated in the 1960s, aimed mainly at new plantations of non-native conifers, which were being planted on bare, deforested hills and mountains. The original layouts were often highly regular, with rectangular compartments, vertical fence lines and horizontal upper margins where the trees were planted up to a contour line.

To overcome the artificial appearance of these forests a number of design principles were adopted from landscape architectural practice, for example considering the shape of the forest, its scale and proportion, the degree of diversity of different species, the unity of the forest with its surroundings, how it related to landform and so on (these are presented in detail below).

Skylines

Skyline notches can have a strong visual impact even when seen in the distant background.

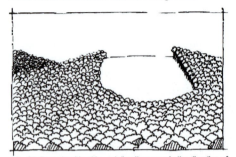

(a) *Skylines should not be cut directly across in the direction of the principal viewpoint, as the coupe edges will remain clearly visible for many years.*

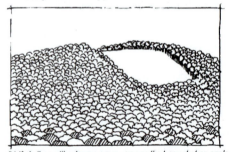

(b) *If skyline cutting is necessary, arrange the harvest at an angle to the main viewpoint.*

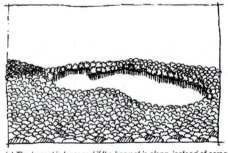

(c) *The impact is lessened if the harvest is along, instead of across, the skyline.*

Figure 3.6
A sample of applied design from the Tasmanian handbook, demonstrating design in relation to a skyline. Source: Forestry Practices Board.

By the 1980s many of the early forests were ready for harvest and the rate of new planting declined, so the focus shifted to designing the patterns of felling, and the opportunity was taken to use the process of gradually harvesting these forests to completely redesign them, especially if they had not had much design input at planting. At this point a series of detailed guidelines on forest landscape design was published, aimed not only at the state forest sector but also at private forest owners. These guidelines describe what standards are expected and how to achieve them. They set out the way the design principles should be applied to aspects of the forest layout but as they need to be interpreted to fit each unique landscape area, they avoid being too prescriptive. The main process presented later in this chapter and applied in the rest of this book is based around the proactive design approach.

The idea of proactive design has also been tested and developed in British Columbia into a complementary method by which to ensure that the VQOs were being met. A programme of training was undertaken to support this and a manual published to demonstrate the process. The design approach also expanded the scale from the single cutblock to the entire forest, looking into the future, so that anticipated changes over time and any unforeseen problems caused by cumulative effects could be tested.

Elsewhere, similar design approaches have been developed over the years. In the 1970s–1980s, a method was established in New Zealand and some regional guidance booklets were produced by Clive Anstey and Steve Thompson. In France, researchers at CEMAGREF, a government forest research agency, also looked at the British model and developed similar guidance for French foresters. Expansion of forestry in Denmark led to guidelines being produced on forest location and design, while researchers in Sweden have also produced a book which seeks to integrate ecology and aesthetics. In Latvia, the design approach was also developed, based on the use of a simpler landscape character system and aimed largely at new afforestation, partly in response to the need to replace the Soviet system and also to reflect the change in land ownership and the abandonment of farmland to natural forest regeneration.

In countries where the forest area comprises many small landowners, the kind of broad-scale assessment and design models described above are less easily applied. In Finland, guidance for forest owners considers three key objectives: paying attention to the regional characteristics of the Finnish forest landscape, harmonising forest management measures with the long-range panoramic views of the landscape and preserving the pleasant impression of the close range, or feature view. The design of felling coupes is also considered.

The great beauty of this landscape lies in the subtlety of the land-formation.

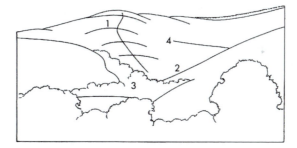

ANALYSIS
The strongly modelled shoulder (1) flanks the amphitheatre-shaped valley (2). The drift of broadleaves (3) accentuates the composition whilst the forest road (4) cuts across it.

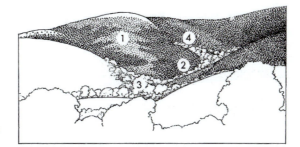

SOLUTION
The modelling of the shoulder could be accentuated by larch (1). Broadleaves would accentuate the valleys (2), linking them to the existing trees (3). The encouragement of native growth at the sides of the forest road (4). curving down to join the valley broadleaves could permanently heal the road scar.

Figure 3.7
A sample from the first design book prepared by Sylvia Crowe for the British Forestry Commission.
Source: Forestry Commission. © Crown copyright material is reproduced with the permission of the
Controller of HMSO and Queen's Printer for Scotland.

Forest landscape patterns

In Chapter 2, applied landscape ecology was used to help understand forests as dynamic ecosystems. Such ecosystems were shown to comprise patterns of elements that arise and change through various natural and human-induced agents. These patterns are to some extent controlled or at least steered by topography and climate. It is these patterns, of stands of trees of different successional stages, of open areas, water, rock and so on, that we perceive (both in the scenic and the participatory modes) and which comprise the sensory landscape. If we analyse these patterns we can usually find that certain relationships occur. For example, forests on big mountainsides often grow best and most densely in hollows and valleys while ridges retain only sparse cover or are free of trees. Open wetlands, dense old forest patches or different areas of species composition tend to occupy particular places as a result of topography, such as wet hollows, sheltered moist ravines, or north- or south-facing slopes.

Figure 3.8 This landscape is strongly patterned. The landform is a dominant feature, which is overlaid by forest stands of different types providing contrast of textures and colours. These are related to the landform and to the history of landscape evolution, such as the fire history.

Natural forests usually present a strongly coherent pattern, at least externally, although not always a complex one. They are frequently mysterious, especially internally. The degree of complexity tends to relate to the degree of variety in topography and thus is also to be found in the soils, drainage and microclimate. This variability is important for understanding how to approach the design of forests for two main reasons. Firstly, when working with an existing, largely natural forest, any changes need, for the ecological reasons outlined in the previous chapter, to reflect the natural degree of complexity, especially in the size of patches and also their shape. The second reason is that when designing new forests to be planted on land probably deforested many centuries ago, it is difficult to know how diverse to make them without some understanding of the driving forces for forest patterns. Since both of these tasks involve creative design and the capacity for visualising future landscapes, and since they must also be sustainable, it often makes the most sense to follow Nature's lead, but it is impossible to copy Nature exactly because it is not completely predictable. To quote forest ecologist Jerry Franklin, it is not necessary to be "slaves to Nature".

The aesthetic implications of this are that in order to balance coherence and complexity we must be able to develop a means of analysing forest patterns, both existing and potential, in (primarily) visual terms. Manipulation of patterns to produce desired aesthetic results is what concerns designers, as manipulation of forest patterns and processes is what concerns landscape ecologists. Thus tools to achieve this are needed. Visual design principles are useful examples of analytical tools developed to help designers. These form a substantial design vocabulary but in most cases it is only necessary to be familiar with a selection of them. A complete explanation of these principles can be found in *Elements of Visual Design in the Landscape* by Simon Bell.

Visual design principles

In both Britain and the USA the development of forest landscape management and design has seen the application of visual design principles as analytical tools to help both describe and understand the existing landscape and to help design the future landscape. Due to the early ascendancy of the traditional scenic aesthetic, many of the principles were first developed and applied to the hill or mountainside where new forests were being planted or virgin forests were to be logged. However, these principles can also be successfully applied to the internal landscapes of less variable topography. Certain key principles primarily apply to these circumstances.

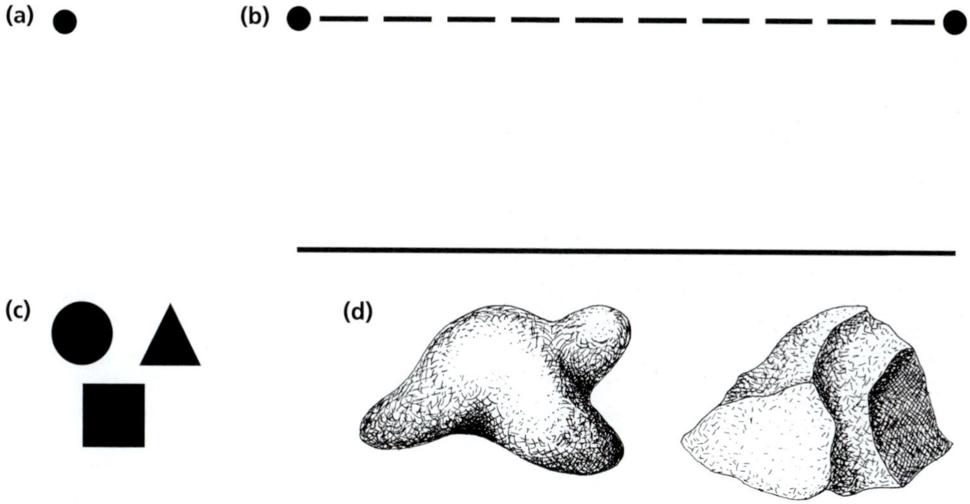

(a) ●

(b) ●— — — — — — — — — — ●

(c)

(d)

Figure 3.9
This diagram
defines the basic
elements of the
visual landscape:
a) point
b) line
c) plane
d) volume

Figure 3.10
This landscape in Colorado, USA,
demonstrates the basic elements
as landscape building blocks. The
green grassy areas are planes,
broken by roads and streams
(lines) and individual trees (points),
the distant mountains are solid
volumes while the valley itself is
an open volume.

81

In this section, these key design principles will be explained, together with examples. The concentration on visual design is a very practical one, because to manipulate patterns successfully we need to be able to understand their spatial characteristics and we cannot do this except through the medium of sight. Also, we tend to "think visually", to conceptualise the world pictorially and to illustrate designs primarily using pictures or other graphic renderings.

Basic visual elements

We can start by breaking down the landscape into constituent parts. These, using conventional geometric principles, can be defined as point, line, plane and volume (no dimension, only position; one dimension; two dimensions and finally three dimensions, respectively). Lines, planes and volumes are the key elements. Lines are very important due to the way our brains process visual images to perceive patterns. Planes are defined by edges, akin to lines in their perceptual importance, while volumes control and define space. Examples of these include streams or roads (lines), fields (planes), mountains (solid volumes) or spaces within the forest (open volumes). The character of these elements and the way they are arranged are what comprise the visual pattern.

Shape

Shape is an extremely important principle. Lines, planes and volumes can all be described by their shape or form (the three-dimensional equivalent of shape). Often it is possible to link the character of a shape with its originating process, such as a meandering river with the hydrological characteristics of water, a rectangular field with land clearance, property boundaries and fence building, and a conical mountain with volcanic eruptions and so on.

In the forest, shapes can be seen externally as defining the patterns of different patches of vegetation, stands of trees of different ages, non-forest elements such as lakes, rocks, bare mountain tops, avalanche slides or fields. These patterns frequently reflect the form of topography (landform). For example, in undulating, gentle landform, simpler, flowing shapes might predominate, while in steeper, more rugged or jagged topography more angular, broken shapes are often more common.

Figure 3.11
These sketches show how shapes of different character can be found:
a) in a landscape of jagged landform, and
b) in a rounded landscape.

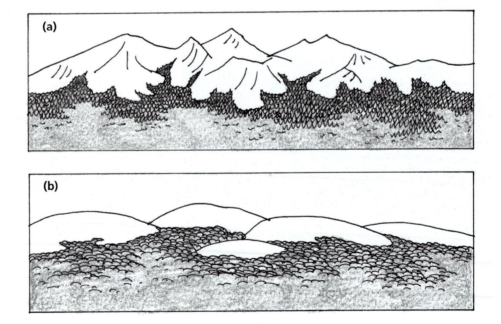

Shapes can usually be defined as lying in a spectrum from the most strongly geometric, such as squares and rectangles, often also symmetrical, to the most organic, being highly irregular, curvilinear and asymmetrical. Many natural shapes also exhibit the phenomenon of fractal geometry, where the same irregular shapes occur at a range of different scales of resolution. In many of the preference studies testing peoples' attitudes to different forested landscapes, organic shapes stand out as being more consistently preferred over geometric ones. There are several reasons for this, some more concerned with ethics than aesthetics (geometrical logging as a sign of unwanted human intrusion into the wild, for example). However, there is a strong case for believing that shapes which are similar tend to be perceived as part of a more complete, harmonious pattern than obviously dissimilar ones. There is evidence from the studies of Gestalt psychology that geometric shapes tend to stand out as "figures" from the background and may disrupt the way we make sense of a scene to the extent that we find it unattractive.

One of the characteristics of many managed forests is the predominance of geometric shapes of compartment boundaries, logged areas or plantations. In Britain, many of the early plantations were deliberately planted in square or rectangular blocks and this was one of the major reasons for the disfavour they found with many members of the public.

Within the forest, geometry can be perceived in straight roads or the square edges of logged blocks. Such simple shapes not only proclaim themselves as unnatural, but fail to retain any sense of mystery, since it is possible to see all the way down a straight road or across the entirety of a flat, square felled area.

Simple geometric shapes can also, unfortunately, arise from the application of rules aimed at protecting the forest, such as the retention of streamside corridors. These often end up as parallel-sided belts dissecting the landscape and as such appear highly unnatural. This tendency to simplify nature is a very human one but often misses the subtle variations that make one landscape different from another and thus can indirectly lead to a standardisation of approach.

Where it is straightforward to describe the shapes found in a forest landscape and to use them as templates for design, it is not appropriate to use such templates in a random or arbitrary fashion. As we have observed, forest patterns tend to occur in certain ways due to the influences of topography, soil and so on. We also tend to look at landscapes in particular ways, drawn by strong lines, edges or shapes that define the structure of the scene. One key way in which we can understand many forest shapes is in relation to landform, through the principle of visual force.

Figure 3.12
An example of a plantation forest laid out in a strongly geometric fashion, with external boundaries following the contour or at right angles to one another. Scottish Borders.

Figure 3.13
These clearcuts are generally geometric in shape and they are laid out in a formal pattern across this natural mountainside in the Rocky Mountain trench in British Columbia, Canada.

Visual force

Strongly defined topography presents itself as a series of edges and lines, such as ridges, skylines, valleys and hollows. As part of the way we make sense of the landscape, it is important to be able to understand how these patterns fit together. Many of these ridges and valleys produce a strong sense of direction and it has been empirically observed that our eyes tend to look for skylines and follow them and associated ridges downwards while seeking valleys and following them upwards. Some of these features are stronger and retain our attention longer than others. This sense of flow or movement in the landform has a definite pattern to it. This is known as visual force.

One of the, often subconscious, effects of, for example, rectangular shapes on mountainsides is that they arbitrarily cut across topographic features such as ridges or valleys. This can induce a state of tension or awkwardness due to the fact that such shapes interrupt, with their powerful presence (often highly contrasting in colour or texture), the direction of the force lines down a ridge or up a valley. Thus, if shapes (whether of plantations or of felled areas) follow the flow of the visual forces, they are more likely to be in harmony with each other, induce less tension and present less of a contrast. This is important for coherence and also for reflecting and building complexity.

However, is there any evidence that shapes defined by the application of a design principle derived from aspects of our perception can also be found in Nature? Happily, many natural patterns show a tendency to be related to landform by being, for example, found more frequently in valleys than on ridges. This is not a universal tendency, but it is frequent enough to be a highly useful tool not only for designing shapes that, although resulting from human activity, take on a natural appearance, but that also reflect those present in the underlying topography, soil, drainage and vegetation patterns, so aiding the application of ecological design principles.

It is a useful technique to define these lines of visual force using photographs or topographic maps. Such an exercise can become a key part of the analysis of local landscape character as part of the integrated forest design process (see Chapter 5) and will be illustrated in several case studies later in the book.

Scale

One of the major attributes of any landscape is its size relative to that of the human figure. Large-scale landscapes that dwarf us tend to induce experiences of the sublime, such as the Grand

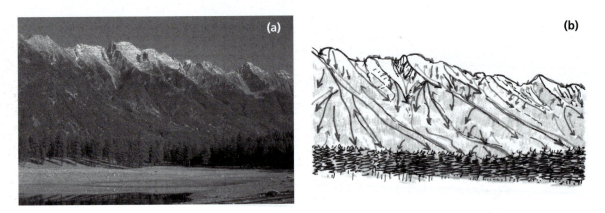

Figure 3.14
a) This example of a mountainous landscape in north-western British Columbia, Canada, has many ridges and valleys. b) This sketch shows how the landform can be analysed using the design principle of visual force. The major features are shown with thicker arrows than the minor features, and the forces rise up the valleys and flow down the ridges.

Figure 3.15
This diagram shows how lines of force can be identified from contours on a map.

Canyon or tall mountains, while spaces and details of a more human scale help create comfortable participatory landscapes such as in urban forests.

Scale can be defined initially by the topography – its general size and how it controls the distances over which we can see. A view from a cliff top where we can see for miles over a plain and an enormous mountain rising abruptly above us are both large-scale landscapes. Likewise, lesser hills and shorter views are smaller in scale, especially where signs of human activity, such as houses and roads, are not dwarfed by the landform. The open volumes created by the enclosure of space within a forest also exhibit scale variation depending on the height of the trees and the distance across the space.

Forest landscape patterns often tend to be related to the scale of the landform. Large-scale landforms tend to demonstrate large-scale expanses of similar forest type, and conversely finer

Figure 3.16
This landscape, in Colorado, USA, demonstrates a large scale: the mountains are high and the distance from the viewpoint to them is wide. Human activities appear small in this scene.

Figure 3.17
This example, from Nova Scotia in Canada, demonstrates a small-scale landscape of small-sized hills. The logging seems to occupy a large proportion of the scene.

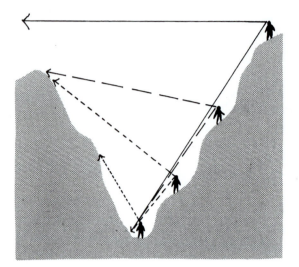

patterns reflect smaller-scale landforms. Topographic scale also tends to vary according to the position of the observer. Down in a valley bottom where the landform encloses views the scale tends to be smaller than that experienced by an observer on a mountain top, able to see a large open expanse of landscape. This variation, too, tends to be reflected in the patterns of the natural forest ecosystem. These scale relationships will prove to be significant when the design of landscape changes due to felling or plantation management are considered, as described in Part 3 of this book.

A related principle to scale is that of proportion. This also involves the relative sizes of landscape elements in relation to that of the composition, though not the human figure. A particularly useful proportioning device is known as the "thirds rule". This device ensures that the proportions of different elements that comprise a landscape pattern are asymmetrically balanced on the rough and ready approximation of one-third to two-

Figure 3.18 This diagram shows how the scale of the landscape depends to some degree on the position of the observer. From low down, enclosed in a valley, the landscape will appear small in scale compared with the wide and distant prospect available from the summit of the mountain peak.

thirds. As an example, consider some proposals for timber harvest. If more than two-thirds of the visible area or landscape unit, as seen by observers (rather than measured on maps), was to be felled, and somewhat less than one-third retained, this would produce a result where the proportion of felled area to retained area appears severely unbalanced. Conversely, felling half and retaining half is too symmetrically balanced, with neither element appearing to be the dominant one. If either one-third is felled with two-thirds retained or vice versa, the asymmetric balance appears far more harmonious. This rule of thumb derives from the classical proportioning principle of the "Golden Section", which itself can be demonstrated to derive from some fundamental mathematical relationships found in natural structures and patterns.

That these theoretical applications of scale and proportions work in practice can be demonstrated from some of the preference studies that have been carried out. However, some aspects of scale and proportion appear at first sight to cause problems. One cause of this emerges from the application of some of the landscape management or scenic planning systems described earlier. These use the concept of "percent alteration". This refers to the proportion of the visible scene that can be altered by, for example, logging, before people feel that too much change has taken place. In British Columbia, for example, this could mean that 15% of the visible portion of a mountainside could be altered. It is argued that this "social scale" should take precedence over other principles. However, it often has unfortunate consequences. One problem is that such consistent and slow rates of change do not necessarily reflect ecological processes and may run counter to the needs of ecological sustainability. A second problem is that the accumulated effects of many small-scale changes can result in a chaotic, "moth-eaten" appearance that is scenically counterproductive. A third problem is that such calculations can only be made from a few set viewpoints, and that if a landscape is seen from many different places, including internally, they fail.

Earlier in this chapter the importance of the cognitive variable of coherence was mentioned. This refers to the sense that the perceived pattern of the landscape is understandable and "hangs together". This factor can be directly related to one of the key principles of good design, that of achieving unity.

Unity

Unity occurs when all the parts of a composition relate to one another in such a way that it is the whole that is the most powerful aspect, rather than any of the individual parts. Similarity of shape, colour and texture all help to achieve unity. Thus, if all the natural forest patterns tend to be organic shapes of a certain degree of irregularity, the addition of strongly geometric shapes may well give too strong a contrast and disturb unity. That is not to say that contrasts always reduce unity. Colours that are opposites (such as red and green) work so strongly together (complementary harmony) that they increase unity.

In the natural forest, there is an inherent unity that arises between the forest cover and the landform due to the interaction of ecological processes such as succession or microsite variation. Thus the forest is to be found as a composition, structure and pattern that reflects the culmination of the various probabilities of the processes occurring. This contrasts with many of the historical forest patterns introduced by human activity such as logging or planting. The geometric shapes produced by much recent timber harvest are the result of limitations on road construction and timber extraction machinery such as cables operating on steep mountainsides. While the topography and forest type obviously influences what is selected for harvest and how it is obtained, the character of the forest patterns has often been ignored. Similarly in the case of plantation forestry, such as in the afforestation of the uplands of Britain, planting was set out in rectangular compartments following straight ownership boundaries and often stopped at a notional "upper planting limit" that followed a contour. Such layouts were planned entirely on flat maps and took no account of the topography, and rarely varied due to changes in soil type. Thus, in both these examples the human-altered patterns created disunity.

In the case of the internal as well as the external landscape, unity can also be achieved through the spatial and structural arrangement of the different elements, such as enclosure, coalescence, interlock, continuity, rhythm and balance. Shapes that are not only organic but enclose space and interlock together (e.g. a jigsaw puzzle) tend to hold together better; as do patterns that pick up flowing, repeated shapes and those that are asymmetrically balanced. These aspects reflect some of the basics of perception of patterns, such as Gestalt psychology. It is also the case that, in the forest, geometric shapes cannot interlock or enclose space and they inevitably show symmetry.

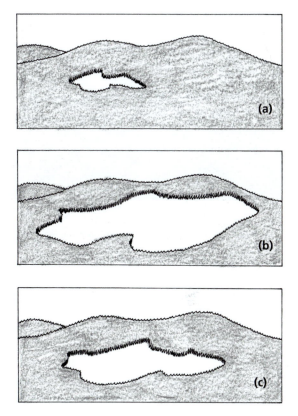

Figure 3.19

This series of sketches shows the application of the "third's rule": a) the felled area of the forest occupies less than 1/3 of the visible landscape and looks out of proportion; b) the felled area occupies around 50% of the view, so that neither the felling nor the remaining forest dominates the scene; c) the felling occupies about 1/3 of the landscape and appears to be much more in proportion than either of the others.

Diversity

Complexity was defined earlier in the chapter as another important cognitive factor. This equates to the design principle of diversity, needed to provide interest and something to explore in the landscape. Diversity balances unity. It is a principle that is subject to the laws of diminishing returns. A scene with little diversity of pattern (low on total number of elements and short on textural or colour variety, for example) can easily bore us. Increases in diversity provide more interesting scenes but eventually too much variety erodes unity, becoming too busy and confusing (losing coherence).

Diversity operates at different scales. Over a large-scale landscape, where the view comprises whole landforms, the diversity is made up of the major components such as forest, mountain, water, fields and so on. At the medium scale, diversity can be read as the different patterns, colours and textures within each component, while at the smallest scale, and particularly within

Figure 3.20
In this landscape in the Arapaho National Forest in Colorado, USA, the forest is naturally unified into the landscape. It lies on the lower slopes, concentrated in the valleys. The shapes are natural and reflect those found in the landform, and the edges are diffuse. The sense of unity is strengthened by the fact that the ecological processes have helped to produce the pattern, which therefore reflects the natural equilibrium between forest and landscape.

Figure 3.21
The natural unity of this forest scene on Vancouver Island in British Columbia, Canada, has been disrupted by the superimposed pattern of logging. This does not reflect ecological patterns and processes, and introduces shapes into the scene that clearly do not seem to belong there.

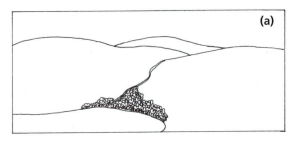

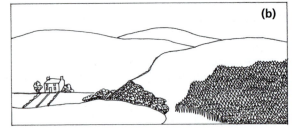

Figure 3.22
This series of diagrams shows the effect of increasing diversity in the landscape. From a simple yet bland scene (a) the landscape becomes more diverse and interesting (b) until too much diversity starts to undermine the unity and visual chaos threatens (c).

forest landscapes, it is the spatial structure and variety of detail, right down to bark, flowers and rock surfaces, that contribute.

This perceptual aspect of diversity may or may not relate to ecological diversity. In the main, ecological diversity follows a similar framework, as explained in Chapter 2. However, some forests can naturally be extremely low on diversity at virtually all scales and it might be expected that people would react somewhat negatively to them. An example might be the vast spruce expanses

Figure 3.23
In this fairly large-scale scene in the Austrian Tyrol the landscape diversity is firstly expressed in the main land-use zones of the fields, forest and alpine area. Within each zone there is also diversity, more so in the fields with small areas of forest, houses and different textures of fields.

Figure 3.24
This area of forest is visually very naturally lacking in diversity: it appears to be composed of a single forest type exhibiting the same colour and texture. This reflects the natural level of ecological diversity, which will be resolved at the stand level while appearing bland from this more distant viewpoint. Coastal forest in the Prince Rupert region of British Columbia.

of the Canadian taiga, where the simplicity of topography and the extreme climate result in a homogeneous pattern of scale well outside that of comfortable human experience. The scale of natural disturbance such as fire is also huge, so that this does not introduce much finer-scale diversity.

However, in many cases the degree of diversity of the forest pattern reflects the scale of variability of the underlying landform, soil and climatic patterns and so attains a degree of harmony and balance with unity.

One of the problems of forest management over landscapes is the tendency to apply simplistic perceptions or rules universally, without adequate sensitivity to the different characteristics of different landscapes. Frequently there is some kind of regional classification of forest types by climate, topography, major soil classification, forest tree species and vegetation. These are useful for the development of policies and prescriptions that are responsive to regional differences. However, at the local level there are also special qualities that help define and identify local landscapes. This is the sense or spirit of place, also known by its Latin name of *genius loci*.

Spirit of place
Spirit of place is that special quality that gives a place its unique sense of identity. There are many possible means by which spirit of place can be generated. Landscape elements such as water, both still and moving, are particularly strong. Reflective lakes or ponds, rapids, cascades or waterfalls all help define the sense of place and also contribute to mystery, that other cognitive variable found to be so important for landscape preferences. Rock formations, groves of old and big trees, historical or archaeological remains and association with art can all contribute to the spirit of place.

One of the problems of past management practice is that spirit of place has not been recognised and, as a result, it has been compromised or obliterated. This is equally true of logging or plantation silviculture, road construction or bridge siting.

Figure 3.25
A landscape with a strong spirit of place, created by the combination of the mountain peak, the blue waters of the small lake and the twisted trees of the forest. This scene is also hidden away amongst the mountains, giving a sense of mystery, which increases its special quality. "Shangri-La", in the Selkirk Mountains of British Columbia, near Invermere, Canada.

Landscape character analysis

As for the role of the landscape ecological analysis described in the previous chapter, a landscape character analysis is also an important first stage in developing any design. Without understanding what makes the landscape what it is and what aesthetic problems may currently exist, it is difficult to look ahead to try to design a potential future landscape. As we have seen, the two drivers that determine the forest landscape are the structure of the underlying landform and the pattern of land use and vegetation overlying it. These are not independent variables, the landform having a profound influence on the other, especially when it is strongly defined. Other processes also contribute to the vegetation pattern and its dynamics, such as ecological succession and disturbance, and human changes such as clearance, felling, plantations, road construction and buildings.

As demonstrated earlier in the chapter, landscape is an experiential as well as a physical concept and therefore its analysis must be considered and presented in ways that reflect as closely as possible the ways people experience it. The external landscapes seen as vistas in the traditional scenic mode of viewing are best analysed using perspective photographs, computer-generated three-dimensional models and, if the technology is available, fly-through simulations of the landscape using virtual reality. Maps are used to help coordinate the results and prepare the landform analysis and vegetation structure or land feature analysis.

There are two levels of landscape character analysis. The first can be described as strategic, the second as detailed. At a large scale (perhaps at a regional level) it is possible to divide the landscape into units that possess a definable character that contrasts with neighbouring areas. These character types are usually built up from a consideration of geology, landform, natural vegetation, land use, building styles and techniques of construction and various cultural aspects. These can form the basis for sets of landscape guidance, both for afforestation (as in parts of Scotland where local guidelines based on these character types have been developed) and for timber harvest and management (as in Tasmania, Australia, where the whole state has been classified). Likewise, in Latvia, a comprehensive landscape classification has been developed, based on three landform types and three land-use types (with some special types), that produces a matrix of nine archetypes that each have design guidance for afforestation associated with them.

At a sub-regional level, within large-scale landscape units based on drainage units, it is also possible to subdivide an area into different zones of landscape character based on topography and forest type. Some of the case studies presented in Part III adopt this approach which helps to take account of local variation.

The detailed level of landscape character analysis is carried out at the design stage as follows:

An important first step is to choose the viewpoints to be used for analysis. These are often selected to fulfil two requirements. Firstly, where the landscape is highly visible from public locations such as roadsides, well-known viewpoints or settlements, these should be used so that the analysis

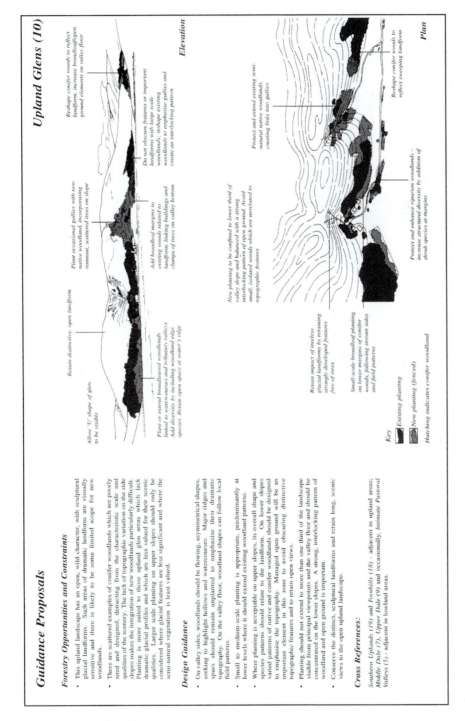

Upland Glens (10)

Elevation

Reshape conifer woods to reflect landform, increase broadleaf/open ground elements on valley floor

Do not obscure features or important landforms with large scale woodlands; reshape existing woodlands to emphasise gullies and create an interlocking pattern

Plant occasional gullies with new native woodland, incorporating remnant, scattered trees on slope

Retain distinctive, open landform

Add broadleaf margins to existing woodlands linked to landform, linking buildings and clumps of trees on valley bottom

Allow 'U' shape of glen to be visible

Plant or extend broadleaved woodlands linked to watercourses and tributary valleys. Add diversity by including woodland edge species. Retain open space at water's edge

Plan

Reshape conifer woods to reflect sweeping landform

Protect and extend existing semi-natural native woodlands, creating links into gullies

New planting to be confined to lower third of valley slope and balanced with a strong interlocking pattern of open ground. Avoid small, isolated woods which are unrelated to topographic features

Protect and enhance riparian woodlands - increase structural diversity by addition of shrub species at margins

Retain impact of treeless glacial landforms by retaining strongly developed features free of trees

Small-scale broadleaf planting on lower margins of conifer woods, following stream sides and field patterns

Key
Existing planting
New planting (fenced)
Hatching indicates conifer woodland

Guidance Proposals

Forestry Opportunities and Constraints

- This upland landscape has an open, wild character, with sculptural glacial landforms. Such areas of dramatic landform are visually sensitive and there is likely to be some limited scope for new woodlands.

- There are scattered examples of conifer woodlands which are poorly sited and designed, detracting from the characteristic scale and qualities of the scenery. The lack of topographic variation on the side slopes makes the integration of new woodlands particularly difficult. Planting is more suited to those upland glen areas which lack dramatic glacial profiles and which are less valued for their scenic qualities. Larger scale planting on upper slopes should only be considered where glacial features are less significant and where the semi-natural vegetation is least valued.

Design Guidance

- On valley sides, woodlands should be flowing, asymmetrical shapes, seeking to highlight hollows and watercourses. Major ridges and spurs should remain unplanted to emphasise their dramatic topography. On the valley floor, woodland shapes can follow local field patterns.

- Small to medium-scale planting is appropriate, predominantly at lower levels where it should extend existing woodland patterns.

- Where planting is acceptable on upper slopes, its overall shape and species patterns should relate to the landform. On lower slopes varied patterns of native and conifer woodlands should be designed to emphasise the topography. Managed open ground will be an important element in this zone to avoid obscuring distinctive topographic features and to retain open views.

- Planting should not extend to more than one third of the landscape visible from principal viewpoints and the valley floor and should be concentrated on the lower slopes. A strong, interlocking pattern of woodland and open ground is important.

- Conserve the distinct, sculptural landforms and retain long, scenic views to the open upland landscape.

Cross References:

Southern Uplands (19) and *Foothills (18)* - adjacent in upland areas; *Middle Dale (7), Upper Dale (9)* and occasionally, *Intimate Pastoral Valleys (5)* - adjacent in lowland areas.

Figure 3.26

An example of locally defined forest design guidance based on landscape character analysis. This illustrates one landscape character type and explains how to incorporate more forest while retaining the essential elements that define the 'landscape character. Dumfries and Galloway Region, Scotland. Source: Forestry Commission.

(a)

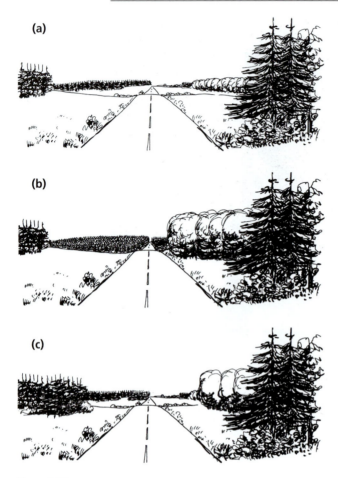

(b)

(c)

Figure 3.27
An example from Latvia showing negative and positive approaches to afforestation in a particular landscape character type: a) shows the original landscape; b) shows how expanding the forest can detract from the character; c) shows how more forest can be added while respecting the character. Source: State Forest Service.

and design can be related to the familiar views that are likely to be of concern to the public. Secondly, views that also enable the design to be undertaken efficiently should be chosen. These may not always be the same as the public viewing points. These "designers" viewpoints should aim to give a good picture of the landscape structures and patterns. Low-level aerial views taken from a helicopter are ideal for this purpose as are photos taken from higher-elevation viewpoints selected for their comprehensiveness. Panoramic photographs should be taken (see Chapter 6 for details on how to do this), followed by analysis based on these or computer simulations of the same views.

Landform analysis

This uses the concept of visual forces in the landform as described above. This normally starts with a topographic map showing contours of the area. Ridges and valleys are then identified and depicted using red arrows for ridges (flowing downwards) and green arrows for valleys (flowing upwards). These arrows are visually weighted to show more or less important features. For example, large-scale dominant ridges or deeply incised valleys are shown as much thicker arrows than small-scale subtle ridges or shallower valleys. This is not a scientific analysis and it requires skill and judgement to know what features to include and how to weight them.

These arrows are then transferred to perspectives either manually to overlays of photographs, or by digitising them and overlaying or draping them on top of a computer digital elevation or terrain model (DEM or DTM). The manual method, while slower and possibly less accurate, has the merit that the action of defining it helps the designer to get to know the structure of the landscape, which is of immense help at the design stage later on. The computer method enables the designer to be completely accurate and to know that all views coordinate together.

Descriptions of the landform are also carried out to determine the origin and type of forms, the scale and diversity of the topography and how it changes from one part to another. For example, in some mountain areas there may be angular, concave, jagged shapes as the result of erosional processes at the upper elevations, while lower down there may be many rounded, convex, smooth shapes arising from depositional processes. Chapter 5 explains in greater detail how to prepare and present this analysis as a key part of the integrated design process.

Land feature analysis

The analysis of the vegetation and land-use pattern is often known as the land feature analysis because it looks for the key features that define the patterns and the sense of place in the

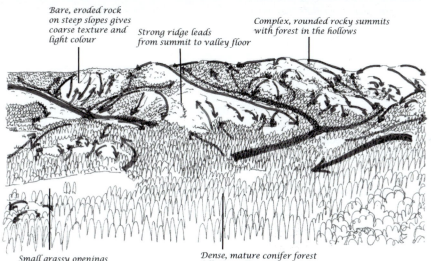

Bare, eroded rock
on steep slopes gives
coarse texture and
light colour

Strong ridge leads
from summit to valley floor

Complex, rounded rocky summits
with forest in the hollows

Small grassy openings
on the valley floor
give diversity

Dense, mature conifer forest
fills the valley and rises into
the side valleys

Figure 3.28 Landscape analysis carried out at the level usual for forest design. This viewpoint, in the Emigrants Gap in the Sierra Nevada in California, incorporates fore-, middle- and background areas. The sketch shows the landform and describes some of the key landscape features of the scene.

landscape. This analysis uses the design vocabulary defined earlier in the chapter to describe and identify the most important contributors to character. These might include the typical shapes, the scale of the landscape, the colours and textures that are contributing to diversity and the unique features that help define spirit of place. This often takes the form of annotated perspective photographs, perhaps with certain features highlighted on the photos. As well as the basic character it is useful to identify any features that are detracting from the sense of unity in the landscape and from its quality and character, particularly past management activities that need to be corrected. A map is often needed to record the aspects of the analysis so that they can be coordinated between the different views.

Internal landscapes are more difficult to analyse because they are multi-sensory and experienced on the move. Video footage of travel routes can be considered and maps of experiences built up that help to determine the spatial and other key landscape elements that contribute most strongly to the experience. A simpler method is to follow trails and other routes and take photos at key locations, which are marked on maps, with notes to record non-visual aesthetic aspects. Virtual

Figure 3.29
This landscape, in the Slocan Valley of British Columbia, is an example of an area where disputes over landscape change have taken place in recent years. People have become accustomed to a scene dominated by healthy, mature green forests on the mountain slopes above their homes. The prospect of significant landscape change, especially made through the agent of logging, worries many people. Managing landscape change is essential and needs to be handled carefully by forest managers. Good design to minimise the potential negative impact is an important tool to help achieve this.

reality models are in development but have drawbacks because the accurate details unique to each landscape cannot be recreated sufficiently to make the landscape like the real thing. This type of analysis is only possible for relatively small-scale areas or for restricted areas such as trails and roads within larger areas, due to the effort involved.

Landscape sensitivity

It is not normally the case that all landscapes are treated equally. Not all have the same qualities, are as visible or seen by as many people so that their sensitivity to change varies. As part of developing the context for forest landscape design it is worth assessing the degree of visual sensitivity of different parts of the landscape. Thus limited resources can be directed more efficiently.

Visual sensitivity of the landscape normally depends on a combination of several factors. These may follow one of two typical combinations. The first has traditionally been used in forestry landscape planning and considers landscape quality, visibility and the number of people who see the landscape (particularly residents and tourist/recreation visitors). The second, now becoming more standardised, especially in Europe – in order to conform with EU directives on environmental impact assessment of a wide range of types of development project from wind farms to power stations – uses visibility, numbers of people, the nature of the viewing experience (whether static, moving, as a tourist or resident, etc.) and landscape value as the factors. These are similar but can be applied in a slightly more sophisticated way.

The visual sensitivity assessment should concentrate on the landscape as a whole, not only the forested portion within it (for example, a plantation set in an open landscape). The landscape to be assessed may be defined by the limits of visibility from the various viewpoints (as defined above) and the forest may only be a small component of it or, if completely forested, it may be the entire landscape. It is an important first step to scope the area for its visibility and where necessary to subdivide it into zones which may become the landscape units of assessment. An example might be a forest lying partly on an escarpment, visible from many places, and partly on the plateau above, hardly visible from anywhere. Landscape sensitivity is likely to be quite different for the visible slope from the plateau, so it would be sensible to subdivide the area. Visibility assessment should consider all possible publicly accessible locations, weighted in terms of the importance as noted below. Photographs should be representative of the character of the area and the forest. They may include the key viewpoints, such as well-known stopping places, but may also be more general, such as a typical view from along a road.

Landscape quality or value

This is the most subjective of all the factors to be assessed, but in some ways it is also easily understood and agreed upon in the case of particular landscapes, beauty spots and the like. At its most basic it can encompass those landscapes designated in some way for their scenic value, such as national parks. In their absence an expert approach using characteristics known from landscape preference research can be adopted.

The US Forest Service scenic Management System uses a similar approach that classifies lands as either A (distinctive), B (typical) or C (indistinctive). Lands classified as A tend to receive higher levels of scenic protection than B or C lands, though other social factors are also weighed.

Landscape visibility

The visibility of the landscape depends mainly on the topography, the presence of elements that block or screen views and the amount of the landscape accessible to potential viewers. The viewpoints scoped within the study area may range from open and unobstructed to those heavily affected and partly screened by trees, buildings or other features. Landform is the major landscape

Table 3.1 Method of combining visual sensitivity factors to create scores

Degree of contribution to sensitivity	Factors affecting sensitivity		
	Visibility	Numbers of people	Nature of the viewing experience
High	Forest is clearly visible from the viewpoint Landscape is open and unobstructed Viewing distance up to 5km	Large numbers of residents High volumes of travellers	Residential viewing Local travellers frequently using the area Recreation and tourism visitors spend time in the area
Medium	Forest is mostly visible from the viewpoint Landscape is partly open and unobstructed Viewing distance from 5–15km	Moderate numbers of residents Moderate numbers of travellers	Residential viewing Local travellers Some recreation and tourism visitors to the area
Low	Forest is partly visible from the viewpoint Landscape is mostly obstructed by objects in the view Viewing distance 15–35km from the site	Small numbers of residents Small numbers of travellers	Some residential viewing Travellers mainly on business Few if any recreation or tourism visitors.

element to screen views. The other factors that affect visibility are the distance to the forest from the viewer and the viewing direction in relation to the lighting direction. Up to 5 km away, a landscape can be considered as foreground and highly visible, possibly dominating views, while from 5 km to 15 km the area will be seen as part of the general landscape. Beyond 15 km, it is more likely to be seen as part of the background and attention is easily diverted from it.

Numbers of viewers

There is usually little or no hard data available on the number of viewers seeing the forest. However, it can be inferred from information on population and from observation of the study area, the strength of the settlement pattern of towns and villages and the number and importance of roads and places used for recreation in a given area. Some viewpoints are used by fewer people since they are on less important roads or at smaller settlements.

Nature of the viewing experience

People who live in a particular area experience the landscape all year round together with its changing moods. They are used to seeing the landscape as it is and may not react favourably to changes taking place. Visitors to the area may see the landscape primarily from the roads, although footpaths also provide a limited but significant type of experience. Travellers see it as a moving experience and may spend greater or lesser times travelling through or around the landscape seeing the development. Local people driving to and from work or local services are likely to be more sensitive than purely business or commercial travellers passing through. Tourists driving to experience the countryside will also be more sensitive.

An assessment using these factors can be prepared for regions of the landscape which can be divided up into those areas that may need special landscape protection, those where especially high design standards should be applied, those where "ordinary care" should be given and those where aesthetics is not an issue. This can be very helpful to managers, particularly if they are worried about the need to devote resources everywhere.

The methodology presented above can be applied across a region or an entire country so as to give a national picture regardless of the breakdown of an area into character zones. This has been applied as a strategic planning tool by the Irish Forestry Board, in order to be able to direct design resources more efficiently. An alternative approach is to base the sensitivity assessment around the landscape character zones, recognising that the landscape quality element of the equation could be refined to take account of local distinctiveness. This approach has been used in Tasmania and has guided the efforts of landscape foresters for some time.

Managing landscape change

One of the problems faced by forest managers in all situations is that people do not generally like to see landscapes change. Perhaps they grew up with a particular landscape, or moved to an area because they liked the scenery. It may also be the case that we instinctively prefer stable, unchanging scenery as an element of security in an otherwise fast-changing world.

However, the issue is not quite as simple as it first appears. We are happy with change when we instigate it ourselves. We also cope happily with gradual change, such as trees growing over the years. We can also deal with cyclical, repeatable changes such as those of the seasons; in fact many people look forward to the different seasons. What seems to provoke the most negative reactions is sudden, unexpected and dramatic change in some aspect of the land that is valued. This is equally true of natural events such as forest fires or windstorms, usually condemned as "disasters", as it is of changes such as forest timber harvest or plantation afforestation. The issue of managing people's expectations of change will be dealt with in Chapter 4.

Forest managers have traditionally reacted to hostile attitudes from members of the public towards the aesthetic changes by avoiding the issue for as long a possible. This means hiding cutting behind retained roadside belts, over the tops of hills, or by planting rows of broadleaves around the outside of conifer plantations. If the issue has been clearcutting, then managers have sometimes resorted to selective felling as a means to extract timber without radically changing the landscape. All these methods, if they work at all, work in the short term. They all miss the main point: that forest landscapes are by their very nature dynamic, and that if they do not change by human actions, they will do so instead through natural events. However, there are some major differences between the dynamics of the natural and the managed forest: the pace and character of change, for example, that have to be addressed.

Managed forests have tended to be run on much shorter cycles than those of natural forests. The rate of change therefore tends to be quicker. The use of clearcutting in tidy geometric shapes produces visual results that are out of sympathy with Nature. Thus, there are some clear design issues to be dealt with as well as those aspects of public participation, consultation and ethics that are covered in Chapter 4.

Forest aesthetics and ecology

As we have pointed out earlier in this chapter, there is a strong relationship between the basic ecology of a forest and how we perceive it. Landform, climate, biogeography, disturbance and succession conspire with cultural land use to shape the forest that we experience visually. How we react to this experience, whether we find it pleasing or not, is influenced both by our biology and cultural conditioning (nature and nurture). Environmental psychologists have suggested that human beings, as common members of a species, possess innate standards of beauty with enormous adaptive implications. Across cultures, humans prefer and assess as most attractive those landscapes that include features that have proved to be beneficial for the biological survival of our ancestors.

The human attraction to "park-like" landscapes of scattered trees with an open, herbaceous understorey has been considered by some to be a reflection of our origins on the East African savanna. This is particularly reflected in surveys of children. Once people become adapted to a particular landscape, expectations change.

It has consistently been demonstrated that people prefer "natural appearing" forests to highly altered ones. Recent research in the southern USA demonstrated that in foreground views, the public clearly prefers forest stands with diverse structure and composition over single aged monocultures with little understorey. Individuals who work directly with the land, farmers and foresters for example, tend to favour "managed" natural landscapes over wild ones.

Cultural expectations about forests are not static. In recent years, we have seen at least four interesting challenges to previously held aesthetics within the USA and to an extent in Europe. First has been the changed way we view marshes and wetlands. Up until a few decades ago, wetlands were viewed as wastelands and swamps that should be drained and put to better use. But gradually, thanks to the work of wetland ecologists, people have come to recognise the ecological value of wetlands, and this has fostered a new aesthetic appreciation of them. Communities now proudly celebrate newly protected or restored wetlands, build boardwalks and viewing points, and recreate in them via canoe or kayak.

The second change is our perception of old growth forests. Again, a few decades ago most people were not attracted to old growth groves. They were seen as dark, dank, drab, dangerous and unexciting. Even the Sierra Club, the leading environmental organisation in the USA, had little interest in old growth conservation until recently, primarily because their focus was on the sublime landscapes of high mountain timberline. But now, old growth has been recognised as ecologically valuable, people go to them for recreation and spiritual renewal, and fight for their conservation.

The third example is the issue of forest fire and other large-scale natural disturbances. The Smokey the Bear campaign, initiated by the US Forest Service in the 1930s, proved very effective in convincing Americans and others that all forest fire is bad. It kills Bambi and her friends, or at least makes them homeless. It leaves an ugly, blackened mess behind. Yet the Yellowstone fire of 1988, and before that the Mt St Helens eruption of 1980, initiated new thinking about the role of large natural disturbances in the forest cycle. It had been predicted that visitation to Yellowstone would plummet in the wake of the fires that burned through most of the park's lodgepole pine forests. Yet visitors flocked to see the burned landscape, the blooming fireweed and the millions of pine seedlings that sprouted as soon as the ash cooled. Mt St Helens is, of course, now a national monument with hundreds of thousands of tourists. Far more visit now than went to see the volcano and its green forest prior to the "devastation".

Lastly, we are on the verge of reconsidering our allegiance to the clipped, weed-free lawn. Numerous studies and popular articles are pointing out that lawns are biological deserts, water wasters and require inordinate levels of environmental poison and fossil fuel energy to maintain them. Already, prairie and wildflower gardens, as well as ornamental grasses, have begun to eat away at the ecological niche presently occupied by lawns.

The implications of these examples for forest design are intriguing. Ecological forestry has a number of characteristics that at face value conflict with pre-existing cultural perceptions. Ecological forestry leaves dead wood behind, in the form of standing dead trees (snags) and large woody debris on the ground. Viewed from a distance, snags left within clearcuts can break the skyline and act as accent points that call unwanted attention. Large-sized wood left on the ground can look "messy". If the scale of harvest increases in order to mimic the scale of natural disturbances (up to thousands of hectares in some cases), this can create serious aesthetic challenges. Even smaller-scale, selective harvest, for example to restore historically open woodland conditions in Ponderosa pine, can result in significant visual change and may be considered unattractive.

The authors believe that cultural perceptions of forests have begun to shift, and will continue to do so as ecological forestry takes root and popular understanding of how forests work increases. While we do not see any inherent conflict between aesthetics and ecological forestry, neither do we see an automatic synergy. Forest managers who make the shift towards ecological forestry, believing that it will be embraced by the public, need to consider aesthetics along with other issues.

Conclusions

Landscape as a concept is a powerful one that we ignore at our peril. Aesthetics is not an optional extra but must be included as a fundamental factor in all forestry planning, design and management. Far from being a subjective issue, landscape aesthetics has received much research attention and some useful results help to guide us in very practical ways. The application of visual design principles has been practised for several decades with good results, as will be demonstrated in case studies later in this book. These principles provide us with professional tools that reduce significantly the subjectivity that can surround debate on the subject.

Chapter 4

Community participation in forest design

Introduction

Chapter 1 stressed the need to involve communities in forest planning, design and management. Most approaches to sustainable forest management identify participation as a critical pillar to support success over the long term. In developed countries there has been a mixed history of community participation in forestry, with a greater emphasis in those countries or regions where forestry plays a central role in the economy, such as Scandinavia. Over the past several decades, interest in forest ecosystems has increased, which has in turn led to growing concern over the way forests are managed. Where economic dependency on forests has decreased, more people tend to support forest protection, and increasingly demand a voice in the way management is carried out. This chapter will explore the concepts, processes and tools for public participation, illustrated with some case studies. The authors are grateful to Max Hislop of the British Forestry Commission for contributing extensively to this chapter. He has researched the subject together with colleagues from the USA and Finland and has developed a valuable toolbox which is sampled in the chapter.

Communities vary widely in their relationship to forests. At one extreme are towns in rural Canada or Sweden where local economies depend on lumber mills fed from nearby forests. At the other extreme are forests near large cities in Britain, the USA or the Netherlands, where the relationship of people to forests is more aesthetic and recreational than economic. Communities in the former tend to support continued harvest of timber, though at harvest rates that are sustainable so as to protect employment, the economic base of the town and uses such as hunting and fishing. City dwellers may make little connection between their daily use of wood or wood products and the forests from which they come, and instead be keen that forests are protected for environmental benefits. The ways in which communities participate in forest planning vary and can depend on all sorts of factors. This chapter examines a number of these and presents some approaches and methods of community participation appropriate to the range of forest landscapes covered by this book. In particular, we focus on forests of the temperate, developed zones that have active rural and urban populations, a high degree of environmental awareness and advanced forest certification schemes.

Public participation in the forest design process, and in forest management in general, is a difficult and challenging task. Professional forest management has a long history of separating local communities from forests. Historically, the first specialised foresters were in effect put in position by landlords to protect increasingly valuable trees and game from encroachments by the local peasantry (see Chapter 1). The legend of Robin Hood may be seen in one interpretation as a peasant revolt against foresters led by disaffected gentry.

In Central Europe, many forests gradually came under the control of state and local governments, which led to foresters working directly for elected or appointed councils who establish working budgets and provide policy oversight. This, combined with long management tenure, has resulted in foresters who tend to be directly in touch with local needs. In Britain, the USA and Canada by contrast, public forests are primarily under the control of national or provincial governments and managed by top-down bureaucracies that have historically been deliberately kept at arms length from local communities, although, as we will demonstrate, this is changing.

Before suggesting techniques and methods for managing public participation in forest design, it is best to consider the reasons why forest managers have tended to avoid becoming involved in participation in the first place.

The first reason arises from psychological profiles of foresters and natural resource professionals, particularly in the USA, which suggest that those attracted to work in forests often tend to prefer hierarchical forms of organisation and do not like interacting with people – working in the forest and interacting with trees and wildlife is a good way of avoiding this. Such professionals also have a scientific training which looks for rational, objective methods of analysis and decision-making. Public participation requires direct engagement with people who lack the technical level of knowledge that professionals typically possess. Lay citizens may rely much more on emotion or aesthetic impressions than on science. This can be exasperating to some forest managers, who may believe they are qualified to make decisions in the public interest by applying the best science or techniques, and should not be second-guessed.

The second reason is that by inviting the public to participate, managers take a risk that the public may not like, or may force alterations on, the preconceived plans or policies that foresters may be already committed to. Managers may believe that they will lose control over the outcome of the process and in effect be forced to adopt a plan they do not like, one they think is wrong scientifically or one that displeases their superiors.

The third reason is that properly done, public participation can be expensive and time consuming. Public forest administration budgets are always tight, and time or money spent managing a public process has to compete with technical analysis or fieldwork.

The final reason foresters have avoided public participation, no matter how well planned and executed, is that public involvement can never be truly comprehensive and result in all issues being laid to rest. In other words, after the time and expense of a good public process it may seem that a working consensus has been reached. Six months or a year down the road, as the plan begins to be implemented, an individual or group that had been left out of the process, or one that formed later on, may unexpectedly appear and present an entirely new challenge to the plan. For example, on a project at Mt Hood National Forest in Oregon, a group of protesters from a Portland middle school arrived at the forest headquarters one day to object to the sale of timber from a project that had been carefully designed and examined during a painstaking seven-year process. One forest officer complained that "These people should have been here to express their concerns when we started this process." It was pointed out to him that the protestors were only six years old when the project was initiated, and could hardly be expected to have attended evening meetings.

Having acknowledged the argument against public involvement, let us now make the case in favour. Our experience is that there are five key advantages to public involvement that apply both to forest design and forest management in general:

1   The public owns the public forest. This fact is of course self-evident, but is all too often forgotten by both managers and the public themselves. Since they own it, they have a right and indeed a civic responsibility to be aware and to engage in important decisions about the fate of their forest. In the USA, Canada and the UK, ownership has usually not translated into direct engagement. This is in large part a question of scale. While one may "own" a share in all of the national forests of the USA for example, this is not an ownership that is a vested property right that can be transferred or liquidated. Elected officials determine public

forest policies, so the argument is often made by forest managers that if the public is unhappy with the direction of forest management, then they need to elect a different Congress or President, since the managers are merely following policy. Also, if one lives in say New York City or London, it may not be very compelling or easy to get directly engaged in a forest design taking place in California or northern Scotland. Thus the local "owners" may have a much greater say in a plan than distant ones, even though they own equal shares. Philosophically, the authors believe that it is appropriate to address national policy questions at the electoral and lobbying levels, and as long as forest designs are done within the larger policy framework it is appropriate to give greater opportunity and weight to local citizens, who after all are the ones most directly affected by the design.

2   Management of forests clearly has both direct and indirect impacts on a range of important public resources, such as water, wildlife, recreational opportunities and scenery. This is why private forests also tend to be highly regulated in advanced countries. Public involvement gives people an opportunity to voice their concerns about potential impacts on valuable public resources.

3   Local citizens may have more direct knowledge of certain aspects of the forest than the forest managers themselves. This can be difficult for managers to accept, but there are many instances where an individual who has hiked, hunted or fished in a particular forest for decades knows a great deal more about certain issues than does a forester who has been on the local job only a few months or years, covers a vast area of country and may rarely leave his desk or the seat of his pickup truck. Forest managers rely much more on remote technologies such as aerial photographs than on field knowledge, while those who use the forest may be more aware of details that are not captured in large-scale databases.

4   Creating opportunities for involvement is the right thing to do. It is neighbourly, and in effect follows a "do unto others as you would have them do unto you" principle.

5   Providing meaningful avenues for public participation in forest planning may be the law. Where it is required, it must be done whether local forest managers like it or not, so it should be done properly.

Principles of participation

Who is meant by the terms "Public"; "Stakeholders"; "Community"?

Forest planners, designers or managers need to engage with one or more of the following:

- The public
- Stakeholders
- The community.

These terms are frequently used interchangeably, but they do not mean the same thing:

- *The public* means relating to the people as a whole and open to all. Political theory often reduces the idea of the public to a set of aggregated individuals. A more accurate concept is that "the public" is plural, and that members of the public often share an identity that is more powerful than those issues that divide them into individuals.
- *Stakeholders* are the people who have a more easily identifiable and vested interest in a particular decision. This can be as individuals or as representatives of a group. Stakeholders includes people who influence a decision, or *can* influence it, as well as those affected by it.
- *The community* means all the people living in one district, or a group of people with shared origins or interests. Thus community is either a geographical or a social concept (see below).

From these definitions it is possible to think of a community as a subset of the stakeholders, and the stakeholders as a subset of the public.

In any approach to community participation it is also important to define what is meant by the term *community*. This will vary from place to place. There are two basic types of community usually defined in the literature: community of place and community of interest.

Community of place

Community of place refers to a well-defined geographic location within which people live, somewhat separated from other communities and possessing an identifiable character. In remoter, less populated areas, communities of place, such as villages or areas of scattered farms or dwellings associated with a place, fit this category. In densely populated regions it may be more difficult to identify a specific community geographic boundary.

Communities of interest

Communities of interest are defined by groups of people who subscribe to common values or interests or belong to a well-defined category that is not always associated with a geographic location. In the realm of forestry, communities of interest might include loggers, hikers, environmentalists and hunters. It is common for one person to be a member of several communities of interest, such as a logger who also hunts, lives locally and depends on the forest for his water supply, in which case values expressed regarding forest management may conflict within the interest of a single individual. In densely populated or urban regions it is often more usual and convenient to identify a range of communities of interest as opposed to communities of place.

It is also possible for a single community of place to comprise many communities of interest. For example, in Scotland the people who live in a particular village may include long-established residents whose parents lived there (locals) and people who moved into the village more recently (incomers). There may be landowners, tenant farmers, estate workers and people who run local businesses such as a bed and breakfast, post office or garage. These may each hold different values in regard to land-use options yet some groups hold more economic power or have different legal rights than others. This demonstrates that simple categories and broad assumptions about what constitutes a community are unlikely to be sufficient.

Communities exist at different levels. The local community of a settlement such as a village represents a first tier, with the most direct interest in the forest and its day-to-day management. A second tier is the regional level, where there is a wider population who pay taxes and elect representatives to local or regional government. They may have strong views on the forest yet be less directly involved, and may know less about practical issues. A third tier includes national or even international communities, often represented by communities of interest, such as environmental or industrial associations. Both may express their involvement in campaigns run through international networks. These may not be insignificant factors affecting a particular forest design, although it is difficult to involve people directly who are at increasing degrees of physical separation from the decision-making process. Some of their views are accommodated through processes such as forest certification or through national legislation and the mechanisms of representative democracy. Their views can be sought, however, especially in terms of the overall balance of forestry policy and how it translates into forest management regulation, through opinion surveys and questionnaires. The development of websites and e-mail make long-distance involvement much more convenient than in the past, although there are still advantages to face-to-face encounters.

The remainder of this chapter concentrates on the practical issues of how to involve local levels of communities in the planning and design of forests. Local communities often feel that regional and national (or even international) interests are well enough represented or have too loud a voice, but that local views are insufficiently taken into account. If there are indigenous minorities who have a stake, this sense of exclusion from decision making may be even more acute.

One of the important principles of community participation is to identify the range and type of stakeholders who constitute or represent the interests or values held within the community. It is prudent to cast the net as widely as possible and to encourage as many people as possible to participate. In a small community it may be simply a case of inviting everybody to a meeting, while in a larger area with more people it may be more appropriate to try to work with representatives of each stakeholder group (some of the methods to use are described below). A stakeholder analysis can be a valuable asset to either draw upon (if it already exists) or to prepare (if it does not).

Identifying stakeholders

Anyone who has worked in a forest for some length of time may feel that they already know who has an interest in management decisions. The collective knowledge of local forest managers, rangers or office staff is valuable and can be used to identify more stakeholders than any one individual is normally able to do. Stakeholder-brainstorming sessions are best held with a range of personnel, including planners, harvesting and forest management staff, rangers and operational staff. A brain-map framework can reduce the time for these sessions and help to encourage the participants to think of stakeholders beyond the usual "local community".

What exactly is "participation"?

Another important principle of participation concerns exactly what is meant by the term and how it is applied in particular circumstances. Arnstein's ladder of participation is an example of the different degrees to which the community may be "allowed" or "encouraged" to participate (see below).

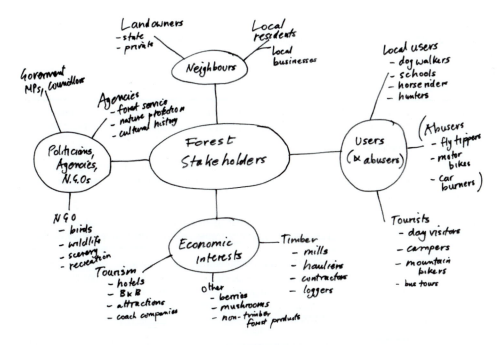

Figure 4.1
An example of a typical "mind map" created to identify stakeholders in a public participation exercise.

Levels of involvement

Any individual may wish to be involved in a particular forest planning process or forestry decision to a greater or lesser degree. Their level of commitment to the process is likely to be determined by what they perceive they are likely to gain or lose, how important the decision is to them and the degree of responsibility they have for the eventual decisions.

The different levels of involvement that people may choose have been classified in various ways. Most classifications follow an ascending scale of involvement, such as Sherry Arnstein's ladder of participation model from the USA. The original ladder represents the different ways in which authorities may try to interact with the public in order to ensure that their decisions are going to be accepted. It was developed primarily to deal with urban renewal or highway projects that had impacts on low-income neighbourhoods.

A second version of the ladder has been developed by the authors to reflect the ways that information, consultation and participation are generally applied in the field of land-use planning, including forestry. As will be seen later in this chapter, different tools can be used to help carry out aspects of community participation, depending on the level chosen. The general aim is to find a position on the rungs of the ladder that can be genuinely classified as participation and which meets specific circumstances. This will depend on the nature of the forest, of the community and of the ownership and management situation. The ladder of participation is a very useful model to help understand the degree of influence participation is intended to have. It can also be viewed as a series of stages which might be tested, step by step, by a forest owner or manager concerned about moving too far or too quickly into unknown territory.

If the forest is owned by the state and is managed on behalf of the public, it is possible to move up the ladder towards some form of citizen control. The Gifford Pinchot National Forest in Washington State is presently experimenting with allowing a coalition of environmental groups to design a timber management project in part of the forest. In effect, forest managers tired of being constantly criticised told the environmentalists: "Let's see if you can do it any better."

Community forests offer additional examples of where this approach is more common. In France and Germany there is a long tradition of commercial forests owned and managed by local communities. In other countries, state or regional governments have sometimes handed management

Table 4.1 Arnstein's original 'Ladder of Participation'

Degree of citizen power	Citizen control	Giving away decision making, resources and control: clear lines of accountability and two-way communication with those giving away the power
	Delegated power	
	Partnership	Two-way communication essential: direct involvement in decision making and action. Clear roles, responsibilities and powers – usually for a shared common goal
Degree of tokenism	Placation	Two-way communication: participants have an active role as shapers of opinions, ideas and outcomes, but final decision remains with the agency
	Consultation	Can be two-way communication: asking opinions, collecting views but final decisions are made by those who are doing the consulting
	Informing	One-way communication: informing the public of their rights, responsibilities and options. Includes provision of feedback of decisions.
Non-participation	Therapy	"If we 'educate' the public they will change their ill-informed attitudes and they will support our plans."
	Manipulation	

Table 4.2 The ladder of participation adapted to reflect forest planning

Degree of citizen power	Community control	The community takes over planning and management functions from the owner or manager, who becomes the agent of the community. Managers offer technical advice and agree to implement the plan on behalf of the community.
	Delegated power	The owner or manager delegates substantial powers to the community and increasingly takes a back seat while objectives and plans are drawn up. The manager may act more as an adviser and facilitator to help the community deal with technical issues of which they have little knowledge.
	Community partnership	The community undertakes substantial activities as part of the development of the plan as partners in the process with owners and managers. Information is drawn from a wide sample of the community to help formulate objectives. Approval of the final plan is sought from the community.
Degree of tokenism	Community involvement	Members of the community are given the opportunity to become involved in a variety of ways, although there is no effort to gain representation of views or to ask the community to draw up objectives.
	Consultation	The owner or manager draws up plans but asks for comments from the community. The owner or manager must show that they are taking comments into account and that they are willing to change their original plans to some degree.
	Persuasion	The owner or manager prepares a plan and tries to gain community acceptance for it by explaining the benefits or reducing concerns about potential negative effects. A selling job.
Non-participation	Information	The owner or manager tells the community what is going to happen without recruiting support or offering the opportunity to comment. May be viewed as an empty public relations exercise by a cynical community.
	Agency control	Plans are prepared by the owner or manager with no actual or intended reference to the community. Still a common situation in countries with strong private property rights related to forestry.

of an area of forest to a community to manage for their own benefit. In British Columbia, Revelstoke Community Forest Corporation was set up in the mid-1990s so that the benefits of forest management could be enjoyed by the local community more directly.

In other cases the state may own the forest but has given a management licence to a private company whose main objectives are to make a profit from timber harvest. The local community may benefit from jobs with the company, but are also likely to want to have some degree of influence over decisions. The company may want to retain overall management control, but if it involves the community in its planning, it is also more likely to retain their support in continuing to manage the land.

Private landowners, especially of large tracts, may feel that private property rights give them a free hand to manage however they wish, but many may also recognise that they have wider responsibilities, and that some form of consultation is advisable. They may be resistant to a greater degree of involvement, although grants or subsidies from the government may have conditions attached that make this a requirement.

Consultation is shown as the lowest point on the ladder that is considered meaningful participation. It requires that the community is consulted fully, given sufficient information and that alternative views are genuinely taken into account before the final plans are approved.

Approaches to participation

Planning for public involvement

Before starting a process of public involvement it is sensible to think through three key issues:

1 The ways that each stakeholder or group of stakeholders desire to be involved in the process;
2 How flexible the process is as the level of involvement changes; and
3 The demand this may put on the available resources.

There is no one ideal approach or practice for developing participation. Ideally, the starting point for the agency or forest manager is to build a *participation strategy* that best fits the location and scale of the project. In addition, it is critically important that the process be designed to facilitate responsible community governance, rather than to pit interest groups against one another. It is valuable at the outset to consider what tools for participation are needed and then to organise the approach around them. The US Forest Service suggests the following set of questions to determine what tools should be adopted:

- What does the public want from us and what do we want from them?
- What information does the public need to give us the information we want?
- What are the social, cultural, economic and geographic characteristics of the stakeholder group or individual and how might those affect the methods and location we choose?
- Will it reach the right people?
- Is it convenient for involvement?

The British Forestry Commission, together with the US Forest Service and Finnish researchers, draw from a toolbox for community participation. It identifies a range of tools that can be used for different levels, such as informing, consulting, involving and working in partnership. Table 4.3 summarises these tools.

A sample of the tools listed, especially from the sections on consulting and involving, will be explored and some of the problems that can occur with participation identified. Some examples of the different tools will also be presented.

Advisory committees

Advisory committees are small groups, usually of between 10 and 20 people, formed from representatives of various stakeholder groups. The committee members meet regularly to discuss issues and bring up ideas. The aim is to find out the stakeholders' views and priorities regarding specific issues rather than set detailed recommendations for action. The committee members should represent a broad range of interests and they can be selected by interviewing potential individuals. The committee should ideally be provided with comprehensive information in order to reduce reliance on experts and technical knowledge. Prior to initiating meetings, background information, minutes and agendas should be prepared and distributed. Several other methods of participation (field trips, presentations, workshops) can be used to encourage participants to explore and analyse issues and to gradually arrive at consensus about the way forward. A third party may be needed to facilitate the process (see section on facilitation below). When setting up advisory committees it is important to define the roles and responsibilities of the parties clearly. The working process of the committee should lead to a final report that gives non-binding recommendations for action.

Strengths:
- Contentious decisions may become more acceptable
- If the committee is truly representative, it takes account of multiple points of view

Table 4.3 Tools for public participation. One tick means it is appropriate but not the best place to use a tool, three ticks means it is an ideal tool for the purpose

Tool sheet	Informing	Consulting	Involving	Partnership
Advertisements	✓✓			
Advisory committee		✓	✓✓✓	✓
Briefings	✓✓	✓		
Citizens' jury		✓✓	✓✓✓	✓
Community issue groups		✓✓	✓✓✓	
Consensus building		✓	✓✓	✓✓✓
Co-view		✓	✓✓	✓✓✓
Delphi			✓✓	✓
Design charette			✓✓✓	✓
Direct observation		✓		
Displays	✓✓	✓		
Electronic democracy	✓	✓✓	✓	
Focus groups		✓✓	✓	
Forest events	✓✓	✓✓		
Forming partnerships				✓✓✓
Forum		✓✓	✓✓	
Interactive displays	✓✓	✓		
Internet survey		✓		
Interviews		✓✓		
Leaflets	✓✓	✓		
Media	✓✓			
Newsletters	✓✓✓			
Newspapers	✓✓			
Nominal group technique		✓✓	✓✓✓	✓
One-to-one contact	✓✓	✓✓	✓	
Open house	✓✓	✓✓		
Open space techniques		✓✓	✓✓	
Participatory appraisal		✓✓	✓✓✓	✓✓
Planning for Real®	✓	✓✓✓	✓✓	
Presentations	✓✓	✓		
Public hearing	✓✓	✓		
Public meeting	✓✓	✓		
Questionnaires		✓✓		
Response cards		✓		
Secondary data	✓			
Shared decision making				✓✓✓
Site visits	✓✓	✓✓✓	✓✓	
Small informal displays	✓✓	✓✓✓	✓	
Staffed displays	✓✓	✓✓		
Surgeries	✓	✓✓	✓	
Surveys		✓		
Task force		✓	✓✓✓	✓
Telephone hotline	✓	✓		
Telephone surveys		✓		
TV and radio	✓✓			
Un-staffed display	✓✓	✓		
Visioning				
Website	✓✓✓	✓		
Working groups		✓✓	✓✓✓	✓✓
Workshops		✓✓	✓✓✓	✓✓

Source: Forestry Commission.

- Participants gain greater understanding of other perspectives, possibly leading toward a compromise/consensus strategy
- Provides the opportunity for detailed analysis on planning issues
- Long time-scale (multiple meetings over months) gives ample opportunity to raise issues and to debate them in more depth.

Weaknesses:
- Can be expensive and take many hours to staff
- Can take many months or even years
- Even then members may not reach a working consensus
- The wider public may not approve committees' recommendations
- The legitimacy of the process is dependent on the attitude and commitment of the agency to listen to committee recommendations
- Participants may have unreal expectations of their true influence over the process
- Large committees can make debate and discussion difficult to manage
- Professional facilitation may be essential, adding more expense.

Citizens' juries

Citizens' juries involve a group of representatives of the community, normally 12–25 persons, who volunteer to spend several days considering a subject in depth, discussing and researching the matters at hand. Juries are organised by independent bodies and report back to the concerned parties. Jurors hear evidence from witnesses who might be experts or members of pressure groups and receive written evidence. They scrutinise the evidence, debate the questions and deliberate their decisions within the group. The commissioning agency is expected to publicise the jury and its report as part of the process of public involvement. An agency, for example, might commission a citizens' jury to contemplate a resolution to a particular, often controversial, dispute then consider its findings when deciding on a policy. A citizens' jury does not replace other forms of consultation or participation, but may provide a new perspective that adds openness and fairness to governmental activities. Citizens' juries are typically non-binding and lack any legal status.

Strengths
- Provides an avenue for the public to identify with the findings and support a recommendation
- New perspectives brought by ordinary people from outside a dispute may highlight new solutions
- Helps to build consensus and share information.

Weaknesses
- Requires considerable resources and time to set up and conduct
- May not produce a consensus even within the jury if the issue is extremely controversial
- Does not normally generate widespread participation.

Community issue groups

Community issue groups have similarities to both focus groups and citizens' juries. Their main aim is to bring new views and external perspectives to the planning process. Community issue groups usually consist of between 8 and 12 participants who meet up to five times or so, over a series of weeks. Meetings build upon the previous discussions, giving attendants time between gatherings to reconsider issues and raise questions. New information can be introduced to the discussion to build up the participants' knowledge of the issue. The discussions are normally taped and analysed later on and reports are produced.

Strengths
- Works more efficiently in creating deliberation compared to focus groups
- Informed discussions offer an opportunity to explore issues in depth
- Opportunity to refine views
- Fairly cost-effective.

Weaknesses
- Time requirements are relatively high
- A relatively low number of participants are involved in groups.

Design charettes

A design charette is a short, intensive session (one or a few days) in which a group of participants explore topics related to a specific problem and redesign project features. It is a process typically used in architecture and urban design, but can be adapted to forest design as well. The actual planning and design may be done by professionals, with community stakeholders invited to review and comment at critical stages. A charette can also contribute to sharing information and increasing participants' understanding of multiple issues. Charettes often begin by distributing background information, followed by a field trip. A group leader presents principles that underpin the planning and design process for participants. The group may be sub-divided to create alternative approaches, or to focus on certain aspects of the design. Ideas are presented, debated, discarded and reiterated, with the group gradually working towards consensus and a final resolution. After an agreed-upon plan is created, a report presenting the whole process and its outcomes is produced for wider distribution and comment. Presentation, graphic images, design standards and implementation strategies produced in a charette provide essential documentation for the planning process. The design charette may generate a prioritised action plan regarding the problems being addressed. It is essential that all participants understand and agree with how the results will be utilised.

Strengths
- Turns the attention of the attendants to solutions and constructive ideas and away from argument
- Helps generate partnerships and positive working relationships
- Allows interactive learning between planning experts and local community representatives
- Helps create deeper understanding of planning issues
- Can be relatively inexpensive and highly productive of ideas compared with more drawn-out methods.

Weaknesses
- Participants may not be representative of broader public
- The effects may not be long-lasting if the design charette is used as a one-off technique
- An experienced leader or facilitator is needed to guide the process
- Requires intensive preparatory work
- Going so quickly to solutions can backfire if important details or issues are overlooked
- Later analysis of results may reveal fatal flaws that are hard to repair.

Nominal group technique

Nominal group technique (sometimes referred as Delbecq groups) can be used to define needs and goals from representatives of different interest groups. It may also help in prioritising ideas and identifying solutions to specific planning questions. The group size should preferably be less than 12 persons. To start with, a few simple questions are asked by a facilitator to generate participant response to the issue. The attendees normally formulate their answers and judgements of alternative ideas independently in written form. All the participants are then asked to read out and explain what they have written. Each idea is discussed more widely and clarified by the participants, and the individual ideas are numbered. Participants then indicate their preferred ideas (for example by voting with sticky dots) and a discussion of these preferred ideas follows. The group tries to reach a common solution to the questions or issues that were originally posed. If there is still lack of consensus, the individual judgements are produced again. The method should lead to a prioritised list of actions or issues.

Strengths
- All participants are likely to contribute owing to the small group size
- As the debate is limited, the participants may express their ideas with minimal fear of being criticised
- The method can help to prioritise between different issues or options
- Can lead to consensus between the participants
- Participants may be from a variety of backgrounds
- Only limited resources are needed.

Weaknesses
- Includes only a very limited number of participants
- Does not generally allow in-depth examination of the issue
- A balanced participation of stakeholders is essential
- Must be combined with other means of involvement where the issues are complicated.

Participatory appraisals

Participatory appraisal is a methodology that creates a cycle of collecting information, reflection and learning. Practitioners design a process based on the needs of the client, then use suitable methods to facilitate analysis and discussion of local issues and perceptions with and by local people. The methodology has evolved rapidly and is continuously modified by users. It can be adapted to work with small groups or entire communities. Each group of participants proceeds gradually from stage to stage, first looking at their perceptions of the current situation, then identifying barriers or gaps and coming up with solutions or issues for change. Participants are able to choose their own level of participation that suits their interests and needs. Many of the methods used are visual which helps to simplify complex issues. Different methods used include brainstorming, institutional analysis diagrams, ranking of priorities or criteria and community mapping.

Strengths
- A highly flexible methodology
- Uses interactive activities involving many stakeholders
- Helps groups to determine their priorities for action
- Can be used in different locations where people naturally come together
- The opinions and concerns of local people have a central role in the process.

Weaknesses
- Requires trained facilitators to guide the process
- May require a long period of time and resources to generate outcomes and reach decisions.

Planning for Real

Planning for Real ® (PfR) is a group involvement technique for soliciting suggestions and opinions from local people who may or may not otherwise be likely to express themselves. The technique, developed originally by the Neighbourhood Initiatives Foundation based at Telford in the UK, is designed to provide a hands-on, non-threatening experience to community members. Participants use a model or a map of the planning area in a workshop setting on which they put options cards or other symbols representing issues, problems or suggestions for actions that they would like to see. The actual equipment can be very simple, and the rules for running the process are also very basic and flexible. It is important that the people lead the process while expert staff are available to answer questions. There are no hidden restrictions on options, although each person can make each suggestion only once, among others. Forests for Real is an adaptation developed by the Fort Augustus Forest District in Scotland, where they have developed options cards specifically related to forestry issues.

Strengths
- Many people can have their interests included in planning activities while preventing a few vocal or articulate people from dominating input to the process
- Can increase the feeling of ownership in any outcomes among the community members
- Increases the level of trust between the parties
- Can be applied to different sized planning areas.

Weaknesses
- Requires a fair amount of preparation time to ensure adequate attendance
- Restraint is necessary from meeting organisers to allow community members to participate fully
- If done poorly can raise expectations beyond the level at which outcomes can be delivered
- Many more than 50 participants in any one session can be unworkable.

Surveys

Surveys can be either formal or informal. Informal surveys tend to reach a self-selected group of people, whereas formal surveys are scientifically assembled and administered to obtain information from a statistically significant sample of population. Surveys are a means to get a general sense of an average response from a specific section of population or the whole population of a particular area. They can provide information on public opinion about particular issues and public concerns related to planning and find out what information people would like to receive. Surveys can be carried out in person, by mail, by phone or by the Internet. The design of surveys and structuring of questions need to be conducted carefully to avoid errors in information gathering. Benefits from surveys – whether conducted by interviewers or completed by respondents – may be improved if local groups are involved in the whole process, including design of the questions, administration of the surveys and analysis of the results. Some common types of surveys are: Internet surveys, response cards, telephone surveys, interviews, questionnaires and one-to-one contact.

Strengths
- Can provide a good cross-section of public opinion and people's views about given issues in the area
- Can be targeted to special groups of population
- If properly designed, the results can be statistically valid
- Informal public opinion surveys are relatively inexpensive
- Helps to gauge the wider public view, which may not be represented by well-organised stakeholders.

Weaknesses
- Can be expensive, both time and labour intensive
- Normally requires professional skills to design and implement

- Surveys do not provide much opportunity for interactions
- Interviews may give false impressions if not conducted in proper way
- May raise false expectations within communities unless the purpose of survey is made clear
- Sample must be taken with care, since the phrasing of questions is crucial.

Task force

A task force group consists of experts or relevant stakeholders; the composition depends on the issues under discussion. A task force can be formed when a specific outcome or policy recommendation is to be developed. It may review the participation process, receive community input and exercise other functions depending on its mandate. A collaborative task force group is assigned a specific task, with a time limit for reaching a conclusion and resolving a problem, subject to ratification by official decision makers. The personnel of the agency usually appoint task force members and a facilitator guides discussion to cover all issues that the participants see important. It may use other group work methods such as brainstorming in order to seek solutions to the specific problems. The sponsoring agency can provide technical support depending on the issues addressed. The members of the task force should have credibility with the public and represent various views. It is also important that the members are independent. Academic institutions may sometimes take part in organising a task force with a local agency.

Strengths
- Findings are likely to have fairly high credibility if diverse interests of the stakeholder groups are presented
- Offers an opportunity to form compromise
- Can produce high-quality proposals and recommendations
- In a collaborative task force, a great depth of discussion is expected.

Weaknesses
- The costs may be quite high
- A skilled facilitator is needed
- Time and labour intensive
- No guarantee that consensus will be reached
- Participants must make an extensive commitment to the process
- The results may be too general to draw any firm conclusions.

Working groups

Working groups offer an effective participation means for organised stakeholders. They help participants to familiarise themselves with the planning issues and keep them and their groups constantly in touch with the process. The size of a working group is normally fairly small, ideally between 5 and 12 people. Before forming a group, the stakeholders have to be carefully assessed. The purpose and role of the group in planning should be clearly identified before starting the group work. At the first meeting, the inclusiveness and the tasks and common rules of the group should be agreed upon and the planning process and aims should be explained to group. An important goal of the group work is to clarify different opinions and try to build consensus. Working groups try to arrive at solutions through dialogue and consensus rather than voting on issues.

Strengths
- Members are usually well informed and have expertise on certain topics or issues
- Provides a good chance for in-depth interaction and negotiations

- Members' knowledge about the planning topics and objectives of other interest groups increases
- Group working gives forestry officials immediate feedback during the whole process.

Weaknesses
- The members of the working group may have to commit plenty of time and energy to the work, favouring organisations with paid staff
- Usually only a few individuals constitute a working group, thus it is not broadly representative.

Workshops

Workshops include a wide range of different group work methods. The participants of a workshop usually formulate, assess and resolve problems related to a pre-defined topic. Workshops enable conversation on and exploration of issues at hand. They can be arranged as a one-off event or several times to ensure effective participation. Alternatively, participants may be divided into smaller groups to increase the intensity of the group work. It is recommended that some background information is provided for attendees before the meeting or that they have a presentation giving a clear overview of the issues. Several group-working methods can be used, for example brain-storming and nominal group technique. At the end of the process, the participants usually reflect what they have achieved in the workshop. "Information exchange workshops" are small groups usually targeted at representatives of different groups. "SWOT workshops" explore strengths, weaknesses, opportunities and threats related to a given issue. "Initiatives workshops" further develop the ideas created in SWOT, consider details on factors and produce an outline work programme for a particular issue. "Action planning workshops" are arranged when there is a need to hold a session specifically for interested parties who will have a role in the implementation process.

Strengths
- Excellent for identifying issues and analysing alternatives
- Fosters public ownership in solving the problems
- A direct form of participation that is likely to promote communication between participants in the future
- Can raise the level of awareness of topics by attendees
- Less formal compared to public meetings or committees.

Weaknesses
- Labour and set-up requirements are usually rather extensive
- Multiple trained facilitators may be needed for large sessions
- Workshops have to be planned and structured carefully, although over-planning can lead to suspicions of manipulation by participants.

Before moving on to some case studies in public participation, it is worth focusing more closely on some popular methods that tend to be used most often.

Surveys and discussion groups

In order to gather information about values and perceptions from the majority of community members who are unwilling or unable to attend workshops or other events, it is necessary to use other approaches. Surveys can be done by knocking on doors (in small communities), over the telephone, by making an appointment followed by a visit, by stopping people randomly in the street or in other public places, by post or via the Internet. Postal questionnaires typically have a low response rate, while face-to-face or telephone interviews are more successful but time-consuming. Surveys need to provide enough background information to help people understand the context of the subsequent questions, and questions must be posed in such a way that meaningful responses can be obtained without leading the interviewee or respondent. One advantage of surveys is that, depending on the overall size of the sample, it is possible to do statistically valid analyses of the results. Surveys can be designed so that the sample is representative of the demographic structure of the local population, with the results extrapolated to the population as a whole. Planners can use the results of such a survey to balance or give a context for the outputs of the more direct phases of participation.

The second method is to hold discussions with groups of common interest or location. Examples include: school classes, older people's clubs, young mother's groups, unemployed people, company employees and so on, all groups that are unlikely to attend workshops and meetings. The facilitator and an assistant arrange to meet the group, are able to present some background information and then, using some questions, start an open discussion. Notes are taken and sometimes sessions are recorded, the results used to inform the rest of the process in the knowledge that a broad set of opinions and information have been included. These group discussions are sometimes known as focus groups. In some more sophisticated processes these focus groups are held early, and the content of questionnaires developed from the issues raised at such groups.

These methods, if they are to work, require some special skills to be found in the social sciences. It is not to be expected that every land manager will be able to practise such exercises without expert help. However, with some training and experience it is possible for key people to learn how to do these techniques.

Large, all-inclusive public meetings are relied upon too often as a way of engaging people, and can be one of the more inefficient and misleading techniques. Public meetings tend be organised with a group of planners or managers behind a table at one end of a large hall, with an audience of concerned citizens facing them, setting up a tribunal atmosphere that breeds mistrust. They tend to reinforce polarisation and often provoke open conflict and grandstanding. They can be difficult to manage since groups have time to plan disruptive tactics beforehand, especially those opposed to the plans or project being presented.

Large meetings work best during the opening stages of the planning process, to inform people of the project process and to ask for their involvement, or to identify their issues. Ideally there should be no plans or even outline ideas at this stage. Instead, the focus should be on pre-empting rumours and to prevent misinformation from taking hold in the community. If possible, these meetings should be chaired by an independent person or respected member of the community instead of the landowner or a member of the forestry company. This approach demonstrates an open agenda with no preconceived plans, and helps to build trust at the outset.

Public meetings can also be used for scoping, where they help identify the main issues in the community. If a scoping report is prepared and circulated back to those who attended, the openness of the approach can be demonstrated and trust further developed.

Facilitation

In order for the participatory process to be successful and for the results to be accepted by all interested groups, it is important that no single stakeholder dominates the process or controls all the key information. In bigger and more complex and controversial projects, it is advantageous for the party initiating the plan – the forest manager, the landowner or community interest group – to appoint an independent facilitator to organise the process, run the meetings and workshops, take notes and report back to the community. A good facilitator has a mix of attributes, including negotiation, mediation, meeting management, communication and interpersonal skills as well as a general understanding of planning and design processes. They must be able to gain and keep the trust of the different stakeholders throughout the participation process in part by speaking "truth to power" when needed.

Examples of participation

Sutherland's Brook, Nova Scotia

Stora Forest Products (now StoraEnso) set out to involve the community in developing a harvest plan for an area in Nova Scotia, Canada, and employed several of the tools outlined above. These included an advisory committee (called here an "objectives committee"), a working group of members wider than the objectives committee, an intensive workshop session lasting several days and a design charette (done within the workshop session).

The objectives committee was composed of representatives from different stakeholder groups interested in the area. They were set up initially to ensure that as wide a range of interests as possible was considered. This was particularly useful given that the area is remote, population is low and there is not a readily identifiable community associated with the forest. Important communities of interest lay some distance away, yet had a legitimate interest in the plans. The objective committee was so called because its principal task was indeed to help formulate objectives for the plan and to ensure that as wide a range of values as possible was taken into account. A facilitator from the Canadian Forest Service helped to identify potential members of the committee together with local forest managers and the Provincial Forest Service.

The workshop involved a wide range of stakeholders invited to participate in a five-day intensive session during which a facilitator led the group through the full planning and design process described later in this book. This allowed a deeper understanding of the issues to be developed and encouraged intensive debate. The design charette was held once the analysis was completed. Workshop participants were divided into three groups, each developing a different design. These were then evaluated, with one chosen for further development. The professional planners within the company then iterated this plan and later presented it to the objectives committee for comment and approval. This committee was consulted at intervals over several years as the plan began to be implemented to ensure that there was continued support from the community.

Objectives committees can be the principal element of participation. Once the objectives are agreed to, professional planners can create a first draft plan, presenting this to the committee for critique and comment. It must receive approval by the objectives committee before it can be moved to the next level, such as the Provincial Forest Service. A wider public can be kept informed via their representatives, who are expected to show their wider constituency information and ideas arising from the plan. This approach relies on a limited number of people as a sounding board and external indicator of the suitability of plans drawn up by the forest managers. It is a good method where the practicality of directly involving a wider public is challenging. The results of the project are presented in a case study in Chapter 9.

Strathdon, Scotland

Strathdon, Scotland, is a remote Highland community where public meetings and a newsletter were the main tools used to inform and gain feedback from the local community. Interviews and questionnaires of a sample of the local population were also combined with workshops facilitated by a professional planner. The use of interviews and questionnaires is valuable because any community has some willing and able to participate actively in planning, but always has others either unable or unwilling to do so. In contentious situations, unrepresentative input can generate serious problems. This can be avoided by recognising that an otherwise "silent majority" can be reached in other ways. Thus there is the need to use both "active" and "passive" modes of participation. In the former, active members of the community can become directly engaged and take part in a range of events, such as workshops. The latter group can be reached in other ways, such as by members of the planning team visiting directly with them. This need not be intrusive. Many people cannot afford the time, are physically incapable of attending events through illness, age, lack of transport or because they have to look after children, for example, yet are keen to

Figure 4.2 Some of the materials produced as part of the Strathdon project in Scotland.

make their opinions known. Sometimes public events are dominated by articulate and educated people who get listened to because they know how to present themselves and their arguments with confidence in front of others. Good facilitators know that other people may have important views, yet may find expressing themselves difficult. In Strathdon, where the population is small, planners were able to gather information from quite a large local sample, and to compare the aspirations revealed through this method with the results of the workshops attended by the more active members.

In Strathdon, the local residents have a lot of knowledge about the forest (often of a specific or detailed sort), so workshops were also a good way for planners to ensure that their work was as accurate and as site specific as possible. Moreover, Strathdon (and other places, e.g. Sutherland's Brook) includes local experts such as bird-watchers, hunters, gamekeepers, trappers or avid recreationists who know the local ecology and are willing to share their knowledge.

At Strathdon, the approach involved workshops and also engagement with different residents through interviews as well as some projects where children and other residents were encouraged to express their feelings for the landscape through art activities. This enabled both active and passive participants to take part from all sections of the community. The workshops were held over two consecutive days, and included professional planners who could work directly with community representatives, exchanging information and building confidence. Workshop sessions split into smaller groups that worked on the same aspects, resulting in different findings that could be compared. The facilitator took all the material and synthesised it into a draft plan which was presented to the community at a public meeting, and circulated widely throughout the village.

Little Applegate Watershed, Oregon, USA

In the 72,000-acre (29,000-ha) Little Applegate Watershed of Southern Oregon's Siskiyou Mountains, a strategic public process engaged the local community on its own turf. Complex ecological assessment findings were translated into lay language and graphic images. Local residents and activists who have first-hand, detailed knowledge of the landscape were encouraged to bring their skills to the table in a collaborative effort. A new understanding of the land that had been made possible through landscape ecology, conservation biology and restoration was joined to a public involvement process. This marriage allowed the design of a long-term landscape structure and pattern that, if followed, is expected to retain native biodiversity, restore aquatic conditions, harness the agent of fire and retain economic benefits associated with farming, logging and wildcrafting over many years.

The Little Applegate flows into the mainstem Applegate River near the unincorporated town of Ruch, Oregon. The Applegate in turn flows into the Rogue River, which carves a spectacular white-water canyon through the Siskiyou Mountains before entering the Pacific Ocean near Gold Beach, Oregon. This is a rugged, twisted landscape with small farms, orchards, vineyards and ranches hugging narrow valley floors separated by steep mountains. South-facing slopes include grassland, chaparral and pine-oak woodlands, vegetation types more common in California than in Oregon. North-facing slopes have mixed conifer forests, including remnant patches of old growth Douglas fir and pine. Sub-alpine parklands decorate the highest peaks and ridgelines. Federal lands in the Applegate watershed (two-thirds of the total land area) were designated as one of nine "Adaptive Management Areas" in the Northwest Forest Plan of 1993. This is a regional plan covering millions of acres west of the Cascade Mountains in Washington, Oregon, and northern California. It was the nation's first regional-scale biodiversity conservation plan.

Adaptive Management Areas (AMAs) were established to provide for social as well as ecological experimentation. They were intended to serve as laboratories where federal agencies and local citizens could collaborate to plan and carry out land management work. This is in contrast to the more formal, legalistic and generally non-collaborative approach that had been used on Federal lands elsewhere in the region. Citizens crossed political lines to join together and form the Applegate Partnership several years before the Northwest Forest Plan was hatched. This is a "bottom-up" organisation that includes loggers, ranchers, aging-hippie environmentalists, organic farmers and trust-funders among others. They share advocacy for progressive land management approaches, including thinning overly dense forests to reduce fire hazards and restoration of salmon habitat in local streams.

To its credit, the Rogue River National Forest wanted full and open community participation. They wanted it done thoroughly and without favouring one group or opinion over another. Since this was to be a very long-range plan (200 years) for a large watershed that included public lands and the private property of several hundred residents and timber companies, there was nothing to be gained by crafting a design in a closet. A critical mass was needed to support the selected design concept. If the community was to support it, then they must believe they contributed to its making. They needed to see their own values and ideas reflected in it. To accomplish this, a strategic involvement approach was crafted.

Public involvement can also be seen as a branch of public relations, and one definition of public relations is: *the planned effort to influence opinion through socially responsible and acceptable performance, based on mutually satisfactory, two-way communication.* This definition formed a foundation for the public strategy. But the process was not initiated simply to have an open-ended dialogue. On the contrary, there was an agenda; the crafting of a long-range landscape plan for an entire watershed within the existing policy framework. Thus there was a conscious effort to influence local opinion regarding the nature of this plan, not in order to sell a pre-conceived design, but to get everyone to understand a complicated design process and its potential for solving some of the difficult ecological issues of the area. There was nothing sneaky or manipulative in this approach. In the end, it was critical for citizens to have trust in the process and take pride in the results. Part of the intent was to build a process that would be repeatable in nearby watersheds, so building local support and trust was also essential as a reference point.

The strategy defined the varied stakeholders concerned with the Little Applegate Watershed. The notion that there is such a thing as a "general public", at least in terms of how a particular project is presented and advanced, was rejected full stop. Instead potential participants were defined by common characteristics, interests, concerns or recognisable demographic features. Individuals often fit into multiple categories: logger, local neighbour and environmentalist, for example. Thus communication was targeted at neighbours, local environmental organisations, elected officials, wildcrafters, the timber industry and so forth. In all, 21 stakeholder categories were identified, including personnel of the very agencies that were sponsoring the project. If this seems bizarre, consider that on another project, one of the authors was asked to serve as "a liaison" between two separate departments in the same agency!

For each group, the following information was generated:

1   Key contact names
2   Membership

3   Objectives for what was expected from them (i.e. active support, expression that they were heard, etc.)
4   Tools for communicating. These included newsletters, attendance at their meetings, avenues for participation, etc.
5   Key messages to get across (known sometimes derisively as "talking points" in politics)
6   Measures for success. These are based on responses expected from the varied publics expressing their satisfaction with the process and willingness to participate in the future.

A useful starting point was a detailed survey of community values done by a local organisation: the Rogue Institute for Ecology and Economy. This survey exposed many common themes that linked conservatives and liberals, young and old. Nearly everyone wanted to restore salmon and steelhead runs. They wanted to retain natural-resource-based economic opportunities at a sustainable level. They were concerned about wildfire and the health of the surrounding forests. And they all wanted to maintain the rural character of the area, along with a sense of community.

The process for crafting the strategy was long and detailed, with a specific budget for outreach. More potential tools were provided than the client could ultimately make use of. The key tools that proved most effective were:

1   Establishing a design task force with trusted representatives of key interests
2   Well publicised meetings in local homes and the Grange Hall (a key community meeting space)
3   A series of articles written for a local publication that reaches every household and interest, and is a trusted source of news
4   Clear communication of the "decision space" available to local managers. In other words, if particular interests had an agenda outside of the ability of this project to solve, they were advised as to where and how to take that agenda to be more effective
5   A true openness to local knowledge and ideas, and clear acknowledgement of various contributions.

*Figure 4.3
A public
participation
workshop taking
place on a forest
planning project
in the USA.*

In the end, budget and time constraints led to less than 75% of the overall strategy being implemented. But the results were encouraging. The long-range design for the landscape generated widespread support. The community developed a better understanding of the complex ecology they must live with, and they were given new tools for making improvements. The effort initiated from the top met the local interest at the bottom. The actual design aspects of the Little Applegate project are presented in Chapter 8.

Other methods of participation

As well as the techniques for engaging communities in forest planning described above, it is also possible to use other, less direct means. Some people are not very good at verbal communication to express their feelings and values and yet they can express themselves through art or music. Art workshops can be held and the resulting products used to inform the more traditional planning work. Often art can present aspects of values that are difficult to capture by other means – aesthetic, spiritual, mythological or tribal, for example. The use of materials taken from the forest may encapsulate a powerful and unique sense of place; stories and poems or music may also express feelings about the forest in general or special places in particular. The very act of making art can help to engage people and raise awareness.

Some people may prefer to engage in the forest more directly, perhaps by planting trees, making trails, clearing rubbish or making benches or signs. Such activities demonstrate care, a direct relationship with the landscape and present a means for urbanised people who perhaps feel somewhat alienated from the forest to renew their contact with it.

In forests close to cities where there are communities with little tradition of direct engagement with the land, participation through any of the techniques described above is problematic. In these circumstances it may be necessary to make an extra effort to bring people out into the forest so that they can gain some first-hand experience as a precursor to engaging in further participation.

There are societies where traditional planning systems have specifically precluded participation by communities, and there are residual low expectations of being able to influence decisions or a lack of experience in local democracy. In parts of the former Soviet Union where the forest planning system was highly centralised for decades, and no role was provided for people to contribute to any areas of local democracy, there is a need for capacity building ahead of any sustained attempts at community involvement in projects. There may also be a degree of cynicism and suspicion about the new readiness of centralised systems to embrace local participatory democracy. Often it is relatively small and emerging NGOs which act as the facilitating organisation to help build trust among both the community and the forest administration.

Problems of participatory planning

While participation by local communities in forest planning is to be encouraged and generally leads to better results, there are several ways things can go wrong.

If there are expectations or suspicions among sections of the community that the forest managers already have firm plans and that they only intend to carry out participatory planning as a cosmetic exercise, then the process is unlikely to be successful unless such suspicions can be refuted early in the process. This means that the forest managers bear a responsibility for generating trust before anything concrete is attempted. Such a situation can be avoided by allowing a trusted third party, perhaps a government agency or NGO, to act as the enabler or facilitator. But in point of fact many public forest managers are still reluctant to share crucial information, let alone decision-making authority. The culture of many natural resource agencies is still that of a top-down hierarchy where promotions, pay and prestige are tightly controlled, and the local manager who opens up to the community risks censure from above.

If the public anticipate that highly controversial proposals, such as logging old growth, are likely to be on the agenda, there may be a risk of staged confrontation. In these cases formal public meetings should be avoided or postponed because they are subject to being hijacked by special interest groups who pack the meeting with their supporters in order to set off a well-organised campaign, making use of the insatiable media appetite for controversy.

In some communities different sectored interests may simply be impossible to reconcile, and attempting a participatory approach may only serve to widen the gulf separating the groups and lead to severe problems in the community. An example is where a significant section of the community depends on and wants to retain jobs through logging and milling while another section wants to see all logging cease in favour of conservation, recreation and tourism. It may be futile to expect these groups to work together if the conflict has been running for a long while, and there may be little by way of compromise that managers can offer. It may be best in this circumstance to focus outreach efforts towards the middle part of the political spectrum, assuming a functional middle exists or could be nurtured.

If the broadest range of community interests and representatives is not sought when planning the participatory process, it is possible for one active section of any community to exert undue influence. If the extent of community participation is near the top of the ladder, such as "community control", or decisions are made by the community to be carried out by the agency, then the group involved can become a "quasi-agency" and themselves be seen as merely informing the rest of the community about what they have decided.

In many cases it may be difficult to motivate the community to take part. While planners, community leaders and agencies may view participatory planning as a good thing, the community may feel they have better things to do than spend winter evenings going to endless meetings. This may be a serious problem where there are multiple participatory processes at the same time or stacked one after the other in short order, leading to "consultation fatigue". If the local experience has been that previous processes failed to make any noticeable difference in outcomes, there may be a high level of cynicism and apathy to overcome. A further obstacle is where planning is in advance of implementation by several years, and thus not likely to affect people's lives in the immediate future. It is ironic but perfectly understandable that the most contentious issues that guarantee a large turnout to meetings are least amenable to a participatory process, whereas those with less time pressure, allowing full analysis and with scope for full participation, may not be seen as a priority by community members.

Conclusions

Public participation in forest planning and management is an increasingly important activity. While there are issues of practicality and costs that may deter forest owners or managers, the benefits, not only to the community but also to the owners or managers themselves, can be significant. There are many approaches to participation that are suitable, depending on the country, its political system, the forest ownership and management structure and the types of communities found in the area. The authors have direct experience of many of the approaches described in this chapter, and believe that there are some key factors that contribute to successful participation and which are common to most circumstances:

1   Project initiation. Someone must initiate a participatory plan, be it one or more agencies or members of the community. The source of the initiation may affect the quality of the process and the degree of participation.
2   Early, informal meetings or a series of small conversations can gauge interest, describe and develop the scope of the plan and identify the resources that will be needed. Basic information about the planning area and community(ies) involved should be shared, sources and availability of information ascertained and any gaps identified so that specific survey and analysis can be carried out.

3   Once the planning process is established, a local community liaison should be recruited who can start raising awareness, identifying people to interview and arrange further meetings. Such a person can provide a valuable bridge between managers, staff and the local community, and establish open lines of communication.

4   A professional facilitator or planner should be given the main responsibility for guiding the public involvement, perhaps under a steering group comprising members of the community and the relevant agencies. This person should coordinate participation activities with the information assembly and the process of creating plan(s).

5   Only a qualified person should be made responsible for carrying out interviews and questionnaires. If local talent is used, training should be given if needed. Together with the local liaison person, the interviews and questionnaires should be completed and the results written in terms that the community can understand. The results should be publicised throughout the community.

6   Technical information should be assembled and presented in such a way that it can be understood by non-technical people. This is where models, such as those used in Planning for Real, are especially valuable.

7   Planning workshops should be set up at convenient times where technical experts can exchange information, views, ideas and values with each other in small sub-groups. The workshops should follow a set structure, perhaps answering a set of pre-defined questions. One person in each sub-group should be responsible for recording the information as it is assembled.

8   The project leader/facilitator should be responsible for taking the material from advisory committee meetings, workshops, interviews and questionnaires, combine this with technical information and blend this together into draft alternatives. These should be shared with the whole community for comment, but within a structured framework that includes technical analysis of pros and cons.

9   Multiple iterations of plans are normally required, at each step narrowing the focus and building understanding and support. It is crucial that an otherwise well-managed public process not be sabotaged at the very end when the budget gets tight and managers grow impatient to reach a final draft.

There are many potential future developments that could make participatory planning better integrated into all kinds of land-use planning, including forest design. A key example is growing use of geographic information systems (GIS), interactive websites and three-dimensional imagery such as Google Earth, which enable local communities to digest complex data and plan options better. However, the keys to a successful public involvement process are openness, honesty, patience and perseverance by the land manager. All the techniques and methods will not help if there is no intention to listen and incorporate the needs of communities into forest management.

Part II

The process, techniques and implementation of forest design

Chapter 5
The process of forest design

Introduction

In the previous section the chapters covered a number of key issues that should be incorporated into the design and management of sustainable forests, particularly if a truly comprehensive solution is to be developed. Due to the inherent complexity encountered in forest design, we have found it helpful to follow a tried-and-tested process conscientiously in order to ensure that nothing is missed, or that we have not focused on the wrong information.

The process to be presented in this chapter has common elements that should be followed whatever the landscape scenario or application. This process can be done as a strictly technical exercise (i.e. with expert staff), but should ideally include public participation along the lines discussed in Chapter 4. Ultimately, unless the interested public have been meaningfully engaged, they are less likely to support a forest design solution, no matter how technically sound it might be.

Since the design process can be applied widely in different countries, there is inevitably a range of different terminologies that applies from place to place. Aspects of the process that apply more frequently to different countries will be highlighted and different terms will be explained. Readers should then select the variant that applies to their own circumstances.

There is one exception to the principle that the process should be applied more or less the same way everywhere, and that is the stage and process by which management and design objectives are set. This is because initial objectives are often derived from existing laws and policies, or reflect ownership priorities. Since these can vary a great deal, local managers need to work within the framework that applies to their circumstances.

This chapter presents the generic or standard-model forest landscape design process. In Part 3, Forest Design Application, we demonstrate how to tailor the process to three different design scenarios: forests where restoration is the overriding objective (Chapter 8), managed natural forests (Chapter 9) and plantations (Chapter 10). A number of different case studies in each chapter illustrate in some detail how the process has been applied in each of these circumstances. Every landscape and every community differ, and there is no "cookbook" formula that can be simply overlaid to save time or money in design, but the process can and should be used much as we illustrate it in this chapter.

Managing the design process

Designing sustainable forest landscapes is, by its very nature, a complex undertaking that normally involves many natural resource disciplines, issues and interests. As such, there are several aspects that should be thought through before initiating a project. Any forest landscape design process should strive for a creative synthesis of four key elements: the client, the designer or design team, the ecology of the land and the community. We addressed land and community to some extent in Chapters 2 and 4, respectively. In the following section, our focus will be on the client and the design team.

The role of the client

One of the first questions that must be asked and answered before beginning a forest design project is: who is the client? This seems simple at first. Obviously the client is the person or legal entity that owns or manages the forest and pays the bills. The client is the one who presumably wants a forest design done for a particular area. However, we have found that in many cases defining the client is not so simple. In the case of publicly owned forests, there are usually overlapping authorities. The person initiating a forest design project may be a resource planner or forester who works for a public agency. They may answer to a district or area ranger or forest district manager, who in turn answers to a forest or regional supervisor, and so forth. The client in this case may technically be the agency charged with administering the public forest, but ultimately the designer needs to identify the precise person or board (e.g. a Forest Leadership Team) that makes the decision regarding the plan. Where communities take an active role in forest management they may be the client, operating through the type of objectives committee illustrated in Chapter 4. The true client is the one who has the authority to approve or disapprove a given plan.

The key roles of the client are to articulate the objectives of the forest design project and to take responsibility for the outcome. Again, this sounds simple and straightforward, but in practice can be frustrating. One of the authors had a client who was more or less dragged into the forest design process at the urging of a junior staff member. She stated that while she supported the idea of forest design, she wanted to know in advance what the outcome would be before she agreed to initiate the process. The author pointed out that any design process is a search for a solution to a stated problem, thus knowing the outcome in advance is a logical impossibility. If one truly knows the outcome then one does not need to spend the effort at design. Design is an objective-driven process. Any design must be measured by how well it is able to meet the objectives set or agreed by the client. Thus the key role for the client is to articulate what these objectives are. Objective setting is dealt with in detail below.

Taking responsibility for the design means that the client has ownership of it. The designer is simply a creative person who knows how to apply the design process to help craft a design for the client. Once it is completed, it must be the responsibility of the client and their subordinates, not the designer, to implement and monitor it. Again, this sounds straightforward, but too often public land managers have given only half-hearted effort and support to forest design projects. The result may be a very good design, supported by the design team and community but not by the person in charge. This is a recipe for frustration and a waste of resources.

The role of the designer

The forest designer is usually an expert brought into the team from elsewhere in the organisation or from outside it, if there are no staff designers. The designer's role is to guide the process towards an outcome that meets the objectives as established by the client. Designers are not solitary geniuses who take the basic materials and then work in isolation, emerging some weeks later with a grand synthesis. In some cases, the designer's hand may never touch the pen, marker or digitiser. They may simply guide the hands of others. In other cases they may hold the instrument, drawing the lines while others peer over their shoulders guiding them. The key point is that forest design is by necessity a group process, because forest management is for the most part a complex communal endeavour.

The authors' experience is that the designer should bring three things to the table. Firstly, the ability to visualise and project landscape patterns at fairly large scales. One design objective might be to have a well-connected old growth forest network. A possible alternative might consider retaining or restoring an old growth forest corridor that follows a major valley up and over a low pass. The designer can help give this form either on a map or in a perspective view. Secondly, the designer must be an active listener. He/she must be able to interpret and incorporate the apparently confusing or conflicting objectives inherent in forest management, and then help a coherent form to emerge from potential chaos. Thirdly, the designer must bring a sense of courage or confidence to the process. This is not arrogance, but rather the confidence that there is in fact a solution that can be reached through design. Many natural resource professionals, particularly in the USA, have lost confidence in their ability to achieve comprehensive solutions. In other cases, rather than make the effort to gain a broad-based, creative plan they invoke their economic or political power to force an unwanted plan onto an unwilling community, at the expense of the ecosystem.

The role of the design teams

Forest design is, in nearly every case, a group process. As such it relies on teams of professionals, often combined with lay citizens or activists through a participation process. As stated earlier in this book, we subscribe to the principle that all ecosystems are too complex to understand fully. Take any 5,000 hectares of forest and you will find enormous complexity across a whole range of aspects, including soils, hydrology, geology, geomorphology, aquatic ecology, terrestrial ecology, game management, recreation, aesthetics, silviculture, transportation, forest engineering and so on and so on. It is vital to obtain as much expertise as time and budgets will allow, in order to achieve the best solution.

Our experience is that most forest designs can be successfully developed by a core team that includes the following:

1 A planning forester
2 A silviculturist or operational forester
3 An ecologist, preferably with broad knowledge of wildlife and plant communities
4 An earth scientist (geologist, geomorphologist, hydrologist, soils expert, terrain engineer), particularly if the forest area encompasses steep or unstable ground
5 An aquatic specialist, particularly where stream, wetland or lake density is high
6 A designer, preferably trained in landscape architecture, and with a good understanding of natural resource management.

While this core team should be capable of developing a design, they often need additional assistance from other disciplines. These are usually referred to as the consulting or technical team, and may include one or more of the following:

7 A public involvement and/or communications specialist
8 An engineer or logging systems expert
9 A geographic information system analyst and computer technical support
10 A social scientist or resource economist
11 A recreation planner
12 A botanist or specific wildlife specialist where sensitive species are present
13 An archaeologist, anthropologist or landscape historian
14 A graphic designer or desktop publishing specialist
15 A writer/editor if a detailed report is envisioned.

Managing the team

One of the core team members, perhaps a planning forester, should serve as the overall project manager, or in some cases a person whose only responsibility is to manage the project is assigned to it. The project manager must be "process and outcome" orientated. Their most important job is to develop a budget and schedule, and see to it that everyone on the team gets their work done on time and in appropriate formats. Good project management is more art than science, and is only learned through experience. It requires both big-picture thinking and attention to detail. The ability and confidence to change course or improvise when the situation calls for it, but also to hold a steady course when others may waiver, is a very important quality. The project manager must have leadership skills.

The project manager must also have the natural or learned ability to grasp the whole rather than to get lost in the parts. Modern natural resource management has followed the path of science towards focusing on increasing detail, often at the expense of an awareness of large-scale patterns and processes. Inventories are often conducted at widely varying scales with very different purposes in mind. Vegetation maps may be derived from remote satellite imagery, while aquatic habitat is mapped from detailed field notes taken during stream surveys. Forest design, conducted at the landscape scale, cannot make use of overly fine-grained data. Conversely, very coarse data can be misleading or so inaccurate as to lead to poor decisions. There is an informal, often intuitive period early in the process where the project manager and team must perform "information triage", organising data and maps into piles marked 1) essential, 2) informative but not essential, and 3) neither informative nor essential.

The team will undoubtedly include both data junkies and generalists. This is not a bad thing. In fact it is a mistake to think that the entire team should be "big picture" thinkers. This can result in a rather sloppy solution that is not well grounded or validated. One or two "data hounds" on a team can force a more rigorous process. A whole team of them will usually result in hopeless gridlock. The key to successful team management (besides having a talented and motivated team) is to provide a clear sense of direction, along with suitable benchmarks that indicate progress.

Start with an outline

The forest design process is fairly simple, but working through it is demanding. Let us assume that a planning area has been identified as a target for design. Ideally, the owner or manager assigns a project manager to organise and lead the effort. The first task then, is to create an outline of the project, starting at the desired end point. In other words, once the design is done, what would one have in hand? If the manager wants or needs an operational design that spells out in detail harvest units, roads, trails and areas to be restored over a 50-year period, then the design process must be set up so as to lead to that end. If, on the other hand, the manager only wants a broad concept for managing a given area, the general pattern or shape of landscape structure, then this is a different challenge. It is important to set the target at the point of greatest utility, and within the available budget.

Once the project manager knows what the finishing point is, he or she can then work backwards to the starting point. This is best done by posing and answering questions, such as: to get a broad design concept, what do we need to have beforehand? Answer: clear objectives and good ecological analysis. To have clear objectives what do we need? Probably a series of facilitated meetings with all the key players in the room, as well as outreach to the local community. How do we ensure good analysis? Make sure the base information is up to date and well documented.

Successful project management includes balancing the available budget, the desired end products, the time available and the team's abilities. It is rare to have a budget truly large enough to produce all the products (maps, report, illustrations, etc.) that are desired at a high level of quality. A project work plan, or flowchart, is an essential tool. Typically, this is drafted to correspond with the basic steps of the design process, articulated below, with interim deadlines and critical paths used to set up the overall elements of the process, how they relate to each other and who is responsible for producing them. All project managers have some experience in developing work plans, and may have their favoured methods of doing do.

The steps of the process

Over the years that the authors and others have been developing, applying and refining the process of forest design, a logical set of steps has become established. The major developments that unified the approaches that had evolved in the USA and the UK happened in the early 1990s, when the landscape ecological analysis (see Chapter 2) and the landscape character analysis (see Chapter 3) were brought together into a unified process. This achieved a high degree of synergy, and the majority of the developments that have occurred through different projects since then have been refinements, testing and widening the sphere of application.

The following are the essential steps of the process:

1 Project initiation: define the planning area, establish the project context, appoint the team and create the project work plan
2 Set initial design objectives and/or identify key issues based on policies, pre-existing plans or previous experience of the planning area (or delay objective setting to step 5)
3 Field inventory/survey or assemble existing inventories and data
4 Analysis of the forest using the inventoried information
5 Review and refine objectives (or set objectives if the alternative route is used)
6 Develop one or more broad design concepts
7 Test concept(s) against objectives and carry out critique
8 Refine selected concept in response to critique
9 Re-evaluate the refined design against the objectives and analysis and refine it further
10 Documentation and approval
11 Implementation
12 Monitoring, adaptive management, revision.

Note that there is no public involvement identified as an "essential step". This is because public involvement, done correctly, happens at all stages of the project. Selecting and defining the area to be planned can be done in response to community concerns. Initial objectives and issue identification should be at least supplemented by outreach. Many local communities have ecological knowledge or data to share, gained from hunting, gathering, living, working or recreating in an area. Communities can also help develop design concepts, establish evaluation criteria and make recommendations as to the selection of alternatives. In some cases they may be willing to help monitor and report back, thus also helping with adaptive management. The keys to public involvement, as noted in Chapter 4, are to understand to what extent the community is a "partner" in the project, to what level they will be allowed to affect the outcome and to create an effective strategy for participation.

Figure 5.1 The design process: the main steps and the organisation of them. These slightly simplify the more detailed description in the text.

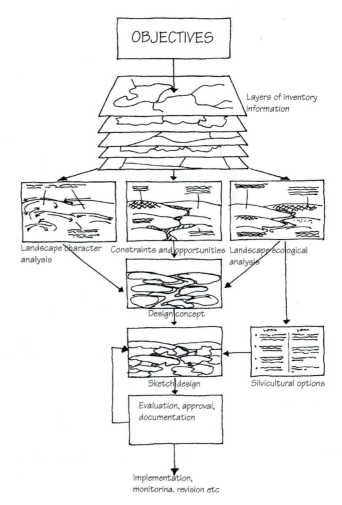

OBJECTIVES

Layers of inventory information

Landscape character analysis   Constraints and opportunities   Landscape ecological analysis

Design concept

Sketch design

Silvicultural options

Evaluation, approval, documentation

Implementation, monitoring, revision etc

Step 1: Project initiation

As a general rule, forest landscape design is most effective on forests between 1,000 ha and 100,000 ha. It is certainly possible to design smaller areas (especially in the plantation context), but at smaller scales landscape ecological analysis loses its effectiveness. Areas larger than 100,000 ha can be designed, but are best subdivided into sections due to the likely problems of overly complex information and design strategies becoming too broad-brush. It is also important at this initial stage not only to set the boundaries of the area to be designed, but also to nest it within the broader landscape context.

The following are the main options for determining the appropriate design unit:

- natural topographic and hydrological units, such as watersheds, catchments or complete landform units such as hill ranges
- natural ecological units based on geology, soils or forest type (useful in rolling terrain where landform and drainage structures are not particularly strong)
- ownership or jurisdictional boundaries such as forest licences, community forests, private land units, national boundaries or other administrative units. (These may present problems for design where administrative boundaries are unsympathetic to ecology or landscape character, but this can be addressed by overlapping the design at the edges.) It is always sensible in these circumstances to analyse an area somewhat larger than the design unit, in a strategic way, before working in more detail inside the boundary.

Often the eventual unit may comprise a mix of all three determinants: a hydrological boundary may encompass most of one or more vegetation units but be cut off in one place by an

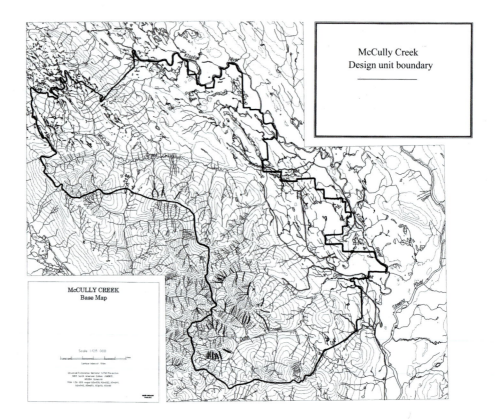

Figure 5.2 An example of a design unit whose boundaries are derived from a combination of topographic, hydrological and land ownership features. Source: British Columbia Ministry of Forests.

administrative or ownership boundary. This is a fact of life and rarely prevents a satisfactory design from being prepared. From time to time land-use practices close to the area may interrupt the habitat of wildlife, for example, so that co-operation with neighbours becomes important. In other places, notably Scandinavia, small-scale ownership patterns need to be ignored during design until a level of detail is reached that can be applied within individual ownerships. The case studies presented in the chapters comprising Part III illustrate these points well.

The context description should consider:

- geological and topographical structure, including drainage
- main ecological classifications
- social factors such as settlement and communication, water supplies, cultural and historical data
- economic factors such as timber production, tourism, retirement housing, sporting/hunting, etc. that occur in the area.

A core project team, including a project manager, as described above, needs to be appointed at the outset as part of the project initiation. Once the team is up and running, the project manager should draft the work plan, normally with the assistance or at least input of the core team. The work plan should list all the main tasks, state who is responsible for them, when they are to be done and what the expected product will be.

Step 2: Set initial objectives and identify key issues

All designs are driven by objectives. There is no point in doing a design unless there is something we want to achieve that is different than what already exists. In most cases, a good deal is already known about an area that has been selected for a design. In some cases public policies contain implied or explicit objectives. In other cases an area may have been under intensive management for some time, and the owners have clear notions about what they want. This pre-existing set of ideas is what we mean by an *initial* set of objectives. They represent the policy and planning framework that guides management decisions. In the USA, most publicly owned forests are managed within a set of laws, and usually have a governing land-use plan or strategy already in place. Thus, an area selected for a design will have already been designated as commercial forest, and there will be some set of stated goals or objectives. In Canada, where forests are managed at the provincial level, the government may have pre-determined that a given area is part of a timber licence, thus some level of harvest is expected. The area may also have been mapped as key wildlife habitat for a range of species, or as a scenically sensitive zone, or perhaps as a municipal watershed. In the UK, national forestry policy objectives and environmental guidelines, as well as the requirements of certification, tend to provide a framework within which objectives are formulated.

By studying existing policy documents to see what objectives they set, either explicitly or implicitly, the design team can understand right away what information is likely to be most important, and thus make data gathering cost effective. There is a risk that the initial objectives may be interpreted too rigidly, which can have the disadvantage of reducing the flexibility of the team by setting aims that may ultimately prove unsustainable, unpopular or unrealistic. Conversely, in projects that are left entirely open-ended, the alternative is to collect an exhaustive array of information, carry out a thorough and objective analysis and then, armed with findings, consider what are the appropriate objectives for a given area. These could range all the way from hands-off wilderness designation to intensive timber harvest. As is the case with most natural resource management projects, the key is to strike a good balance between a too rigid or too open approach at the outset.

In most management plans for working forests, initial objectives are often related to achieving a given output, such as timber volumes, return on capital and so forth. These are also important objectives for forest design projects, but since a broader approach to sustainability is implied, there must also be additional objectives that relate more directly to the state of the forest landscape. In other words, rather than simply focusing on outputs, initial objectives for a forest design ought to include at least general statements about what the forest should look and be like over time.

Initial objectives arise from multiple sources. At first it might seem obvious that the landowner or manager is the main stakeholder and, since they are probably paying the bills, they have the final say. Nowadays, however, there are many interests which desire a say in how forests are managed, as was discussed in Chapter 4. In areas where there is multiple property ownership or overlapping claims, it is very important to consider the objectives of all stakeholders.

Large-scale private landowners or licence holders increasingly need to include objectives that are not income related, such as watershed protection, wildlife habitat and maintenance of scenic quality. In many cases, these objectives are required under sustainable certification criteria. The team may draw upon these as a starting point for the initial objectives.

Communities may view themselves as owners of public land (as tax payers) or as stakeholders in public or private land where they have freedom to roam, rights to obtain fresh water from the forest, hunting rights or where the forest is an important part of the scenery viewed from their homes or communities. Local native or indigenous people may also express particular values or have treaty/first nation rights associated with their culture, such as trapping, subsistence gathering or protection of sacred sites.

The issues advanced by various stakeholders may or may not conflict or coincide. They normally represent extrinsic values. Intrinsic values are difficult to express but could be deemed to include all the higher-order requirements set out in international treaties. These may not hold much sway with logging companies and citizens of mill towns but must be taken into account for the needs of non-local residents and future generations. They are usually expressed through government policies and programmes associated with meeting key targets such as habitat or species protection, or ensuring reforestation and fire control.

In most cases, it is useful to draw up a public outreach strategy that includes an early discussion with all or most stakeholder interests to identify key issues that should be addressed in the plan. The outreach can be done through the application of one or more of the methods described in Chapter 4. This can help prevent conflict and can also demonstrate where interests overlap or coincide. If a full, open and unconstrained discussion can take place, preferably facilitated by the leader of the design team or someone trained in this role, this stage can greatly ease the progress of the rest of the plan. Often, managers are afraid that the public will want all sorts of what they consider to be ridiculous, expensive and impractical things to be incorporated. They are pleasantly surprised in most cases that this is not so, and they can feel more confident that they can produce a plan that will be agreed by all stakeholders. In fact, in many of the plans we have worked on, a very similar list of issues is developed, perhaps because there may be a finite number of issues that can be identified regarding forests (for example, to protect biodiversity, maintain game species, keep an attractive appearance at key use areas and so forth).

In order for issues to become operative, they have to be translated into objectives. Initially phrases such as "to produce", "to sustain", "to protect", "to enhance" frequently occur and, in the main, these express values that nobody can reasonably object to. However, phrases such as "to protect water quality", "to maintain habitat values for black bear" or "to produce a sustainable supply of timber" need to be expressed rather more fully, in order to provide planners and designers with an indicator of how they know they are meeting these objectives in the design. This is where some more detailed knowledge about an area is needed, and where a combination of expert and "lay" opinion can generate statements that reflect more meaningful and measurable criteria. These criteria may sometimes have to be qualitative descriptions (such as for landscape protection or enhancement) while at other times they can relate to measurable standards (such as water quality), minimum areas of habitat or expected volumes of timber. Some of these indicators can be measured from analysis of the design (such as timber volumes, patch size, riparian corridor width or percentage of old growth), while others may need to be measured over time on site

Table 5.1 Example of method of presenting objectives

Resource value	Objective	Criteria of objective being met
Timber production	To produce a sustainable supply of timber over the period of the plan	To produce on average 20,000 cubic metres of timber every year
Scenic quality	To maintain the attractiveness of the landscape	To ensure that the shape, scale and distribution of cutblocks reflect the landform and landscape diversity
Ecological integrity	To enhance the degree of connectivity of the landscape	To manage the forest towards 60% in a late successional stage at any time, connected both across and up/down slope

(water quality, habitat use by wildlife) or through simulation (landscape change, successional stage change, timber volume production).

Presentation of objectives can be in text form, or, more clearly, in tabular form using a three or more column table where one column states the resource value, the second the main objective and the third a more detailed description of how to meet the objective (see examples in case studies). Additional columns can be used later to develop a monitoring system so that the gradual implementation of the plan can be measured in terms of achieving the objectives.

Step 3: Inventory/survey

This is a very important step because it is at this stage that information is collected about key aspects of the forest landscape. This may include collecting data from a wide range of sources, including national, state, provincial or local government agency databases, specially commissioned surveys and also the collation of information from members of local communities (possibly through a community mapping exercise as part of the outreach). It is important to state that, for practical reasons, it is impossible to expect that all possible knowledge about an area can be obtained. We can never know all there is to know about a forest, so most projects rely on what is most readily available, supplemented by cost-effective surveys to fill significant gaps and bolstered by expert professional judgement and opinions.

At the start of this step it is useful to consult a checklist to see what types of information are needed, then to cross check with various sources to see what is available (its scale of resolution, how up to date it is, etc.), after which major gaps can be identified and strategies to fill them developed.

All of the data should be mapped, preferably within a geographic information system (GIS). Where data collected at, say, scales of 1:100 000 is subsequently to be presented and used on maps of 1:20 000 scale, boundaries should be treated with caution and, if possible, refined at the smaller scale. The following is a comprehensive list of all the types of data that might be needed. Not all of it is needed everywhere.

1. Climatic data

At the scale of tens of thousands of hectares there can be quite a variation in climate, particularly continentality, accumulated temperature, rainfall and wind exposure. These may affect growth rates, species choice (for plantations), silvicultural systems, ecosystem types, natural disturbance regimes and successional pathways.

- Temperatures vary, being warmer at lower elevations than at higher ones. Where the number of days in the year that the temperature is above 5°C (the point at which plant growth starts) are added up a measure of accumulated temperature is obtained (day degrees). This is corre-lated to tree growth and from this the occurrence of or suitability for different species. Accumulated temperature can vary across the landscape, especially if there is a large altitude variation.

- Rainfall can be affected by weather patterns and topography (rain shadow effects) and altitude. Across a landscape it can vary a lot and, in conjunction with temperature, have a marked effect on available soil moisture and thus on tree growth and species suitability.
- Wind can affect tree growth or act as a disturbance agent, affecting forest structure and composition. This exposure depends on the local wind climate (this can be built up using anemometer readings), the exposure to winds that typically blow from prevailing direction and altitude. When the storm pattern is also known a measure of wind damage risk can be built up to help managers identify both the likely disturbance pattern and the risk of storm damage to timber. Local experience is often the best source of data on wind disturbance.

Maps of these factors, resolved to 50 m cells in a GIS or simple sketch maps that indicate broad patterns, can be prepared. In some places climate information is already embedded in ecological site classification systems, so does not need to be prepared separately. Examples of these include the British Columbian Biogeoclimatic Ecosystem Classification (BEC) and the British Ecological Site Classification System (ESC). Examples of their application can be found in the case studies in Chapters 8 and 9.

2. *Physiographic information*

There are likely to be many variations of geology, both solid and superficial, soils and hydrology within the landscape. These may already be mapped and available. The information is of use for a number of purposes:

- Solid geology, mapped as units of rock type together with information on any structural weaknesses or particularly unstable areas, may be important as locations of operational constraint related to timber harvest and/or road engineering. Geological information may also include sources of rock for road construction. A geotechnical engineer may need to make a specific survey and map particular features, especially in mountainous areas.
- Surface geology, including glacial, aeolian or fluvial deposits, can also present issues important to forest management. Some materials may be in unstable positions, where roads cannot be constructed or where removal of the forest cover could result in loss of stability due to increased moisture accumulation or reduction in root anchoring. They may also indicate flood or debris torrent hazard zones. Active landslide zones and other eroding features such as slow-moving earth flows may affect potential forestry activities and may already be generating sediment into streams and lakes. Maps of the risk of slope failure or sediment yield may be needed in these circumstances, and should be prepared by a qualified hydrologist or geotechnical engineer. Terrain stability maps can be generated using fairly sophisticated GIS models that combine slope steepness and shape to identify different categories of risk.
- Hydrology is particularly important in almost all situations. The entire stream system should be mapped and zoned into source, transport and depositional reaches. Each of these reaches may interact with the adjacent forest in different ways, thus requiring different management strategies. It may be advantageous to subdivide the landscape into a series of sub-catchments, to facilitate monitoring of water quality and quantity before, during and after plan implementation. Streams can also be classified into stretches that are accessible to migratory fish, such as salmon, with natural and man-made barriers to their movement (waterfalls, dams, culverts, irrigation diversions) marked on maps. Streams, lakes, ponds, rivers, bogs, fens and other water bodies should be identified, especially if these act as sources for base flows or as traps for sediment generated upstream. Peak and base flow information may be

available from gauging stations. It is most helpful if data on hydrology is used to generate an overall picture of how the system operates. Is it a rainfall-driven system with very flashy streams? Is it a snowmelt and spring-fed system with very steady flows? How does it generate, transport and store sediments and organic material? Most importantly, are there places in the landscape critical to hydrological functions?

- Soil types are often mapped into units of major varieties such as brown earths, podzols, skeletal (colluvial), gley, etc. (the terminology varies from country to country). These may indicate fertility, drainage status and suitability for different species (especially in plantation forests). They may be arranged in series according to parent material (geological and from sources such as peat deposition). Of particular use in ecological site classification systems are the two measurements of soil moisture and soil nutrients. These can be presented using a grid, set within different climatic zones and related directly to the ecology in terms of climax forest types. Soil erosion potential is a particularly important concern in most hilly and mountainous areas. Granular soils are easily mobilised and transported downstream, whereas sedimentary soils tend to be more stable. Another issue in intensively managed forests may be soil alteration or compaction from past operations, which may now limit future options.

3. Ecological information

Several aspects of ecology are needed for forest design and are used initially for one particular type of analysis:

- Some classification of the landscape into different structural and compositional plant types is critical. This can be recorded according to pre-determined classifications such as the BEC or the British National Vegetation Classification, using certain indicator species. One of the problems in relying on existing information is that the forest inventory may be constructed on the basis of tree species rather than vegetation communities. In a number of cases, a more refined classification will be needed. In addition to vegetation types, successional stages are usually very important. Most ecologists use at least five successional stages, from herb dominated to shrub, small trees, tall trees and finally mature or old growth, which usually has multiple canopy layers. These should conform to whatever is standard terminology in the area being designed (see Chapter 2). Of particular interest in most cases is the presence of "old growth" or other extremely old, undisturbed forests. Special vegetation types or associations, even if in very small patches, should be noted, since these are likely to become design factors.
- In plantation forests or bare areas to be reforested there may be two types of maps produced. Plantations are simple to classify in both type and successional stage, although there may be areas of natural or semi-natural vegetation remaining that might have greater ecological value. Land to be afforested may have semi-natural vegetation, such as the moorlands of Scotland or grassy slopes in New Zealand, which may have habitat value in their existing condition. Where the objective of the design is to recreate a native forest a map of potential forest types can be constructed from the soil and climatic information allied to the current vegetation. This will help assess the possible future ecological functioning because it will present the likely scale and complexity of the forest spatial pattern (see Chapter 8).
- Information on wildlife should include the main species that use the whole landscape, or significant parts of it, both common and rare. Wildlife movement should be mapped, either actual routes, such as deer trails, or general pathways (i.e. riparian corridors, topographic breaks). If there are species that require special management, their location and habitat should be recorded separately. This information may need to be treated confidentially to avoid illegal hunting, harassment or egg collection. Wildlife habitat can also be characterised using guild systems, such as that used in the computer program HABSCAPES in the Pacific Northwest (see Chapter 7), or the Matrixes for Wildlife Habitat Relationships in Oregon and Washington

by the National Habitat Institute. These systems relate wildlife communities to the type and scale of existing habitats within a given region.

- In natural forests it may be possible to uncover the disturbance history – the type, scale, frequency, intensity and resulting character of fires, windstorms, ice and rain events, floods, landslides and insect outbreaks. Plantation woodlands may be susceptible to these same patterns, especially wind, fire and insects that relate to climatic cycles. It is possible to use aspect maps to build up a picture of local microclimatic variation that could lead to differential fire regimes, and detailed aspect/shelter measures to reflect relative windiness across a landscape. Avalanche tracks and forest types prone to particular insects can also be mapped.

4. Cultural information

Human use and various social, historical, cultural and recreational values may have greater or lesser degrees of importance, depending on the location.

- Archaeology, history and cultural heritage include sites, features and historical or cultural associations. These may include protected archaeological sites dating from many thousands of years ago to relatively recent times. Non-protected sites should also be recorded. Cultural sites include places where key historical events took place, where artists worked (painting, poetry, prose featuring the landscape), those associated with famous people, of religious significance, used for traditional purposes by indigenous people and those forming an important setting for any of the above. Cultural sites can also be contemporary, related to the ongoing practices of indigenous people. It may also be important to examine the broader landscape history and its development, perhaps using old maps, aerial photographs or documentary evidence.
- The visual landscape should be recorded as a series of photographs taken from a range of viewpoints, particularly from major travel routes, scenic vista points, settlements and tourism facilities. Maps of visual sensitivity or viewsheds may be available, or could be generated if this comes up as an important issue. Chapter 3 outlines ways of assessing visual landscape sensitivity for use in planning.
- Recreational uses of all types and locations should be recorded. These may include trails and recreational driving routes, water bodies and special settings. Areas used more widely for hunting, berry and mushroom collection or as recreational activities, should also be noted.
- Places within the forest that possess a strong "spirit of place", such as waterfalls, flowery meadows, rock formations, historical sites, enclosed lakes or groves of ancient trees should also be identified.

Some of the above information can best be obtained from local people, and it is often the case that these correspond to the values associated with the landscape that are of particular concern to them. Asking people to map and record this information, a process known as community mapping, is a very good way of including them in the overall design process in a tangible way. However, it may be worth checking the accuracy of this information, since memories can often be misleading, and some people may not be able to place a spot on a map very precisely.

5. Management information

There is a lot of information normally generated by forest managers that enables them to develop practical strategies. Access to the forest, pest control, silvicultural possibilities, fire protection and so on will all have bearing on design options.

- A map showing the growing stock (a stock map or forest cover map) will encapsulate a lot of information such as species, productivity, age, height, canopy closure, etc. (depending on the location).

- Road and trail access is always going to be crucial. Some areas may already have roads that will impose certain constraints just by their location. In non-roaded areas there may be questions of operational feasibility, economics and public acceptance of direct management. A map showing potential or planned road locations should be drawn up based on slopes, stream crossing points, places to avoid (such as important habitats and archaeological sites), terrain stability and forest operability. Qualified forest engineers should be involved in this.
- Harvesting systems information includes the feasibility and flexibility to use different equipment. Steep slopes, soft ground and fragile vegetation may limit machinery, for example reducing the chance to use ground-based equipment such as skidders, forwarders or processors. Cable systems are more costly, and may limit cutblock/coupe layout. On the steepest slopes, where road access is not desirable, helicopter systems may be possible, if timber values are high. It is useful to have a map that delineates zones where different systems are feasible. A more sophisticated approach is to combine the road location, landings and systems to produce a map of small units each loggable from one set up. This is sometimes known as a total chance plan (because it looks at the maximum total chances of getting timber from a particular area – see Chapter 1).
- In places where the climate permits, winter logging is often preferred, since it dramatically reduces ground and soil damage. A harvest system map should identify places that should only be winter logged versus those that can be harvested in dry summer conditions, or with protective mats of brush or slash.
- Certain "no harvest" zones may be set by legislation or operational guidelines, such as riparian areas, lake shores, proximity to archaeological or cultural sites, extremely high landslide-prone areas, etc. These need to be identified early on, in order to be incorporated into the design.
- Silvicultural information includes past management activities, such as previous logging, thinning or respacing, the effects of fire, insects, disease, blow down, different growth rates, regeneration or replacement species choice. In plantation forests, areas may have to be fenced against deer, sheep and other domestic stock, rabbits or people. This may affect flexibility in planting or replanting after harvest.
- Harvest suitability maps may include ones showing ideal felling dates according to economic criteria, such as the time stands reach maximum net present value at a test discount rate. Alternatively, stand volumes may be used, measures of thriftiness or the time the stand reaches the age of maximum mean annual increment, or even a minimum age that stands must reach before they can legally be cut. These guide managers on the best time to fell trees to get the best financial/volume return, though they should not be treated inflexibly, or there is no point to the design. Operability is a concept based on a combination of stand merchantability (volume/tree size/tree quality/proportion of rot in the stand) and harvest cost. In mountainous circumstances there may be an operability limit at certain upper elevations, which limits the extent of logging. Once again this should not be treated too rigidly.
- Fire risk maps are very important in certain areas, such as interior forests of the western USA and Canada. These often synthesise a number of factors, including access, vegetation types, built structures, prevailing winds, lightning frequency and terrain. They are essential where prescribed fire is a likely option within the design.

Field trip

In addition to mapped data, we cannot over-emphasise the importance of scheduling an early team field trip. Where the scale of forest design is usually large, an aerial tour is usually more cost effective, and also more revealing with regard to landscape pattern and structure issues. By touring as a team (possibly including members of the public), a certain synergy can take place right from the beginning, where the various resource specialists are able to demonstrate findings or observations over a range of issues. In some cases, an initial field trip can be combined with a public workshop, and a good spirit of camaraderie established at the outset of the project, building goodwill that may come in handy later on, as the inevitable tradeoffs are debated.

Step 4: Analysis

Once the inventory has been completed, the team should be ready to move on to the analysis stage. There are three parallel types of analysis that are most helpful: landscape ecology, landform/landscape character and constraints/opportunities. By definition, any sustainable forest design has to have a strong landscape ecological basis, except possibly where the forest is of plantation origin in a much altered landscape and is not of large extent. Where the forest is highly visible, or has complex topography, an analysis of landscape character is likely also to be a driving force. Since all forest designs occur within a social framework, social, economic, practical and legal parameters that establish constraints and opportunities should be mapped and analysed.

Landscape ecological analysis

At the inventory stage a number of layers of useful information were identified. At this stage this information is used within a structured process of analysis, designed to avoid too much confusion or entanglement in excessively fine details. This process is presented as a series of steps embedded within the main design process. It is important to remember that while the output of the ecological analysis may be one or more desired future conditions (DFCs), this is not the end result of design or planning. A DFC is only one element that must be woven into the main concept or strategy, as will become clear later on. The process should examine the landscape according to the approach described in Chapter 2.

The ecological analysis is best carried out by the entire core team working together in one setting. It can also be done along with interested stakeholders as part of a facilitated session.

1. Present landscape structure

This first stage describes the current landscape using the terminology and classification presented in Chapter 2. It will be illustrated here using simplified material from the "Leoland" project mentioned in the Introduction. It is useful to consider each of the following categories, map them, describe them briefly and also to visualise them through photographs or sketches. The classes of landscape structure include:

- *Matrix* This is the most connected part of the landscape, which is also likely to occupy the largest total area. In some cases there may be several competing matrices, or none at all (where the landscape is all smaller patches). Typical matrix types include mature forest, plantation forest or open woodland. These can be further broken into sub-categories, perhaps relative to an ecological site classification. Plant communities other than forests may also be matrices, depending on location and scale.
- *Patches* These are discrete areas, homogeneous within the landscape, that contrast in structure and composition with their surroundings. It is useful to categorise them into at least two sub-types: forest patches (different species and/or successional stages from the matrix) and non-forest patches (water, rock, tundra, grass, fields, quarries, bogs, human settlements, etc.).
- *Corridors* These are linear patches usually associated with movement along them that is dependent on the vegetation character, or landform (includes forest belts, hedgerows, waterways, tracks, power lines, roads, ridges, etc.).
- *Pathways* These are routes used for defined movement but without the associated structure (might be a route across a landscape that covers many vegetation types).

As noted in Chapter 2, some landscapes may better fit the "landscape continuum model" rather than the "matrix-patch-corridor model" described above. In the former, the contrast between patch types may not be clear, and most key flows may not be much affected by subtle differences in landscape structure. The team may conclude through this analysis that landscape continuum is a better description to use.

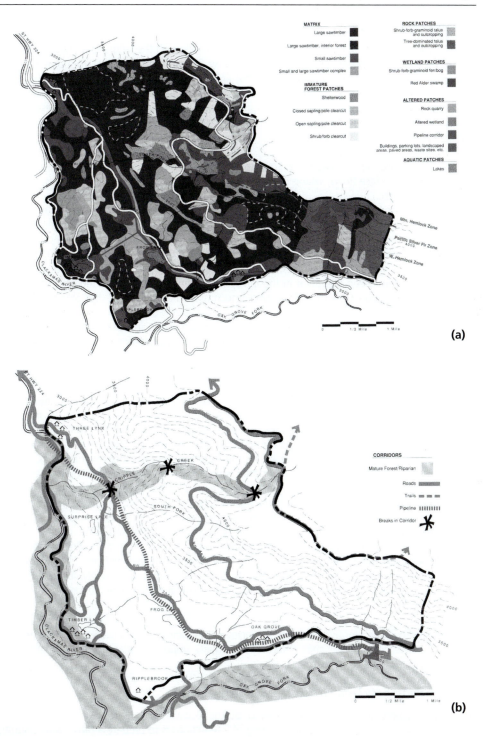

Figure 5.3
Maps of landscape structure: a) an example of a map showing the landscape classified into matrix and patch elements; b) an example of a map showing corridors. Source: US Forest Service.

2. Landscape flows

As described in Chapter 2, flows are elements that move through the landscape. They can include both living and non-living, be on the ground, in the air, in water or under the ground. It is a good idea to differentiate between regular flows, that are a part of the day-to-day functioning of the forest ecosystem, and those which are periodic disturbance agents. There is not necessarily a clearly defined boundary between these, so it becomes a pragmatic choice by the design team for each individual project. As a rough guide the main flows that are usually taken into account in most projects are:

- *Water*, considering the places it is collected, how it moves (in the stream system, as surface flow or sub-surface flow), where it is stored and released (in swamps, bogs, ponds, lakes, as snowpacks, etc.).
- *Large terrestrial fauna* (e.g. bears, wolves, deer, elk, moose, caribou, pine marten, goats, beaver, otter) that move around the forest, possibly seasonally, using different landscape structures or corridors.
- It is usually important to include *rare, sensitive or keystone species as important flows to consider*, since specific strategies to maintain or restore their populations are often required. These may have been identified in the initial objectives, determined during step two.
- *Aquatic fauna*, especially migratory species such as salmon and trout, but also resident fish and amphibians.
- *Birds*, especially raptors such as eagles, osprey, hawks, falcons and owls; waterfowl including divers; ptarmigan, grouse and capercailzie; woodpeckers and other obligates of forests; groups of passerines that relate to particular parts of the landscape. These may be grouped as collections of interior forest, forest edge, alpine, water, etc. species. They may include seasonal migrants as well as year-round residents.
- It is optional to include other wildlife flows that may be characteristic of the area. Remember that the analysis mostly involves landscape-scale functioning, so normally fauna that only inhabit discrete stands or patches are excluded. However, if movement from one patch to another requires certain conditions for certain important species, they should be included.
- *People* who use the landscape for recreation (hunting, hiking, horse riding, cross country skiing, snowmobiling, climbing, water-based activities, berry and mushroom picking, enjoying the scenery) or for work (logging, game management, farming, trapping, commercial berry or mushroom production, ecotourism). The patterns of these two main categories may be very different. Also, where there are indigenous peoples with specific cultural and subsistence relationships to the landscape, these should be as carefully identified as possible.
- In landscapes where expansion of natural forest is part of the project, the flow of seed into the open parts yet to be colonised may be important for determining the likely pattern of the spread of regeneration out from the existing seed sources. Conversely, if the forest has an urban or agricultural boundary, there may be a flow of unwelcome weeds into the forest at its edge.

Generally, flows are mapped as directional arrows that indicate broad routes, or more detailed corridors. Converging arrows may indicate particularly important places where attention to connectivity is likely to be needed.

3. Interaction between structures and flows

This step is where a deeper understanding of the ecosystem functioning is developed. A constructive way to organise an understanding of interactions is to prepare a table, with columns representing the flows and the rows the matrix, patch types and corridors.

The goal is to note the relationship between each structure and each flow. This can be done in one of two ways. Firstly, a description of the interactions can be entered into the boxes. In this

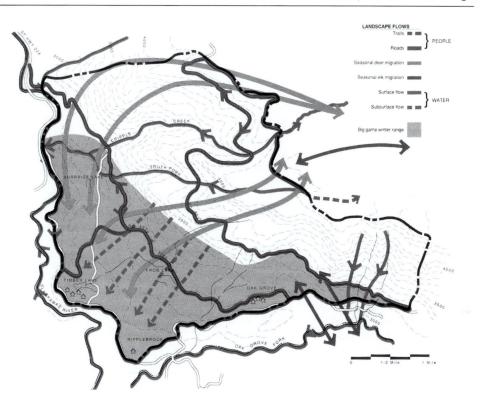

LANDSCAPE FLOWS

Trails ⎱ PEOPLE
Roads ⎰

Seasonal deer migration

Seasonal elk migration

Surface flow ⎱ WATER
Subsurface flow ⎰

Big game winter range

Figure 5.4
A map showing
landscape flows.
Source: US
Forest Service.

way more comprehensive information on the habitat requirements of the flows can be built up. However, this can be a time-consuming process and it may be difficult to see the patterns emerging (although it has been the method used in most of the case studies presented in Part III). The second method, in order to keep the process simple, is to mark either a positive (+), neutral (N) or negative (–) relationship in each box. By dividing each box into two parts (with a diagonal line), it is also possible to note these relationships in two directions: flow to structure and structure to flow. For example, the flow of Northern Spotted Owl is positively influenced by the old growth forest structure or matrix (+). But the owl may have little or no effect on the old growth (N). Some interactions will be very significant, for example where the structure is the main habitat for a particular animal, and could merit a double plus (++). These relationships should consider multiple interactions such as foraging, browsing, denning, nesting, mating, roosting, hibernating, bathing, hunting, travelling, etc. for wildlife. Water may interact by being collected, intercepted, transpired, evaporated, stored, released, cleaned or filtered.

Once completed the team should have a very comprehensive understanding of landscape ecology. However, to be of use, some further analysis is needed. From the table it should be possible to identify the structures that are most significant for particular groups of flows. Some of these may be non-forest, or they may relate to particular successional stages (such as early successional or late successional with lesser use in between). By associating the map of structure with the degree of importance it will become clear if some of the structures are inadequately represented in the current landscape or are almost completely lacking. These would become targets for restoration. Structures that are absolutely critical in that they must remain in one place (that is they cannot be replaced by being developed in another part of the landscape) and must stay intact, can be considered as "critical natural capital". Structures that result from succession and thus keep developing and changing, yet always need to be represented in the landscape, can be thought of as "constant natural assets". There may also be structures that negatively impact on

141

Table 5.2 The method of describing the interaction between structures and flows

Structure \ Flow	Water	Bear	Elk	Salmon
Old growth forest	Snowpack retention, evapotranspiration, generally retains water	Denning, hibernation, berries	Food, cover	No interaction
Late successional forest	As above but less so	As above but less so because there are fewer denning sites	Food and cover – less food than the old growth due to less understorey	No interaction
Lake	Storage	Fishing	Cooling and avoiding insects in summer Drinking water	Habitat but not spawning
Wetland	Storage, filtering of sediments	Berries from shrubs	Major habitat in summer	Clean water flows into streams
Rock area	Early snow melt and run-off, some sediment washes from here into streams	Berries from shrubs growing among the rocks	Some browse, minerals and breezes clear flies in summer	No interaction

Table 5.3 The simpler method of noting the nature of the relationship

Structure \ Flow	Water	Bear	Elk	Salmon
Old growth forest	++	++	++	N
Late successional forest	+	+	+	N
Lake	++	+	+	++
Wetland	++	+	++	+
Rock area	-	+	+	N

landscape functioning, such as man-made features that cause erosion or prevent wildlife movement in a critical way. These are "liabilities" and should be targeted for remedial action.

4. Disturbance and succession

This step introduces an awareness of the natural dynamics of the landscape. The concept of disturbance, where the landscape structures are affected by exogenous factors, and subsequent ecological succession were defined in Chapter 2. At this stage of the analysis, all the main types of disturbances should be considered and their likely effects described. The most typical types of disturbance are fire, windstorms, avalanches, snow break, ice storms, floods, landslides, insects, pathogens, mammals and people.

A good way of developing and presenting the analysis is by a table, possibly split into different sections according to any major ecological dimensions in the landscape (for example, forest type or zone). For each disturbance, consider the following: scale, intensity, frequency, duration and the resulting structure of the landscape.

It is possible to create one or more maps showing the pattern of risk and of different structures arising through disturbance. One major use of this information is to develop models of possible silvicultural approaches based on emulating the effects of disturbance, for example in the shape, size and resulting structure of cutblocks/felling coupes (this will be demonstrated in several of the case studies in Part III). These models may be defined in relation to the different zones identified for the disturbance analysis and possibly refined into different aspects or soil moisture classes, for example to subdivide the fire regime according to the places where vegetation is expected to be normally drier or wetter, thus affecting the rate of spread of a fire.

Forest succession stages should be identified and described for each of the major forest zones. It is useful to create one or more sequences describing the vegetation composition and the expected duration of each successional stage (by using the Oliver and Larson model). Intimately connected with this is the effect of disturbance on succession. This may reset the clock back to stand initiation, or it may help to move it forward. Presenting successional pathways clearly can be a problem. It may be acceptable to describe them, or to develop a diagram showing the effects of each stage. These are used to help determine the progress of implementing the design so as to ensure that all successional stages are represented and that forest rotation lengths are harmonised as far as possible to the natural cycles. In plantation forests where trees are grown on much shorter rotations, the comparison of a natural with a plantation cycle can help show where some parts of the forest should be retained in order to supply late successional elements if these are of particular ecological value, as well as showing where planted stands contribute to ecological functioning, even when the species are non-native.

5. Linkage with the surrounding landscape

No area is a true island (even most islands!), and there can be many important links with the surrounding landscape (or nearby islands) that need to be explained and incorporated into the analysis. The hydrological system, migratory routes for wildlife, territories for some far-ranging species, forest structural elements that extend beyond the boundaries and human uses are all

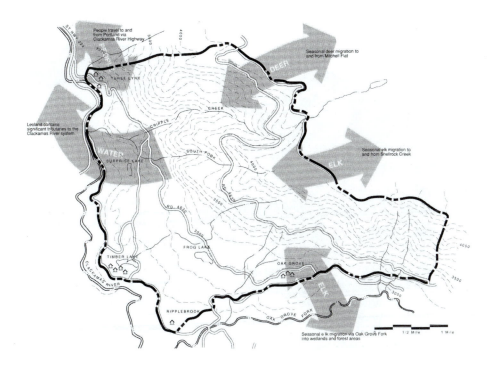

Figure 5.5
A map showing linkages from the design unit to the wider landscape.
Source: US Forest Service.

143

typical examples. It may be the case that important sections of connectivity run through the area, or that problems with neighbouring areas affect the functioning of the plan area.

6. *Desired future condition*

This final, and most important, step of the landscape ecological analysis is where all the previous steps lead to. It is a description of what constitutes the ideal ecological landscape, and is used as a key step towards the eventual design strategy. It can be quite difficult to develop this, and there may actually be more than one DFC, especially in a heavily altered or plantation forest where there is not necessarily a "natural state" that can be considered. In these circumstances it is better to concentrate on ecological functioning and the degree to which the managed landscape can address the different functions (remember that this is not the point in the process where practicalities or economics are introduced to constrain or justify different possibilities). In the case of afforestation or reforestation, the desired DFC is a distant potential goal once the forest has proceeded through an entire generation and natural or quasi-natural processes have been introduced. In the case of logging or silviculture in a natural landscape the first intervention starts the process.

A useful first step towards the DFC is to consider the elements defined as critical natural capital, and delineate these as more or less permanent features in the landscape. It may be necessary, for ecological functioning, to maintain physical linkages between these, in which case a kind of forest ecosystem network is established. The rest of the forest can then be broken down into management units of different shapes, sizes and silvicultural models that reflect the natural disturbance regimes. In practice this means describing a range of different intensities and levels of silvicultural intervention with accompanying within-unit retained components. This should be described and a map prepared showing these different zones. It may not be possible to incorporate the DFC completely in the final design due to practical, economic, social or legal factors, but at least managers will be able to assess the potential losses in functioning arising from deviating from the ideal.

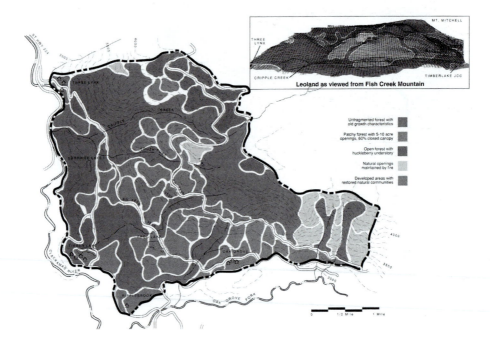

Figure 5.6
A map depicting the desired future condition for a landscape; the inset perspective shows it more three-dimensionally.
Source: US Forest Service.

In places where three-dimensional design is to be used (see design steps below), it is often useful to be able to visualise the DFC as the different zones on the perspective views used in the landscape character analysis (see next section).

Landscape character analysis

Chapter 3 presented the main principles behind the aesthetic aspects of forest landscapes and described in general terms the content of a landscape character analysis. The practical preparation of this analysis, especially in visually significant landscapes (defined according to a landscape sensitivity analysis as described in Chapter 3), should be thorough and carried out carefully. A design is normally there to be uncovered: it just awaits good analysis to help the designer to unlock it. Consider the following:

1   The bigger the landscape unit under design, the more likely it is that it can be broken down into a number of different sub-units or character zones defined by landform type, forest cover/landscape pattern type and zone of visibility. This zonation can be very useful because it may mean that the eventual design is different from one part of the landscape to another. A general description of each of these zones accompanied by a map should be prepared.

2   The landform analysis is an important stage in any design, except where the landscape is nearly flat. This analysis relies on a detailed topographic map, over which is drawn a comprehensive pattern of "lines of force". These are drawn as arrows flowing down ridges and rising up valleys. The best technique for this is to start with the valleys and develop the

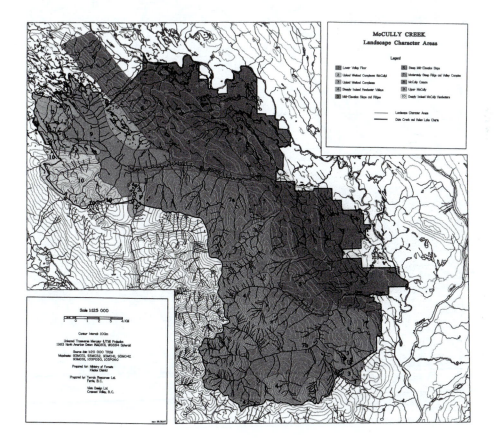

*Figure 5.7
The design unit divided into zones of different landscape character.
Source: British Columbia Ministry of Forests.*

145

complete pattern, first the major features, then the minor ones. Show them as green arrows pointing up the valleys, hollows or concavities. Expect there to be many more arrows in broken, complex topography than softer, more rounded landform. Often the arrows connect to produce a dendritic pattern with a distinct hierarchy to it. Once the valleys are completed, commence to define the ridges using red arrows flowing down. If there are definite summits, it is likely that a radial pattern proceeding from that point will develop, possibly branching as subsidiary ridges develop from the main ones. There is likely to be a strong intermeshing of the two sets of arrows, although it is possible for some landform, such as a series or simple, rounded domes, to be dominated by downward arrows expressing the concavities with few upward valleys, and vice versa, where a simple shape might be dissected by a number of valleys with little in the way of definite ridges in between. Knowing when the analysis has proceeded far enough is to some extent a matter of experience, skill and judgement. You will know if you have produced an inadequate analysis because it will be difficult to use it for design. If on the other hand, it is overdone it may be too busy to detect a meaningful pattern.

Figure 5.8 Preparing a landform analysis from the contours should be done in four steps. Source: British Columbia Ministry of Forests.

Step 1. Major ridge lines identified.

Step 2. Minor ridges added.

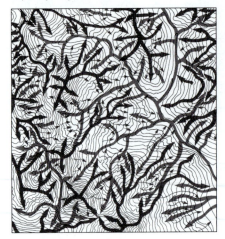

Step 3. Major hollows identified.

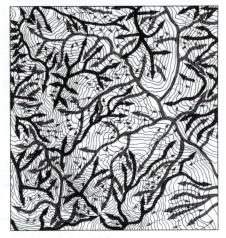

Step 4. Minor hollows added.

This analysis must also be carried out in perspective when the landscape is unusually important visually. If this is to be carried out over a photographic base it pays to work between map and photo as you proceed. This is firstly so that there is a good match between plan and perspective, and secondly so that landform features hidden beneath the forest canopy can be interpolated. Where computer-aided design is used, the mapped arrows are digitised and overlaid on the digital elevation/terrain model (DEM/DTM) so that the computer-generated views correspond to the original photographs.

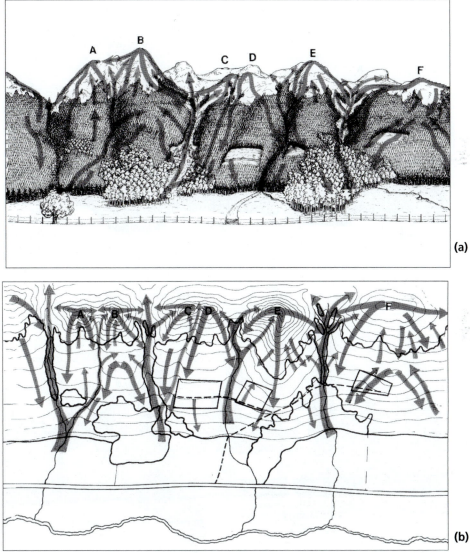

(a)

(b)

Figure 5.9
A landform analysis presented a) on a perspective of the landscape (in this case a sketch, but it could be a photograph or a computer-generated perspective) and b) on plan. Source: British Columbia Ministry of Forests.

The description of the landform is also very important, because this will help define the type of slopes to be used eventually in the design. Differentiate between rounded/irregular, jagged/smooth/broken/coarsely textured or simple/complex varieties. Consider also the scale of the landform, its diversity in terms of the different types in the landscape, and the overall degree of unity present. If there are unique landforms that convey a sense of place these should also be identified.

3   The land feature analysis refers to the elements lying over the topography, together with landform aspects such as rock outcrops, scree (talus) and other elements that contribute to

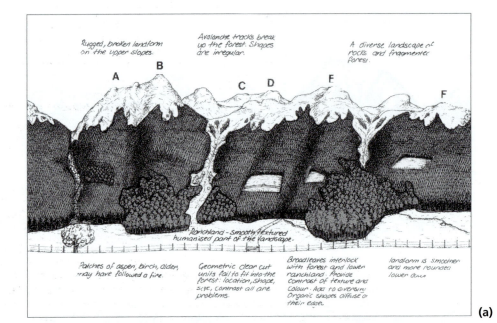

(a)

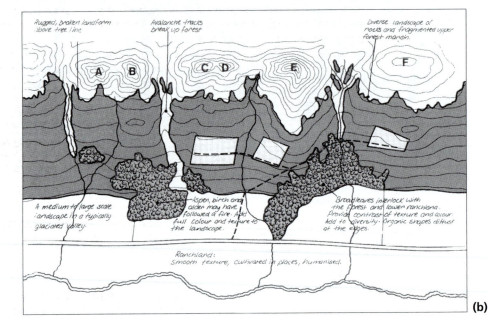

(b)

Figure 5.10 The land feature analysis: a) in perspective and b) on plan. Source: British Columbia Ministry of Forests.

landscape diversity. Describe the different patterns in terms of their shape, scale, interlock, relationship to landform; consider the textures and colours, especially in relation to seasonal changes; look for disharmonious patterns, perhaps caused by previous human intervention or the presence of intrusive elements. These descriptions should feature as a number of annotations to the panoramic photographs as well as a general statement that encapsulates the overall sense of unity, diversity and spirit of place. A map that identifies the major features should be prepared, so that elements visible from different viewpoints can be integrated correctly into the design.

Constraints and opportunities analysis

This third type of analysis is used to evaluate the implications of all the aggregate environmental, practical, social, economic and other functional issues that are likely to affect the strategic direction of the design. Synthesising both constraints and opportunities produces a fuller picture, especially where the plan aims to meet a number of specific objectives. Where forest planning is limited to harvesting timber or afforesting as much land as possible, any factors that reduce the possibilities of maximising timber production are often seen as constraints. In the modern drive to achieve broader sustainability, however, a constraint on one objective (such as having to leave stands of timber on wet or boggy ground uncut) may provide an opportunity towards achieving another objective, such as water quality protection or ecological functioning. Thus a constraint can also be an opportunity.

To carry out a constraints and opportunities analysis, there are two linked tasks. Firstly, the various factors that affect the possibilities of meeting the objectives should be identified, and their influences recorded. The best way of doing this is a kind of "brainstorming" session with the design or planning team. A table should be prepared which is filled in at this session and worked up afterwards. There are two ways of approaching this. The first way is to prepare a three-column table in which the rows correspond to the objectives. For example, the objective of timber production is taken and all the factors constraining the ability to meet that objective are listed in one column, and those factors providing opportunities are listed in the next (see the example below). The second way is to list all the factors (e.g. road construction, terrain stability, etc.) in the first column and then to identify the effects of each factor in constraining or providing opportunities to meet the objective (see example below). The first method helps the team follow through the objectives while the second relates somewhat better to the inventory information layers.

The second main task is to sieve out of the inventory data the significant information related to the factors already identified, where it can be spatially recorded. This means taking only that information which is going to make a significant difference. For example, terrain stability might be an important factor in steep, mountainous areas. However, of all the classes of stability, it is only likely to be the most extreme that seriously constrain activities such as road construction or the ability to use clearcutting as the preferred method of harvest. Intermediate instability risks may reduce the freedom of action, while moderate and lower risks can be effectively ignored. The same applies to steepness, wetness, windiness, soil fertility, etc.

If the inventory is stored on a GIS, then it is relatively straightforward to isolate certain factors and record these on a composite map. However, in many circumstances there is too much information to record on one map. Thus, it may make sense to sub-divide the analysis. One sub-analysis may be forest engineering, based on the inventory layers of steepness, slope/terrain stability, stream classes, harvest systems and road location opportunities (see case study in Chapter 9). Another sub-type is social and cultural information such as historical or archaeological features, recreation, traditional uses, hunting, spiritual uses, etc. A third might relate to the economies of timber harvest (especially in plantation forests), merchantability and so on.

As for the landscape ecological and character analyses, the constraints and opportunities analysis should be presented in perspective, so that the physical location of many of the relevant factors can be seen in three dimensions and better understood.

Table 5.4 Example of constraints and opportunities organised around objectives

Objective	Constraint	Opportunity
Timber production	Inoperable areas reduce available volume Riparian areas reduce available volume Visual sensitivity constrains the scale and phasing of clearcutting	To carry out commercial thinning in younger stands To replant some stands with more productive varieties
Scenic quality	Economics of harvesting demand large-scale cutting Road construction may create scars Use of cable systems reduces flexibility of cutblock shape	To retain extensive areas on prominent rocky slopes To use the inoperable areas to control the scale of landscape change To use the broadleaves to provide visual diversity
Ecological integrity	Logging may disrupt connectivity unless cutblocks are well distributed across the landscape in space and time Road construction may create breaks in stream and riparian corridors, creating barriers to movement for some flows	To use inoperable areas to maintain connectivity To link riparian, inoperable and other retained areas into a forest ecosystem network
Water quality	Timber among wetland complexes will be difficult to harvest without causing damage	To develop wide buffer zones along streams and around waterbodies

Table 5.5 Example of constraints and opportunities organised around factors

Factor	Constraint	Opportunity
Terrain stability	Areas of high mass wasting hazard are off limits to harvesting Road construction is impossible across such areas, limiting access to some places	To use these areas as part of the ecological connectivity areas To incorporate them into the proportion of the landscape that will appear intact over the long term
Operability	Low operability reduces available volume	Inoperable areas can be linked together to form part of the forest ecosystem network
Forest structure	The age structure is unevenly distributed across the landscape due to past disturbance events, causing harvest access to be expensive	The spatial distribution will enable cutblocks to be distributed across the landscape and reduce visual impact
Slopes	Steep slopes will require the use of cable logging systems Road construction will result in visible cut slopes	To avoid logging the steepest, most visible slopes that will be expensive to harvest

Step 5: Review objectives

Now that the crucial analysis step has been completed, it is usually time to take a breather, sit back and review the objectives in the light of what has been revealed. Also, if the decision has been made to wait until now to set the objectives, the assembled information and analysis may help identify what values are associated with the landscape.

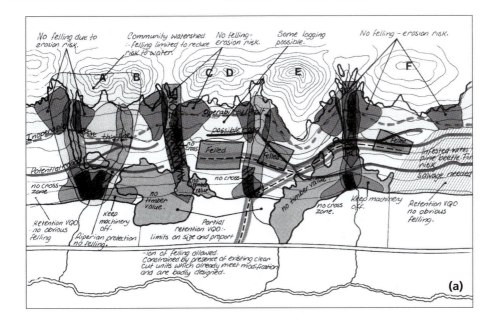

(a)

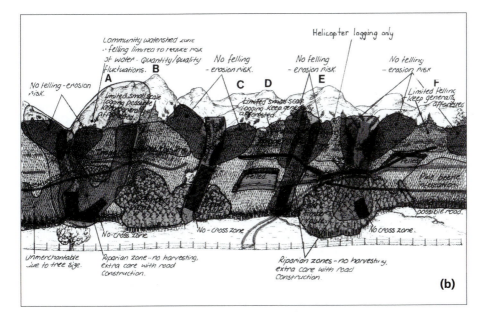

(b)

*Figure 5.11
Constraints and
opportunities
analysis: a)
shows the
significant
features sieved
out of the
inventory
information and
presented on a
plan; b) shows
the same
information in
perspective.
Source: British
Columbia
Ministry of
Forests.*

At this stage the objectives might have to be considerably revised if the analyses revealed unexpected issues that were previously unknown and which preclude harvesting access or planting on areas hitherto deemed operable. If this is the case, it may be necessary for the team to go back and to report their findings to key decision makers in order to obtain approval for any radical revisions.

If the objectives are to be set at this point, then it is possible to reach a consensus on what values are present. We can assume that everybody has been working from the same information base and has come to understand the complex interaction of factors that affect the possibilities.

This is particularly helpful where multiple stakeholders have been involved in all of the steps of the analysis, and can see that their original positions may need to be revised. The logger can see all the factors that constrain timber outputs. The environmentalist can see that key ecological capital has been accounted for and the local resident can be sure that his concerns have been included.

Once the objectives have been refined, then the team is ready to carry out the next step, that of developing the design strategy or concept.

Step 6: Design strategy or concept

One of our ecologist colleagues refers to this step as "when the magic happens". Design can be somewhat mysterious to our more science-based colleagues, since it involves creative responses to complex information. The design step is a particularly important one, but also one of the more difficult, especially when significant alteration to the existing condition of the forest is contemplated.

In the development of a design for a largely natural forest, it a good first step to integrate the main outcomes of the analyses, especially the ecological and landscape character. If the landscape has been subdivided into a number of different character zones, for example, it is a good idea to use these as the basis and to summarise the main structures, flows, dynamics and important ecological elements together with an interpretation of the desired future conditions for each zone. The next step is to consider which values in the list of objectives are particularly associated with each zone. This is useful because each objective is not achievable over every hectare/acre of the landscape, at least not at the same moment in time. The main constraints and opportunities that apply to each zone can also be examined and from this a statement defining the direction of landscape development can be formulated. This statement should include a description of the type of silviculture to be adopted, the elements to be protected or retained, inaccessible places, the shape and scale of management units and the expected rate of change. These statements provide the designers with a blueprint for the next step. Chapters 8, 9 and 10 explain this in greater detail, through the case studies.

When developing a strategy or concept for restoration of a native forest that has been heavily altered, it is necessary to develop a new vision, to see beyond the layout and the structure as it is now to the forest as it could and should be. This is where and why a high degree of creativity is needed. The desired future condition arising from the ecological analysis should be an important guide, as should the guidance on shape, scale, unity, etc. provided by landscape character. What is also needed is a route by which to reach the final goal; that is the type of management that will enable, for example, a fragmented, patchy landscape to become restored to an intact matrix. In this case the zones may or may not reflect the character zones as described earlier. They may, instead, be defined by the degree to which restoration is necessary, or where greater efforts are needed. Chapter 8 explores this in more detail.

In the case of restoration of native forest in deforested areas, the concept can be inspired by examples of existing native forest as well as by the ecological patterns defined by analysis. Zones can be defined based on topographic variation, forest type predictions and visible areas. Brief descriptions of the desired character can then be related to each of these zones for use by the designer. See Chapter 9 for this issue as a case study.

A similar approach is needed in the design of new plantations. Here, however, the zones are likely to be split between the productive element and the open space/native forest component. Ownership boundaries may often pose a severe constraint, as may neighbouring land uses. Chapter 10 considers this type of design.

A concept for the felling and replanting of plantation forests may pose the greatest challenge. This is because these types frequently require major restructuring following felling. As in the case of the fragmented natural forest in need of restoration, it is necessary to be able to see the landscape as it could be instead of how it is. A good starting point, therefore, can be to consider what the forest should be like and then work out how to get there. Since one of the aims of restructuring even-aged plantations is to develop greater age structure, it is often not desirable

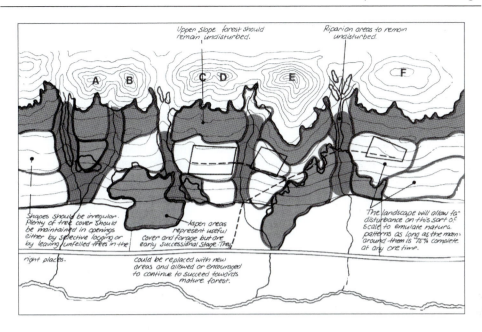

*Figure 5.12
The design
strategy
presented on an
annotated map.
Source: British
Columbia
Ministry of
Forests.*

to remove the entire plantation all at once and to start again. Thus, the concept also has to consider how to develop this at the same time. This is tackled in Chapter 10. Zones where different mixes of objectives are to be concentrated also help to guide the designer.

The design strategy or concept should be presented as a zoned map together with brief statements associated with each zone, in terms of how the objectives are to be met and what overall pattern and structure are to be developed.

Step 7: Sketch design

Once the concept or design strategy has been developed and evaluated in general terms against the objectives, it is time to develop the design in more detail. This step is an interactive process, where the draft design becomes ever more refined. The term sketch design refers both to the process, literally of sketching design ideas, and to the product, a sketch of how the forest might appear either at the end of several stages of implementation or at various intermediate stages along the way. Depending on the visual prominence and sensitivity of the area undergoing design, greater or lesser emphasis may be given to perspective or plan views as the main medium for design and presentation. There may be more than one option of a design concept to be developed or several options for a single concept.

Any forest design is likely to contain one or more of the following elements as major components:

- External margins: i.e. the outer boundary of a plantation, the operable limit of a harvest plan, the edge of a watershed
- Species patterns resulting from afforestation or replanting after logging
- Areas where different silvicultural systems or approaches are to be applied
- Felling coupe/cutblock or prescribed fire unit layout
- Reserve areas, riparian areas, special habitats, open meadows, berry patches and other non-forest parts of the landscape
- Forest roads and pathways
- Detailed design of edges
- Key linkage zones where special attention is needed.

The general principle is to work from the broader, bigger scale down to the more detailed level. The following examples show how this works.

For an afforestation project, the first step is to design the external margin to establish the correct balance between forest and non-forest (in an open or agricultural landscape) and to ensure that it is well unified with the landscape setting. The next step is to develop the pattern of species or forest types to be planted within this boundary. Each of these should be located according to the concept and their precise shape and scale worked out in detail so that a sense of unity and balance is achieved. Thirdly, the non-forest elements, such as open spaces, non-forest habitats and other unplanted areas, should be designed. If roads are to be built during afforestation the location and alignments should be incorporated next. Finally, detailed aspects such as edge structure along the external margin, junctions of species and internal edges should be developed.

Figure 5.13
This series of sketches shows the stages of afforestation design:
a) the first step is the external margin and the balance of forest to non-forest or other land use in the landscape;
b) within the external margin the different forest stand types or species are designed, with their shape, scale and balance in the landscape carefully considered;
c) adding the final detail of small-scale non-forest features into the design.

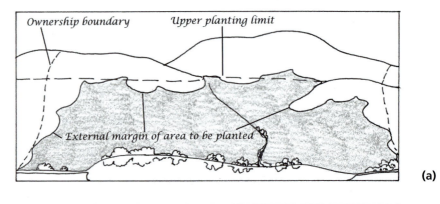

(a)

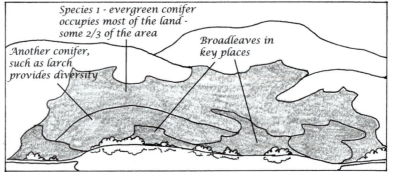

(b)

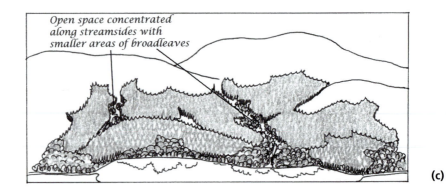

(c)

For a harvest design in an existing forest the key is usually to balance units that will be cut with those that will not. In part this is done by determining the boundaries of operable and non-operable areas, but also by identifying important ecological capital to be retained. This is sometimes referred to as a "forest ecosystem network", or FEN. This should result initially in large-scale shapes that are integrated into the wider forest landscape. Following this, a network to access the operable area should be designed, following engineering analysis and ensuring that impacts are kept as low as possible. The broader operable units are then subdivided into felling coupes/cutblocks or silvicultural units, which should reflect underlying landform and the major natural shapes found in the landscape. Conservation units may also be subdivided to facilitate restoration, prescribed fire or other activities. Finally, edges and any sub-unit divisions should be designed.

The main manual techniques for sketch design are the use of acetate or tracing paper laid over maps and/or photographs. Computer-aided design is also possible. The main outputs at this

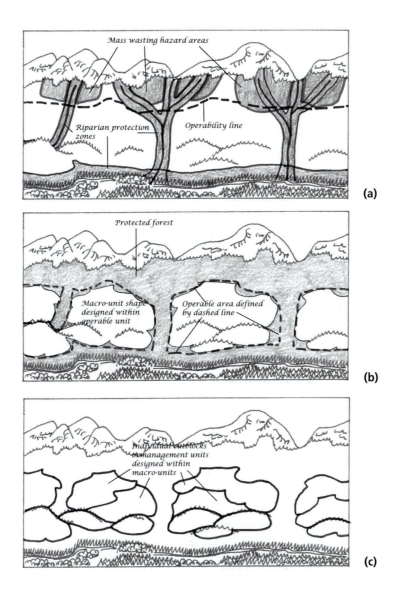

(a)

(b)

(c)

Figure 5.14 Harvest design presented as a series of stages: a) shows the definition of the operable/non-operable areas; b) shows the larger units designed from the operable forest and c) shows these larger units subdivided into coupes, cutblocks or management units.

stage are maps showing each of the elements and their boundaries, colour coded for species, felling period, silvicultural system or reserve category. The resolution of the map should be such that the boundary lines of each unit of design can be transferred to the ground with an adequate degree of accuracy. In other words, these are not indicative boundary lines. Road lines should also be shown, once again as accurately as possible. In visually sensitive locations the sketch design should be presented in three dimensions so that stakeholders can evaluate its anticipated effects on the landscape. There are various high- and low-tech visualisation methods that can be used for this purpose (see Chapter 6).

Step 8: Evaluate the design against the objectives, analyse and refine it

As every iteration of the design develops, and especially once it has been more or less finalised, it is necessary to evaluate it against the objectives. This is best done using the criteria that were developed during the stage of setting objectives. If the design concept was properly evaluated, then there should be few problems in meeting the objectives, although more refined analysis, especially the quantification of key indicators, may be necessary or desirable. This may include timber production over time, habitat and ecosystem measures, as well as checks to see that the design fulfils the criteria of certification. Qualitative assessment of visual design quality, recreation potential, protection of cultural heritage and other aspects is also required. Many of these factors may involve complex evaluative mechanisms. Rather than explain them in depth here, Chapter 7 provides an opportunity to present a range of methods appropriate to different circumstances.

The sketch design should also be evaluated against the various analyses that have been carried out. It is very important that the design should take full account of the range of opportunities and constraints. In particular, the practical aspects should receive close attention in order to ensure that the design is workable. It may not be possible for the designer to carry out all of the evaluation alone. Other experts from among the design team should be given the task of thoroughly assessing the design in terms of particular elements of the analysis. For example, logging engineers should ensure that all cutblocks are practically feasible using the anticipated road layout and harvest equipment envisaged for the project. Ecologists should check that those sensitive habitats have been adequately protected and any necessary corridors retained.

There is clearly some overlap between evaluating how the sketch design meets the objectives and how it responds to the analyses. An effective way to organise the evaluation is for the designer to present the sketch design (or several options) to the design team and clients/stakeholders, concentrating on how the design meets the objectives and where key constraints and opportunities have influenced this. The clients/stakeholders should then have a chance to question the designer, and to identify a preferred option, or to ask for revisions or other alternatives to be developed if they are unsatisfied. Following this public evaluation, technical experts (either from the team or an outside peer group) should be asked to carry out a more detailed assessment. All the results of the evaluation should then be presented to the designer who is responsible for revising the sketch design or for one particular option to be developed further. Depending on the degree of revision needed, it may be appropriate to reconvene the client/stakeholders for another presentation or it may be left to the designer to finish the design and to prepare the final demonstration. It is typical for there to be several revisions before a final consensus is reached.

The key point is that one should not expect to get the design right at the first pass. Constructive critique and multiple iterations are key steps of good design. On the other hand, one must guard against simply changing the design to satisfy unsubstantiated objections or prejudices, however strongly they are put forward. Flexibility is important, but so is the need to be able to defend the design solution. Be reasonable but not rigid.

Step 9: Documentation and approval

The purpose of this step is to prepare the final content of the design in an appropriate format and to submit it for final approval to the clients/stakeholders and, where necessary, to whatever authorities exist in the country concerned. This could mean the national or state forest service as well as other governmental agencies who issue licences or permits for one or more aspects of forestry plans and operations. If such agencies are represented in the stakeholder group and have been involved in the evaluation of the design, then final approval steps should be a formality.

Documentation of the design should serve two related functions. Firstly, it should record the final design intentions as a set of maps that specify what is to be done where, when and how. This becomes in many cases the basis for a formal agreement or contract between, for example, a logging company and a forest service or owner. Such maps should be extremely clear, as accurate as necessary and conform to whatever legal requirements exist. The second function of documentation is to communicate the elements of the design to all the clients or stakeholders, to any other consultees or regulatory authorities not represented in the stakeholder group and to the managers who are to be responsible for implementing the design. This is particularly important when a forest ecosystem design is expected to cover a long period of time, since staff change and new people need to know what reasoning lies behind some of the prescriptions they are being asked to carry out.

The main elements of documentation are as follows:

- Limited text to describe the background, context, main inventory information, the analysis, concept and any specific aspects of the final design.
- A series of maps. The main inventory information may be better placed in an appendix. Major maps include context and setting, showing the location and surrounding land use, access, river systems, etc. Landscape character maps should depict the landform analysis together with descriptions of character zones and, where appropriate, visual problems. Ecological analysis maps should show landscape structures, flows, zones of different disturbance types and desired future conditions. Constraints and opportunities should also be included. The initial design concept map is particularly important and should be clearly presented. The sketch design maps will vary depending on the type of project. These may include layout of new planting showing outline, species, open spaces and roads; felling maps showing coupes/cutblocks keyed with roads/silvicultural system, felling period, retained and inoperable areas; replanting maps showing revised layouts, species and open spaces.
- Tables are useful to present the objectives, landscape zone descriptions, interactions between landscape structures and flows, disturbance regimes, opportunities and constraints and harvest or silvicultural schedules. A table demonstrating how the objectives have been achieved in the plan is also valuable.
- Charts can be used to show species proportions, timber volumes, age class distributions, proportions of retained stands and open space, especially projected over time to demonstrate

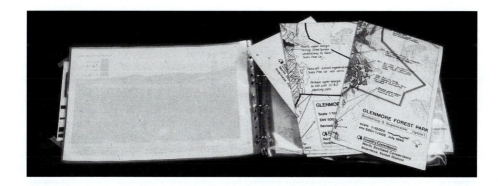

Figure 5.15
The documentation of a forest design plan, held in a ring binder for easy access and presentation.

how well the objectives and the requirements of sustainability are met during the duration of the plan and beyond.

- Photographic simulations or illustrations, including both aerial and perspective, and illustrations to show the landscape character, the basic design and the expected appearance of the forest over time. Photographs can also be useful to show the various landscape structures of the ecological analysis to supplement the descriptions.

Step 10: Implementation

For most forest designs, implementation is through road construction, logging, silvicultural activity and restoration, such as controlled burning, weeding or planting. Since these plans cover a long period of time it is usual for work to proceed in phases. Depending on the map scale used for planning, it may be appropriate to take each element (coupe or cutblock, etc.) of each phase and enlarge the scale to show greater detail. Usually there are aspects of detailed layout and design to be prepared such as final road layout, setting out of boundaries to take account of local features not recorded on the planning maps, small-scale elements of biodiversity potential and so on. Thus, the staff responsible for planning and design must hand the implementation to field staff who should have been involved in the development of the plan, especially its evaluation and testing for practicality. Each management unit (planting area, coupe/cutblock, etc.) should be planned in detail including many operational details such as site production measures, before any work starts on site. Chapter 7 explains this stage of the process in greater detail.

Step 11: Monitoring and revision

Since most plans proceed by phases it is appropriate to pause after the conclusion of each major phase and to review progress. If new information has become available, perhaps because of new research, its implications on the continued implementation of the plan should be considered. Adaptive management is a particularly valuable approach where special monitoring systems, surveys and even experiments are set up as part of the implementation, so that the most important assumptions contained in the plan can be tested. For example, implementing a silvicultural prescription based on natural disturbance may be expected to produce habitat for certain species. Surveys to test if this is a success will help in proposing whether to repeat such prescriptions or to modify them. Thus, the plan is in a more or less constant state of testing and potential revision. See Chapter 7 for a more detailed description of monitoring and adaptive management.

It may be appropriate for regulatory authorities only to approve one phase at a time, so that a formal review stage prior to obtaining the next phased approval should be an integral part of the process. This also enables changed stakeholders views, policy developments and regulatory changes to be incorporated.

Conclusions

In this chapter we have presented the process of forest design as it has evolved over the past 20 or so years. We believe this to be a time-tested process. Every project is different, but the basic steps we have presented are almost always needed. What varies are the details, such as the composition of the team, the size of the area, the amount of detailed information available to work with, the interest of the public in participating, the time and budget available and so forth. We strongly recommend that anyone contemplating the undertaking of a forest design project follow the steps we have laid out as closely as possible, and avoid missing some out in order to try to save some time and money, without careful consideration of what might be missed out. Sometimes some steps are not needed – for example the detailed ecological analysis may not be possible or as relevant in some plantation forests. Following this advice will not inhibit creativity, but will help avoid pitfalls, and will increase the odds of a successful outcome being achieved.

Chapter 6
Design and visualisation techniques

Introduction

In Chapter 5 a comprehensive overview of the design process was presented. At each stage of this process there are a number of tasks, many of which require special materials and techniques. These techniques fulfill one or more of the following functions:

- Recording and presenting information
- Undertaking and presenting analysis
- Developing the design
- Presenting and communicating the design.

For each of these functions different materials, equipment and skills are required. Some of these functions can be carried out manually, while others require sophisticated computer software. There are cost/benefit implications to all of these. The degree of simplicity or sophistication also depends on the sensitivity of the area and the degree of public participation during the design process.

Basic materials and equipment

Forest planning and design often encompasses large tracts of land, and all the maps, overlays and design drawings inevitably take up a lot of space, even if a lot of work is carried out digitally nowadays. The first basic requirement is for a space large enough to accommodate at least one large, flat surface, a drawing board, a desk, pin-up space on walls and a computer workstation. Storage can also be a problem. Maps can be rolled up or kept in special chests, either hanging vertically or laid flat in drawers. Design documentation is frequently presented in landscape format at A3 or 11″ × 17″ format, so shelf space for these is also required.

Good light is important, preferably natural but supplemented by satisfactory artificial provision. A large moveable desk lamp over the drawing board is also ideal. If the design is to be completed manually, the majority of the work will take place at the drawing board, which should have height and tilt adjustment so as to provide comfortable working conditions. Materials needed include tracing paper, coloured pencils and felt tip pens, drafting tape and the usual office consumables. Otherwise a computer workstation with a large screen and a large format printer or plotter are also needed.

Inventory information

The type of inventory information to be collected or made available was described in Chapter 5. This must be presented at a suitable map scale and should ideally be at the same resolution. Problems can occur if information sources have been developed at different scales, since the accuracy of lines defining features presented at 1:100 000 is likely to be less than that of those of 1:20 000 or 1:10 000. All information should be keyed with a suitable legend and use the same map base. Occasionally some information is available in different map projections, which prevents accurate overlaying of information. If possible these should be converted and corrected.

It is preferable if map information is completed and stored in a geographic information system (GIS). Not only does this permit efficient storage but also allows maps to be presented at different scales, enables various analyses to be performed and very clear presentation maps to be produced. Typically, databases accompany the GIS layers so that it is easy to find out what each shape or polygon represents without having to refer to reports or data print-outs. More will be said about GIS later in this chapter.

Aerial photographs are very valuable tools. These can be of four varieties:

1   Oblique aerial photographs can be very useful for showing the character of an area. Oblique panoramas taken from, for example, a helicopter, at a few hundred metres above the ground, are particularly valuable. The best way to obtain these is either to take the helicopter door off beforehand or to take the pictures through a window that slides open. The helicopter should describe a slow, tight circle, during which a series of overlapping photos can be taken (further details of photography are described below). The altitude and position from which the photos were taken can be recorded from the helicopter's instruments (altimeter and global positioning system).

2   Vertical aerial photographs in stereo, in true or false colour and black and white, are useful because, with a stereoscope, the three-dimensional nature of the forest can be seen. This enables the topographic variation to be made clear, although exaggerated, and also the structure of the vegetation, particularly the forest canopy. These photographs are normally obtainable from commercial or government sources. It may be useful to obtain earlier sets, so as to see how the landscape has changed. In the case of British plantation forests, it can be useful to see what the area was like before the forest was actually planted, using photographs taken shortly after the Second World War.

3   Orthographically corrected vertical aerial photographs have been treated to remove the distortions present in ordinary vertical aerial photographs. This enables them to be used like maps. Sometimes contours are overlaid on them. These can be used to interpret vegetation, for example, as part of the landscape ecological analysis. The orthographic correction is nowadays usually carried out on a computer and so, if the resolution is acceptable, these photographs can be used within a GIS and as a base to create fresh maps.

4   Satellite photos are becoming more readily available at high resolution and in colour. For forest design of large remote areas they provide excellent contextual information and may be as useful as some aerial photographs. They are usually digitally formatted and can be orthographically corrected and so can also be used in a GIS. It is to be expected that these will become more commonly used in the future.

Photography

Where the landscape is visually sensitive, or wherever it is helpful to convey the character of the landscape, a comprehensive set of photographs is needed. These may include single-frame examples, perhaps of details of individual landscape features (such as the elements defined in the landscape ecological analysis) or panoramas.

Panoramic photographs are widely used to convey a complete scene. They may vary from 60° right up to 360°, in other words the complete view as seen from a single point. Normally an angle of 90°–150° is sufficient. These photographs can be taken in one of several ways:

- A series of overlapping frames using a single lens reflex (SLR) camera, either digital or by using negative film. This remains the most common, simple and flexible method. The photos are taken from a fixed point. The focal length is normally 50 mm as this ensures that the view is as close as possible to what the human eye sees. Zoom lenses can be used to focus on detail but also foreshorten the view. Each frame is taken with around 20% overlap of the previous frame. This allows them to be joined together accurately. If need be, a tripod can ensure a stable, level trajectory; otherwise success is possible with a handheld camera. It is important to take account of the lighting levels and, if necessary, to adjust the exposure to ensure a good balance. Some modern automatic cameras do not permit a manual override of the exposure so there may be problems where the sky or an area of water or snow present a strong contrast in brightness to the forest. Wide-angle shots should be avoided: they cannot be joined into a panorama as they are too distorted.
- Digital cameras are now common and even if not to SLR standard are capable of taking high-resolution photographs. Their advantage lies in being able to download the image directly onto a computer and, by using special software that can adjust brightness, colour and contrast, to join each frame of a panorama together before saving the final photograph.
- The photos, if printed on conventional photographic paper from negatives, have to be joined together by overlapping them and slicing through the overlap using a straight edge and sharp knife. The two edges are matched and secured by adhesive tape from behind. The completed panorama can be trimmed top and bottom to tidy any stepping from frame to frame. The photos can be taken either in landscape or portrait format and it is important to include sky, foreground and external parts of the landscape to ensure that the whole section of the landscape is present in the photograph. All of the above also applies to oblique aerial panoramas.
- Panoramic cameras can be used to eliminate the need for joining points together. It is possible to obtain special cameras that have a rotating lens. This moves from right to left and gradually exposes a long section of film. Some excellent models were produced in the former Soviet Union and can still be found. Modern automatic cameras that produce panoramic photographs from a single shot are also available.

Photographs are used in a variety of ways for description and analysis. How they are used depends on the facilities available to the designer. Appropriate graphic techniques for visualisation are described later in the chapter.

- The digital panoramic photographs joined up as described before can be used in both colour and greyscale format. The colour photographs can be put in a word processed report (at A3 or 11″ × 17″ format) and text annotations added using the word processing package. Greyscale versions can be overlaid with symbols, cross-hatchings and text in different colours using a computer graphics package. The greyscale version presents a good background for other colours which would be confusing on a colour photograph.
- If prints from negative film are used then photocopies are made for analysis purposes. Modern high-quality laser copiers produce excellent copies, either in colour or black and white (monochrome or greyscale). The colour copies can be used in a report with annotations pointing out key features. The monochrome versions provide the base for analysis as described above.
- In the absence of good-quality photocopying facilities a sketch of the photograph may be used as a base for visual analysis, design and presentation. This type of sketch is simple to

produce. Lay a sheet of acetate over the photograph and, using a fine black felt tip pen, trace the outlines of the main features and add in as much detail as necessary. The resulting tracing can then be copied on an ordinary office photocopier.

Geographic information systems

Geographic information systems, or GIS, are excellent tools for a number of aspects of forest design. Depending on the type and comprehensiveness of the package, varying degrees of analysis and modelling are possible. The level of skill required to make the most out of sophisticated packages can be quite high. The subject of GIS is vast and so there is only room for a summary of the main aspects to be covered here.

GIS systems come in two forms: raster and vector. Raster systems use information collected and presented as small cells or squares. The degree of resolution depends on the size of these squares compared with the scale of the landscape. If the resolution is below 10 metres for a project of several thousand hectares, then the zig-zag edges of the shapes will not be very obvious. Raster systems, while not as accurate at depicting shapes and measuring areas, have advantages for data interpretation and analysis. For example, it is possible to collect site information such as soils from a limited number of sample points. These can then be interpolated to fill in the gaps and produce a complete map. Maps showing changes of slope, aspect, elevation, climatic information and any other pattern that changes gradually are ideally prepared and presented in raster format. Analysis involving combining the effects of several variables to produce a resultant map is also best carried out in raster format. The application of Ecosystem Site Classification briefly described in Chapters 5 and 8 works well in raster format. Several factors are combined according to a set of decision rules and a map of, for example, species suitability can be produced.

Vector systems use information presented as lines or polygons with precise geo-referenced locations and areas. The term vector comes from the fact that each line segment proceeds in a specific direction or vector from one point to another. Information captured in this way is ideal for area measurements of features on the ground, such as patches of different types of forest, water bodies, linear features such as roads or tracks and so on. Each feature can be linked to a database where information describing them is held, such as species, age, height, timber volume, growth rate, canopy closure and so on. It is, however, difficult to recombine vector information and produce a completely new map in the way raster systems can. The essential patterns of lines and polygons cannot be altered without specific work by the operator. Another disadvantage is the need for each polygon to represent a single homogeneous entity, separated by a hard line from the next entity. Forest landscapes are not necessarily so clearly defined, so interpretation of the information is crucial. Vector systems are ideal for analysing the results of plans in terms of areas and volumes, especially when such statistics are needed for evaluation purposes.

It is usually the case that both raster and vector systems are needed for full analytical or modelling potential. Many systems, such as ESRI's "ARCInfo" or its easy to use version "ARCView", allow both types of analysis. How much is possible depends on having all the modules that go with it, such as "3D Analyst", "Spatial Analyst" or the "Forester" extension (allowing lots of special forestry analytical operations and outputs) as well as the basic mapping and database capabilities.

Design methods

In Chapter 5 the creation of the design concept and the stage of sketch design were briefly described. Since the design stages are crucial to the success of the entire process, it is important to describe something of the process undertaken by a designer.

The term sketch design, as explained in Chapter 5, refers to the process of sketching out ideas or options and to the end product, a sketch of the proposed landscape. In forest design there are two main areas of design: planting layout for newly planted areas and designing a set of coupes or management units with subsequent replanting or regeneration and associated reserve areas.

The sketch design process for new planting involves taking the existing unforested landscape and designing the areas to be planted with different species of trees or left open. Coupe design starts with the existing forest and subdivides it into a complete pattern of coupes linked to reserve areas. These are two fundamentally different tasks but both take as a starting point the analyses carried out in plan and perspective.

Map-based design

In places where the landform is not very significant or where the visual sensitivity is low, it is acceptable to design on a map base. The technique is to use tracing paper laid over the map and to sketch out the proposed shapes on it. The underlying map should be one of the analyses, so that the developing design is built around the main factors. If the base map also has contours present then it normally makes sense to relate the shapes to these because they are likely to reflect site types and drainage.

Start with rough general shapes, using a pencil so that errors can be erased, and do not be worried about just playing with shapes to start with. Keep the wrist loose and relaxed so that the shapes are flowing and not stiff and too detailed. As elements of the design start to firm up use a felt tip pen or heavier pencil line to select the final line or lay another layer of tracing paper over the top and draw the chosen version of the lines on it. Then remove the earlier layer and continue to refine the lines, making sure that the final sketch is accurately registered to the boundaries and fixed landscape details such as streams.

Use this first version on tracing paper for the initial discussions with other team members – this keeps the process flexible and avoids the sense that lines have become fixed or frozen and that it is too much effort to change them. Fixing final lines should be kept until the last possible time. It is not indecision to keep options open at this stage. If more than one design is needed, the conceptual stage should suggest what these are and then the sketch design stage develops them further by repeating the process just described. Eventually the design of the final option needs to be fixed, when a final, accurate tracing is made carefully annotated with species, felling dates or silvicultural systems.

An important aspect that needs to be considered is the level of detail in the design. It must be stressed that, although the map scale may be 1:10 000 or 1:20 000, the process just described is design, not planning, and the lines created are intended to be laid out on the ground to within a few metres of accuracy. Therefore, the degree of detail of the shapes needs to reflect the various influences affecting the design and the subtleties of the land as well as the characteristics of the natural patterns being emulated. However, it is equally the case that since the maps are small-scale, it is not possible to design patches of trees much below 0.5 hectares in area as the thickness of the lines drawn on the map are equivalent to several metres in width.

The final sketch design can then be transferred to the GIS or by hand onto a map base and colour coded for species, felling phase or silvicultural system, etc. Reaching this stage does not mean that the design is fixed for ever, but it does imply that the design is well worked up and ready for detailed checking and evaluation. If, following this checking it needs to be revised, whether to a major or a minor degree, it is important not to become defensive about the original design and the effort that went into it. Remember that design is a team effort and the final, agreed and approved design is the product of the entire team, not just the actual person who drew the lines on the map.

Perspective-based design

Where landform is a major influence and where the landscape is visually significant it is necessary to base the sketch design process not on maps but on perspective. This is because, owing to distortions in perspective, it is impossible to predict how a map-based shape will appear in perspective. Three-dimensional landform and the effects of perspective depending on the viewer position mean that planar map shapes can appear extremely distorted and, in any case, may only be partly visible. A map is a flat representation of a three-dimensional world and no amount of skill in reading contours to understand the topography can help.

It has been standard practice in Britain to design forest shapes, whether for new planting or for felling and replanting, in perspective, using photographs as the base. The landform analysis using the principle of lines of visual force, as described in Chapter 3, becomes the main underlay for this approach. Starting with the photograph, or a monochrome photocopy upon which the lines of force have been drawn, tracing paper is overlaid and shapes of species, coupes or management units are sketched out in exactly the same way as described for the map-based design. The shapes are designed to respond to the landform and to reflect the other main features of landscape character such as the types of shapes found in the topography (angular, rounded, etc.), in the vegetation or both.

Designing shapes in response to landform, using the guidance provided by visual forces, requires some skill. This tends to act on two levels. The first level is to guide the general location of shapes on the landform, for example identifying knolls and hilltops to be capped with a shape and valleys to be occupied with a shape, other shapes being located on the slopes between the caps and valleys. The second level is in the detailed shape and interlock of the lines forming the boundary of the shapes. It is best to think of the forces as acting to distort a straight line drawn across the hill or mountainside. The upward lines of force push and distort the line up into the valleys or gullies and the downward lines of force have the reverse effect. If the analysis has been carried out correctly there is a ranking of relevant strength to the lines of force so that the stronger, major forces exert more pressure than the weaker, minor ones.

As well as considering the shapes of the different elements being designed, it is important to consider their scale and proportion. The scale of the landscape increases with elevation, so it is usual to find larger-scale shapes on the upper slopes and smaller ones lower down. The "thirds rule" of proportion is useful here too. Other aspects to consider are ensuring that there is no unnatural symmetry introduced into the landscape by designing too-similar shapes or ones which rise up into hollows at the same height and whose peaks are the same distance apart. The balance of species and felling periods across the landscape also needs to be considered. The proportions of different species used can also reflect "thirds rule" as can the proportion of the area felled at any one time phase. It is also important to balance the distribution of phases, so that there are some at low elevation, some higher up and a general distribution across the landscape as opposed to starting at one end and working along a slope. Of course, there are other considerations in deciding on the species or the phasing, such as the soil types, maturity of the forest, its merchant-ability and the need to develop roads to follow the phases. For this reason it can be helpful to have the main opportunities and constraints affecting the design also depicted in perspective as overlays to the photograph, so that the design can take them into account as it develops. The case studies in Chapter 10 give examples of this.

If the area to be designed is seen in sections from a number of viewpoints the design also has to be done in sections. It must be recognised that due to perspective effects, shapes that look fine from one view may look different and somewhat foreshortened from another. This is normal but it can cause problems to the inexperienced designer. This is why it is best to draw the most directly visible sections of the design from each viewpoint and then to extend and connect them to the neighbouring views, which should overlap to some extent. It is not necessary to worry too much about whether the design will work from all views. If the aim of the designer is to try to hide all the cutblocks from view, then it is likely to be impossible from all views. However, when designing shapes to follow landform the design to some extent becomes viewpoint-independent because, as the landform appears different from each viewpoint, so do the forest shapes that are directly related to it.

As for map-based design, the sketch design proceeds to become ever more refined until it is time to fix it. At this point, if not before, it becomes necessary to transfer it from the perspective

Figure 6.1
An example of an oblique aerial photograph taken from a helicopter. This gives a very good idea of the forest landscape
and the spatial relationships of the different parts of it. Eastern Lakes, Cornerbrook, Newfoundland, Canada.

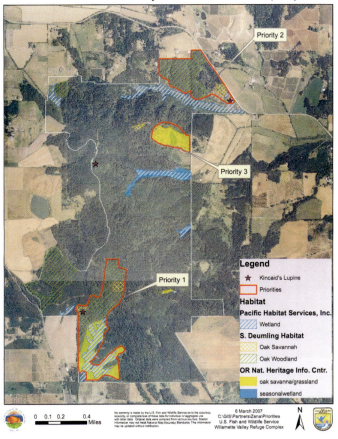

Figure 6.3
Satellite photographs are now available with fine resolution
so that a lot of detail can be seen on them. They are good
for working at the larger landscape planning scales. This
one shows part of central Russia to a fine resolution.

Figure 6.2
A vertical aerial photograph which shows all the detail of the forest and
other features. If it is ortho-corrected it can be used as a map and
information can be digitised from it into a GIS. This example shows the
photograph being used as a map with management information overlaid.
Source: Ecotrust and Trout Mountain Forestry.

(a)

(b)

Figure 6.4
A panorama made by joining photographs using computer software: a) the original photograph, and b) a greyscale version, either photocopy or print of a scan, used for analysis.

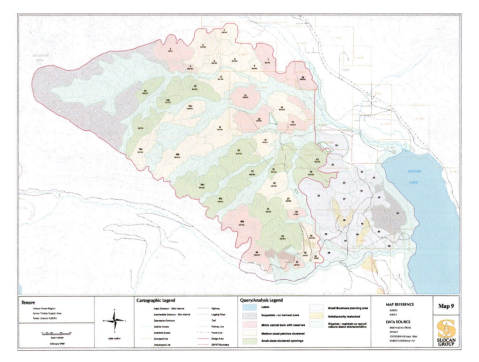

Figure 6.5
An output from a raster-based GIS, in this case a map with a lot of background detail showing a pattern of cutblocks. Each of these polygons is linked in the GIS to a database containing details of area, species, silvicultural treatment, phasing and so on, making the system very flexible. Any changes that take place in the map are reflected in the database and vice versa. It is also more convenient than large paper copies.

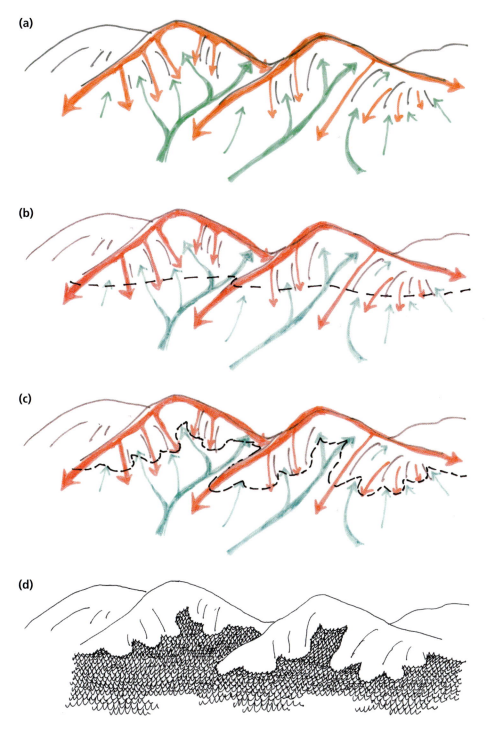

Figure 6.6

The process of designing to reflect landform: a) shows a sketch of a mountainside with the main lines of visual force drawn on it; b) shows the approximate line of an upper planting limit following the contour. It can be seen that this line conflicts with the way that the lines flow; c) shows how a designed line responds to the effect of the lines of force, being deflected by them. The strength of this deflection depends on the relative strength of the force line; d) shows the result of the design, the upper margin of the proposed forest following a much more naturalistic and organic line that relates strongly to the underlying landform.

(a)

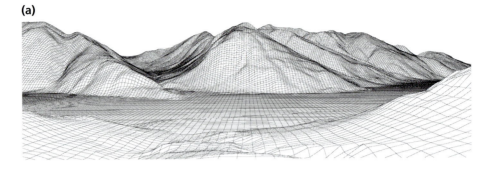

(b)

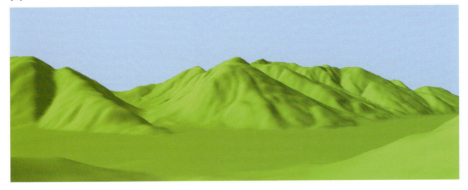

Figure 6.7
Simpler computer graphics: a) shows a wireframe rendering of a digital terrain model;
b) shows the same landform with solid shading, giving a more easily interpreted result.

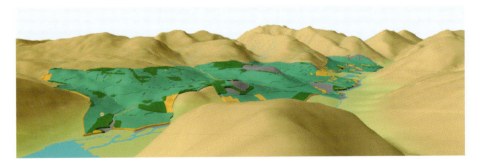

Figure 6.8
An example of simple computer modelling showing areas of different species of trees
as solid layers whose colour represents species, thickness and tree height.

Figure 6.9
Some harvest units rendered on a landscape using photo-editing software. These show different harvest types, including selection cuts. Source: Tom Savage, US Forest Service.

(a)

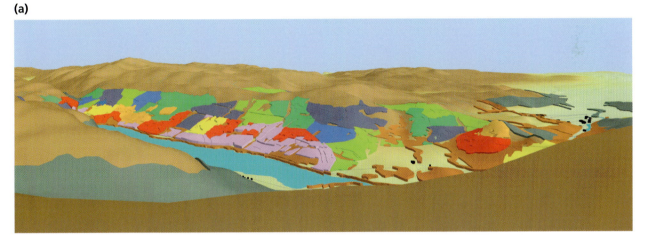

(b)

Figure 6.10
Two examples of computer rendering: a) a felling design where the coupes are shown as slabs ("treeblocks") colour coded for different phases while b) shows the replanting design of the same area using simple but effective tree symbols.

(a)

(b)

Figure 6.11
a) A replanting design rendered more realistically using a more sophisticated modelling
system. This uses scanned images of different species of trees. b) Shows another version of a
design in different weather conditions, illustrating how versatile such rendering systems are.

(a)

(b)

(c)

Figure 7.6 (Continued overleaf)
An example of a time-series for a forest design:
a) shows the original forest;
b) shows the felling plan;
c) to g) show the stages of development at 10-year intervals.

(d)

(e)

(f)

(g)

to the map. This can be difficult if the landscape is monotonous, for example when designing coupes in an even-aged dark green-canopied forest of a landscape where there is a high degree of foreshortening in the perspective. This is where aerial photographs become valuable, especially ortho-corrected ones. The starting point is to identify features on both perspective and aerial photos and to use these as common reference points for the shapes. If the landform analysis has been carried out accurately in plan and perspective so that the lines on the perspective photo reflect those on the contour map, this also provides guidance. Other features, such as roads, can also be used for reference.

If the initially mapped design is put into a GIS with some kind of visualisation option then the computer-generated perspective can be checked against the original design to see where mapping inaccuracies have taken place, so that they can be corrected in the GIS. This difficulty in moving from perspective to plan in design was one of the main motivating factors in the development of three-dimensional computer-aided design for forestry, as described in the next section.

Computer-aided design

In architecture, engineering and increasingly in mainstream landscape design (parks, housing, gardens, etc.), computer-aided design, or CAD, has become ubiquitous. It allows complex three-dimensional designs to be developed and altered at will. In the realm of forest landscape design, CAD has been used for some 15 years and continues to develop. Often, CAD is confused by non-designers as the final rendered output showing a building, structure or park as a realistic three-dimensional projection. This is visualisation for presentation purposes and will be described later in the chapter. CAD, when used by the designer, looks far less realistic and glamorous. CAD enables the forest designer to plan the layout of the forest in three dimensions and/or plan simultaneously. As in the vector GIS described above, the forest layout is reduced to a pattern of lines and polygons representing different elements, for example cutblocks, retained stands, water bodies, forest roads or streams. The designer uses both the existing pattern of stands (obtained from a GIS layer) and creates a new one (of felling coupes, of new planting). Instead of preparing the design or plan and using three-dimensional visualisation to test the results, it is more effective to design in three dimensions from the outset.

CAD systems for forest design start with a digital elevation or terrain model (DEM or DTM). This is a three-dimensional representation of the landform underlying the forest. It can be presented as a deformed grid, the interval between the lines making up the grid being 20–100 metres depending on the size of area and resolution of the data. This DEM/DTM can be obtained ready-made from mapping agencies in many countries, or can be constructed from scratch by digitising contours from a topographic map.

CAD systems are able to view this topography in true perspective from any given viewpoint, terrestrial or aerial. The viewpoint is specified by three coordinates – the easting, northing and height, or X,Y, Z according to the coordinate system in use for the area. In many instances it is also possible to reproduce the exact scene of a photograph, if the precise camera location, focal length and viewing direction are known.

It is possible to overlay linear and polygonal information on the DEM/DTM either imported from GIS or digitised specially, or a combination. Since forests are not flat, but possess height as well as area, it is necessary to be able to represent this. The usual method is to give an average tree height to each forest polygon and to show it as a slab of standard thickness overlaid on the terrain. While not very realistic, this is ideal for design purposes where too much detail is confusing. These forest polygons can be given more realistic in-filling of detail for rendering purposes once the design has been completed (see below). These thick polygons are usually known as "tree-blocks".

The design of new forest patterns proceeds either by the creation of an entirely new layer of forest over the DEM/DTM, or by modifying the existing pattern of treeblocks – splitting them up, changing their shape, removing them and so on. Manipulation of these shapes in perspective as well as plan enables their appearance in the landscape to be tested as an integral part of the design, while the corresponding map view allows the plan shapes to be evaluated, areas calculated and practical aspects to be noted. This technique also removes the laborious process of manually mapping shapes designed over perspective photographs as described above and was the main motivator for developing the system.

The main CAD package currently available is "Tretop", a suite of programs from the Edinburgh-based Envision 3D Ltd. The complete package enables the designer to import or create a DEM/DTM, import or create surface information, carry out design in three dimensions, produce annotated maps and create final visualisation in formats that match panoramic photographs. This can include oblique aerial panoramas described earlier in this chapter, where the viewpoint location was recorded from the helicopter navigation instruments. It is compatible with GIS, so that base files can be imported from ARCView and the final design exported back.

Graphic presentations

There are three main types of graphic presentations used in forest designs:

1 maps
2 perspectives
3 graphs.

Maps
Maps need to be clear, at the most appropriate scale and usually presented in colour with good legends and annotations. If these are to be prepared by hand, paper bases can be coloured in pencil, felt tip pen or tone film. Legends can be handwritten in neat lettering or printed or typed and fixed to the map. If a GIS is to be used it is possible to add legends, annotations, even photographs to help explain each map. The section in Chapter 5 on documentation lists the types of map to be produced.

It is worth thinking carefully about the colours, hatches or other in-fills and symbols to be used. Different maps should be presented using different techniques:

- Inventory maps of physical information can use whatever colours or fills are desired, unless there are standard ones used in particular places. Forest structure and growing stock maps may be shown using naturalistic colours but there is a limit to how many greens can be differentiated, and mixed stands always pose a problem.
- Analysis maps need particular care as the various aspects recorded on them often overlap. This means that they can easily become visually confusing. One approach is to use a range of solid colours for some features and hatches for others. As long as the colours contrast, the hatch can be read over the solid colour. Avoid trying to put too much information onto one map. If necessary, create two or more for the constraints and opportunities analysis, as this is particularly likely to be overloaded.
- Landform/landscape character maps should show the contours clearly, the forest planning or ownership boundary and the pattern of red and green arrows that depict the lines of visual force (see Chapter 3). Symbols can be used to show viewpoints, aesthetic problems in the landscape and to give a sense of the spatial character along roadsides, for example.
- The felling coupe/harvest unit/cutblock map should be clearly colour coded with both silvicultural system and period or phase of operation. Solid colours and various cross-hatches can be used, together with a set of colours for retained, protected and inoperable areas. It is a good idea to use a sequence of colours such as the spectrum (red, orange, yellow, green, blue, indigo, violet) to represent the different phases.

- Planting/restocking/desired future conditions maps should use colours that reflect those of nature as far as possible. This may be species (in the case of plantations), successional stages or forest types.

All maps should have a title, a written and linear scale, a north point, a date of preparation and whatever keys and legends are necessary.

Perspective visualisations

Perspective visualisations are used to communicate aspects of the design as it might appear if it is implemented. Forest design, by its very nature, is concerned with dynamic systems, so there is never a final point when the design has been completed. Visualisations are needed, not only to present the eventual design, but have a role in helping people who are uncomfortable with maps to understand the design process. These visualisations may be either from high-elevation oblique views or from particular viewpoints (for example, those identified during the landscape appraisal). Graphic illustrations can be used alongside the photographs to help people understand the interactions between the factors influencing the design. Where these visualisations are related to various maps, the same colour coding should be used so that there is a direct link between them.

There are several methods of producing visualisations; these will be explained below. The choice lies in the skill, time and resources available to the design team, the sensitivity and actual or potential degree of controversy surrounding the project and the type of stakeholder group involved.

Graphs

Graphs are a useful means of showing some of the quantitative information generated by the design. Bar charts, pie charts and other outputs can be produced from database and spreadsheet programs which may be linked to a GIS used to prepare analysis. Especially useful are comparisons of current and future states or of changes to the forest over time. These might include timber output projections, proportions of different species or forest types, amount of different habitat or recruitment to late successional or old growth forest conditions.

Graphic techniques

There are a number of techniques for producing visualisations of the forest design. These include manually produced or computer-generated varieties.

Manually produced visualisations consist of hand-drawn sketches, generally using the photographic base described earlier in the chapter. There are four basic approaches:

1   Using a base sketch or tracing overlay on the panoramic photo, a line sketch is prepared in pencil or fine felt tip pen. Future planting, felling or other management activities that change the appearance of the forest are transposed from the sketch design shapes and illustrated graphically. This pencil or line work sketch can then be photocopied or scanned into a computer for subsequent use in a report. Colour can also be added to make the sketch a little more realistic and easier to understand.
2   The same type of sketch method can be used, started with pencil outlines to develop the design but subsequently the design is completed in coloured pencils, which are used to create different tones and textures. In the hands of an expert this can be very effective. Details occurring on the original photograph can be reproduced to make the result as lifelike as possible. These can also be scanned into a computer for use in a word processing package.
3   Also starting from a pencil sketch base, the graphic rendering can be accomplished using felt tip pens. These often produce a more vibrant effect compared with the coloured pencils and they scan particularly well.
4   Instead of preparing a sketch base, a black and white photocopy of the photograph can be used, the sketch design being rendered straight onto it using the same techniques described

above. This approach has the advantage that all the detail in the original landscape remains visible and the alterations to the forest are emphasised. If the tones of the original photograph start to show through the superimposed illustration, they can be flattened out with black- or white-coloured pencil, the new design being applied over this base.

These techniques used to be ubiquitous until the advent of computer graphics. They can produce attractive and subtle results but are time consuming and expensive as well as needing a skilled person to create them. They still have a role in certain circumstances – see one of the case studies in Chapter 10.

Computer graphics

As technology improves it is easier to replace hand-produced illustrations with ones created on a computer. These fall into three main categories.

1   Altering photographs using computer graphics packages. One of the main examples is Adobe Photoshop, where the original panoramic photograph is scanned into the system and the design is presented by manipulating the scanned image. Areas of different trees can be added or taken away or substituted with other types of forest or other land-use types. The results, when done well, can be extremely realistic.
2   Simple computer perspective, showing the landscape and forest as a series of wire-frame or solid digital elevation/terrain model and shaded blocks, are quick to produce and effective for showing aspects of the design that do not need to be realistic, such as colour-coded areas of trees to be felled. In some of these visualisations it can be counterproductive to try to be realistic. The point is to display the pattern of basic shapes and phasing of operations so that it can be evaluated. Similar renderings can be used to show soil patterns, variations in habitat value or other physical or ecological aspects of the landscape.
3   Visualisation of the forest as it develops over time through planting, logging, silviculture and normal forest development can be generated to varying degrees of realism using a number of possible packages. Some of these, such as Brendan, part of the Tretop CAD system, use fairly simple symbols in different colours, heights and densities to show the variations in the forest texture. Others fill polygons from the forest design with symbols to reflect the stand composition and structure from the GIS. The symbols may be derived from a library of scanned images or may be designed by the user as part of a bespoke library. "World Construction Set", part of the Virtual Nature Studio suite of programs, is a good example of this approach. A third variety attempts to grow the forest from basic principles. The areas or polygons of each type are specified, including species, age, spacing and so on. Using rules developed to describe tree growth, such systems grow the entire forest. Some trees such as conifers look better than others, especially broadleaves, because the formulae used to define them are rather simpler and more predictable. Different systems require different amounts of computer capacity and take shorter or longer times to produce their output.

Virtual reality

A final type of output that is fast developing is that of virtual reality. This is not yet of the type where users wear special helmets, but they allow people to move around and through the landscape and have a look at whatever part they want from wherever they want. This approach puts more power in the hands of the stakeholders compared with visualisation prepared from fixed vantage points that may or may not have been chosen by stakeholders. The widespread use of virtual worlds in many computer games and the techniques of navigating around such worlds are well developed. This has been used in some fields for public participation, using a mobile wide-angle curved screen. People sit in front of this and the virtual landscape is projected onto it. It can give the effect of moving around the scene and people can request views or routes that are of interest to them.

Forest ecosystem modelling

There is currently a lot of activity in the field of forest ecosystem modelling, both at the stand level and at the landscape scale. Where the forest ecosystem design is based on natural processes or where the ecological functioning over time must be understood, such models offer a means of trying to predict the effectiveness of both natural developments or the effects of human intervention such as the implementation of a forest design. In this section, a selection of approaches will be briefly examined.

Landscape modelling approaches

One type of model has been developed in order to provide tools to help evaluate and predict whether ecological integrity or biodiversity will be maintained (at the landscape scale) by the implementation of certain plans or forest management activities. Many of these tools focus on conserving particular species, while others are used to model the changing balance of habitats over time or to predict the effects of natural disturbances across the landscape. The species approach, by focusing on one or a few species, tends to ignore the rest of the species present in the ecosystem. This may be justified where forest management is explicitly directed at conserva- tion of particular (usually endangered) species, such as the case of the Northern Spotted Owl in the Pacific Northwest of the USA. Often this is not so, in which case the approach may be generalised by adopting species guilds or keystone, umbrella or biodiversity indicator species, such as used in the HABSCAPES program described in Chapter 7. Some examples of approaches of varying sophistication used with different degrees of success are presented below.

Habitat suitability models

Early attempts at developing habitat suitability models were non-spatial because of the lack of GIS or other spatial modelling systems to act as a platform. After years of development these now work well on GIS, which of course is ideal for use in forest design plans. In most of these modelling approaches a number of measurable habitat variables are used to produce a habitat suitability index (HSI) for a particular species or guild of species. These models can either be deductive or inductive in the way they operate. The HSI can therefore be either derived from the habitat variables according to a theoretical rule base or according to correlation of observed factors using a method such as logistic regression.

Such models can be used simply to calculate the total quantity of suitable habitat in a landscape. This is their most basic application and can be used for simple comparison of different design options (see Chapter 7). A more sophisticated level involves examining the spatial arrangement of the resulting habitat, usually by applying landscape indices or metrics. This may include measuring spatial and structural variables such as patch size, edge-to-area ratios, fractal dimension and so on, all within the capacity of GIS tools. HABSCAPES (see Chapter 7) uses this approach.

Metapopulation models

The earliest versions of these models tended to be analytical, with strategic approaches to the landscape whereas the more recent ones, linked to GIS, are more tactical, making them more useful for comparing different landscape-level scenarios. Population Viability Analysis is commonly used for practical applications, mainly to compare different scenarios in terms of the likelihood or risk of extinction of local populations of certain species. GIS-linked metapopulation models use an HSI map (see above) in order to derive a connected graph structure (a set of vertices connected by edges). They then model population dynamics according to this graph structure.

Spatially explicit population models

Spatially explicit population models (SEPMs) provide an alternative method of relating species population dynamics to spatial landscape patterns. No assumptions are needed about the population structure of the species in question. SEPMs use a spatially explicit representation of the landscape (usually raster grid or hexagonal tessellation) upon which to model population dynamics directly. SEPMs may be individual-based rather than population-based models, where the location of each individual of the target species is explicitly modelled. These models assume a static state for the landscape (comparing different landscape patterns arising from different scenarios) unless some element of dynamic modelling is also introduced. There are several ways of doing this.

OWL is a SEPM used by the Bureau of Land Management in Oregon. It applies to the Northern Spotted Owl and is used to help decision making in relation to different management scenarios projected over a 100-year period. Representing the landscape as static over such long time-scales would be inappropriate and inaccurate so the OWL model uses a dynamic landscape. This is included by increasing habitat age over each time step of the population dynamics model, acceptable in landscapes where there are no significant natural disturbances over the time period and where successional processes are the most important mechanisms of landscape change. In landscapes where natural disturbance has a major effect on structure, leading to more complex vegetation dynamics, more sophistication is needed. In this case SEPMs can be linked to another class of model, a vegetation dynamics model.

Forest landscape dynamics models

There are a variety of forms of vegetation dynamics models. Of particular interest in relation to forest ecosystem design is the currently developing set of forest landscape dynamics models (FLDMs). These models are both spatially and temporally explicit simulation models of the changing structure and composition of forest vegetation operating at landscape scales. These provide both a mechanism for generating dynamic landscape representations for use in population dynamics and habitat modelling and also as stand-alone models of tree species and forest habitats. A major development permitting the recent development of FLDMs has been the increasing power of computers, since the simulation of large landscapes at high resolution places heavy demands on processing speed and memory usage.

FLDMs now can be seen as part of the suite of tools for landscape ecology, but their forerunners have mainly been the non-spatial stand-scale models of community dynamics used to explore ecosystem dynamics, succession and natural disturbance. As such they are also part of the family of transition models, because the component stands of the landscape undergo transition from one state to another over time according to different processes, each of which have different probabilities of occurring.

The conceptually simpler landscape-scale transition model construct has led to the development of more management-orientated application, while gap models have been used to explore community dynamics in different regional forest types.

Transition models have now been incorporated into spatially explicit landscape management tools and have been applied in the real world. TELSA (Tool for Exploratory Landscape Scenario Analyses), originally designed for the Interior Columbia River Basin Project in British Columbia, has since been developed as commercial software. When using TELSA it is necessary to define the types of vegetation occurring across the landscape and to estimate the transition probabilities in relation to the different processes taking place. The dynamics are simulated on a vector GIS layer where the landscape structure is broken down into a Voronoi tessellation. In the model disturbances are applied non-mechanistically. They spread across the landscape via adjacent polygons to create a continuous extent, at a rate and to a spatial scale corresponding to the size distribution that has been pre-defined by the user.

Another model tool, SIMPPLLE (SIMulating Patterns and Processes at Landscape ScaLEs), defines the structure of the model and the way it operates but not the nature of the vegetation types or transition probabilities to be used. The vegetation types are defined using the three descriptors of species, size/structure and density. The transition dynamics operate over a database of vegetation patch types based on a vector GIS layer of the original vegetation cover including adjacency information.

A final model to mention here is the GALDR model (Glen Affric Landscape Dynamics Reconstruction) developed in the UK for application in the restoration of reforested areas in Scotland. This is a forest landscape development model (FLDM) for the simulation of forest landscape dynamics, characterised as a stochastic, cohort-based model of natural disturbance and succession, capable of depicting change in species composition and forest structure over large spatial extents and long time-scales. The GALDR model was implemented using SELES (Spatially Explicit Landscape Event Simulator), a modelling support tool developed at Simon Fraser University in British Columbia, Canada. SELES was chosen as the most appropriate development platform because it allows for rapid model prototyping without forgoing a large degree of flexibility in model design. It was also convenient to use in that it is compatible with ARCView GIS.

SELES represents the landscape by a collection of raster layers. Spatial data is divided into two distinct types by SELES: *static layers* have constant values throughout the simulation run and represent aspects of the landscape that can be regarded as unchanging over the time-scale of the model run; *dynamic layers* represent the features of the landscape which change over time. In a typical forest landscape dynamic model, the static layers will represent (relatively) permanent landscape aspects such as topography or underlying geology while the dynamic layers will represent the forest vegetation and other dynamic aspects of the landscape. Together, the static and dynamic layers plus non-spatial variables describe the model *state* – the complete data representation of the landscape within the model.

A SELES model consists of two principal elements: the model *state*, and the set of *landscape events*. The landscape events determine the dynamic behaviour of the model, i.e. how the model state changes over time. Typically, each landscape event will represent a well-defined biotic or physical process in the landscape. For example, a wind landscape event could simulate the effects of a storm and make appropriate changes to the vegetation layers of the model. Landscape events may be defined to occur periodically or episodically. Continuous processes (such as growth) are represented by periodic events occurring every time step (i.e. at the level of the temporal resolution, such as 10 years). The application of this model is presented in greater detail in Chapter 9, where the Glen Affric project for landscape restoration is presented in depth.

Conclusions

Forest ecosystem design, as a practical process, needs tools to enable planners, designers and managers to carry it out effectively and efficiently. It is also necessary to be able to communicate the design intentions to a wide audience, often of non-professional people in local communities. The resources available to the designer will determine the level of sophistication adopted in the process and in this chapter it has been possible to demonstrate the range of available techniques, from the simplest to the most sophisticated. As time goes on the range of tools available increases and the job of the designer is made easier. Hopefully the days of manual transfer of designs from sketches to perspectives or vice versa are over and there should be few places where GIS databases are unavailable, but if this is the case, it is still possible to carry out design but it is just more time-consuming and tedious to do!

There is a difference between preparing a design and presenting a visualisation. Some people have made the mistake of assuming that a visualisation is a design. Good visualisation of a bad design should enable everyone to see that it is a bad design, but poor communication of a good design may hinder the approval process because people cannot understand it properly. What is needed is the creation of the best design that can be achieved under the circumstances, communicated as effectively as possible. The techniques and tools presented here should enable this to happen.

Chapter 7
From design to implementation

Introduction

In Chapter 5 we presented the integrated design process and explained how to carry out and how to manage it step by step. A design might appear excellent on paper, but the real test of its quality will only become apparent when it is implemented. Equally, since the implementation may take many decades to complete (with revisions and unforeseen circumstances having the potential to change the design or to render it obsolete), some aspects of testing have to be carried out by other means, such as modelling. At the time of writing, some aspects of forest ecosystem design, such as predicting timber flows or landscape quality, are better able to be assessed than others, such as habitat provision or water quality protection.

To implement a design on the ground, boundaries of coupes or management units must be laid out, and landscape and silvicultural prescriptions and other operations such as road building or decommissioning accomplished. There is also a change of scale from the landscape level to the coupe or stand, allowing for more detailed planning, perhaps with a further fine-scale design element included.

One of the main problems of forestry is the long time-scale, leading to uncertainty in the nature of external events, the impact of new policies, changing scientific knowledge and shifting societal demands. Therefore it is prudent to adopt a cautious approach to design implementation. Adaptive management is a valuable method for monitoring and evaluating the effects of the design as it is progressively implemented, as well as for changing the plan in accordance with new knowledge. Through this approach, criticisms levelled at designers regarding the lack of perfect knowledge can be overcome without forest management activities being forced to cease altogether.

This chapter presents some approaches for evaluating different aspects of the forest design and then goes on to discuss implementation. A number of the methods are still under development at the time of writing, and some ideas for future development are also suggested.

How does the design meet the objectives?

In Chapter 5, ways of describing how to establish objectives for the forest design were explained, and a later step of the design process (Step 8), evaluating the design against the objectives, was summarised. In this section we take a closer look at this important step.

Objectives are best set in relation to the values associated with the forest as identified by policies, local managers and stakeholders. Ideally, objectives are fleshed out using one or more of the public participation methods presented in Chapter 4. As part of objective setting, a number of criteria should be established, against which to evaluate the design. There are several ways in which these criteria can be developed and presented. Some may refer to outputs from the forest, such as timber, drinking water or recreation experiences. Other criteria may involve characteristics of the state of the forest over time, such as habitat availability, aesthetic quality or soil stability. A third set of criteria relates to how well the design satisfies rules, regulations or certification standards, such as the width of riparian buffer areas, maximum clearcut size or the presence of protection zones around archaeological sites. Clearly, some of these criteria are quantifiable while others are more qualitative in nature. Some of the qualitative criteria may become more quantifiable, or more measurable as methods of evaluation evolve and improve (such as habitat requirements for certain species), while others may remain qualitative, although some qualitative criteria can be calibrated (for example, aesthetic quality tested against public preferences). The following example illustrates how such criteria can be developed. It is important to note that the degree of sophistication applied in this test should match the degree to which the design objectives are developed. There may be a temptation to be too detailed and quantitative than the circumstances warrant.

Conservation of old growth forest is one design objective typical of many plans in the Pacific Northwest of the USA and in British Columbia in Canada. The following questions may arise:

- What sort of old growth?
- Where in the landscape should it be?
- How much should be present at any given time?
- What should the minimum patch size be?
- How should old growth patches relate to or connect with each other?

Initially, the design team may have started with a rather general goal that the forest design must "conserve and continue to create old growth conifer forest in a pattern and distribution adequate to maintain population viability for dependent species." "All right," the designer asks, "what do you mean by that exactly?" The project team should then develop a more refined description of the old growth conservation objective, perhaps as follows:

- At least 15% of the planning area must be in old growth condition at any given time.
- There must always be at least three old growth patches larger than 300 acres/125 hectares each (to provide for interior dependent species).
- The patches should be connected by corridors or "stepping stones", to facilitate the movement of old growth-dependent species.
- Old growth patches should have naturalistic, complex-shaped edges rather than simple, geometric ones.

Where does this more detailed description come from? Is it scientifically based? In fact, as with many such descriptions, it probably emerges from a combination of regulation, research findings and the combined expertise of the design team. Whatever its source, this description is probably good enough to guide the designer. It may need to be revised or refined following the inventory and analysis phase of the design (see Chapter 5) in the light of more detailed knowledge of the plan area.

The design team should follow the same approach for each resource listed under the objectives, making sure that where the questions about a particular value are qualitative then the criteria are also expressed qualitatively, and if quantitative, expressed in quantities.

The next complication arises when objectives and criteria for different resources contradict or conflict with each other. It is the designer's job to try to find one or more design solutions that meet all the objectives, perhaps with some bias in favour of one or the other (ranking objectives in order of importance is helpful here). The major irreconcilable conflicts can be evaluated and the implications assessed by the whole project team and, if need be, referred back to the local managers or stakeholders for a decision. The more values that are included in the objectives and the more complex the interactions between them, the more the creativity of the designer is challenged. Clear criteria thus help the designer by establishing a solid framework.

Qualitative measurements of objectives

A useful way to test forest designs against objectives expressed and described qualitatively is to establish an expert review process. This process takes each objective, establishes a grading system and then rates design alternatives against it. Thus, a design option can be graded from A to F or 1 to 6, or by whatever method suits the particular project area.

The following example illustrates this approach in relation to an objective for stream conservation. Perhaps the initial objective was stated as follows: "The forest design should establish a stand pattern and structure (and associated road or access system) that prevents sedimentation, protects riparian areas and augments summer base water flows." This is an almost standard objective for many forest design projects in the Pacific Northwest and British Columbia. How would the designer know whether any particular design option meets it? Sedimentation is notoriously difficult to measure or predict. There is much argument over what constitutes a true riparian area, or how wide it should be, although there are now regulations in most jurisdictions. Relationships of upland forests to summer base water flows are also little understood. Therefore, since the establishment of precise quantitative criteria may be difficult we should apply intuitive skills and expert knowledge about the ecosystem to this issue. Three alternative design concepts will be compared to see how well they address this objective using qualitative measures.

First of all, a brief analysis of the issues and the setting is needed in order to make this example meaningful. The forest in question has highly variable soils, somewhat variable slopes, a moderately dense network of streams and enough change in elevation for higher areas to receive winter snow, while lower elevations do not.

Some soils have a greater susceptibility to erosion than others. Silty loams are highly susceptible, while sandy soils are not. Slopes over a certain degree of steepness are more erosion-prone than flatter areas. Thus, a design that placed protected or less intensively logged forest stands on steep slopes with silty soils would have a greater chance of keeping sediment out of streams than would one which located intensively logged stands on such areas, particularly if roads are kept off these areas as well.

For the riparian question, there may be some debate over the exact extent of the riparian zone and where the functional as opposed to the legally defined boundary should be drawn. A design that takes a more generous approach to riparian conservation would tend to be superior to a minimalist approach, perhaps one only barely complying with regulations, whereas the functional riparian zone is significantly wider.

For summer base water flows it is often necessary to go with the most intuitive approach. In areas where snowpack retention is believed to be important for releasing water slowly into the hydrological system, then forest patch types that maximise both interception and retention, balanced against evapotranspiration, are likely to result in higher summer base flows. In many areas, these might be forest patches with a fairly open upper canopy, but which provide enough shade to prevent rapid snowmelt. By locating such patch types in important snow elevation zones, particularly where soils are somewhat porous and well connected to shallow aquifers, the design would respond well to this objective. By including such areas in intensive harvest zones, a negative effect on summer base flows might result as a consequence of rapid snowmelt.

Three alternative options for the design are proposed:

1   *Alternative One* includes protection of the areas with the highest erosion potential, has a generous protected riparian area and retains patch types on upper slopes that facilitate snowpack retention, using selection harvest as a means of management to maintain a fairly dense canopy over the area.

2   *Alternative Two* protects the erodible soils and maintains a generous riparian area but allows greater timber harvest on stands with high snowpack retention.

3   *Alternative Three* includes timber harvest on erodible areas, retains minimal riparian zones and ignores snowpack issues altogether.

In grading these, Alternative One would score high, Two would score medium and Three might fail altogether.

Now clearly this example is over-simplified. While Alternative Three might fail the aquatic protection and snowpack retention objective, it may well score more highly at meeting the timber output objective. Alternative One may of course fail the timber output objective and Alternative

(a)

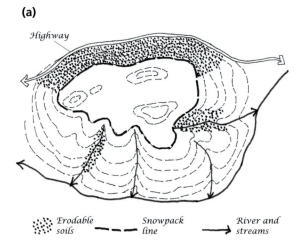

(b)

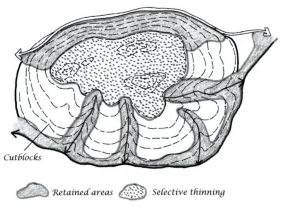

Figure 7.1
This diagram illustrates the scenario used for explanation of the application of qualitative assessment of a forest design plan. a) shows the original area with the streams, areas with erodable soils and the snowpack retention area.

b) shows Alternative One, which includes protection of the areas with highest erosion potential, has a generous protected riparian area and retains patch types by selective harvest on the upper slopes that facilitate good snowpack retention.

(c)

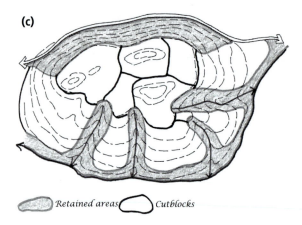

(d)

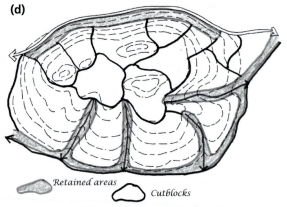

c) Alternative Two protects the erodable soils and maintains generous riparian protection but allows harvest on stands with high snowpack retention.

d) Alternative Three includes timber harvest on erodable areas, keeps minimal riparian protection and ignores snowpack issues.

Two may be more of a compromise all round. Normally, since the forest design process is aimed at getting a plan that meets all the objectives to at least a minimum level of satisfaction, neither Alternative One nor Three would be acceptable. In the real world of forestry, there are trade-offs and unfortunately, few perfect plans. The key point of the above example is that alternative plans can be evaluated qualitatively in order to assist decision makers.

One advantage of qualitative analysis is that it can be done with non-professionals, as long as everyone is relatively well informed about general forest and ecosystem issues. An evaluation team can rate alternative plans over a range of objectives in a consensus fashion. The authors have used this approach on several projects.

The evaluation discussed above can be made clearer by use of a table. This enables the different alternatives to be compared. The temptation to score using numbers and simply add these up should be resisted, since this may give a false assurance of objectivity. An alternative is to use letter grades (A–D, with A the best and D the least). The table might appear as follows:

Table 7.1 Assessment of alternatives against water quality objectives

Alternative	Protection of areas with high erosion potential	Riparian area protection	Protection of areas with high snowpack retention	Timber production potential
1	A	A	A	D
2	A	A	D	B
3	D	B	D	A

Aesthetic quality is another aspect where qualitative measures should be used. In this case, the implementation of each of the design principles described in Chapter 3 can be converted to the letter scale. As an example, three design options may be tested:

1   In Option One the shape of cutblocks are very organic, they reflect landform and landscape scale well, but tend to reduce diversity of colour and texture. The design is well unified and reflects spirit of place well.
2   In Option Two the shapes are not quite as organic and reflect landform less well, but their scale is acceptable, diversity and unity are maintained and spirit of place is adequately reflected.
3   Option Three incorporates good shapes that reflect landform but are somewhat out of scale. Diversity is reasonably well reflected but unity less satisfactorily. Spirit of place is adequate.

The table evaluating each option is as follows:

Table 7.2 Evaluation of visual design quality

	Shape	Landform	Scale	Diversity	Unity	Spirit of place
1	A	A	B	D	B	B
2	C	C	B	A	B	B
3	A	A	D	B	D	C

The judgement here might have to be balanced with other factors. For example, Option Two, with simpler shapes, may be more practical and less costly to lay out and manage. Option Three, with larger scale, may satisfy timber outputs and harvest costs. If visual aesthetic quality is rated as a more important objective than timber, Option One would be the preferred option; if timber flow is more important, Option Three might be preferred. These trade-offs are very common, especially if trying to balance visual aesthetic quality against practicality.

Figure 7.2
This sequence of sketches illustrates designs with different qualities:
a) shows Option 1, with organic shapes to cutblocks which also reflect landform and scale well but reduce diversity of colour and texture. The design is well unified and reflects spirit of place well. b) shows Option 2, where the shapes are not quite as organic and reflect landform less well, although their scale is acceptable, diversity and unity are maintained and spirit of place is adequately reflected. c) shows Option 3, which incorporates good shapes that reflect landform but are somewhat out of scale. Diversity is reasonably well reflected but unity less satisfactory. Spirit of place is adequate.

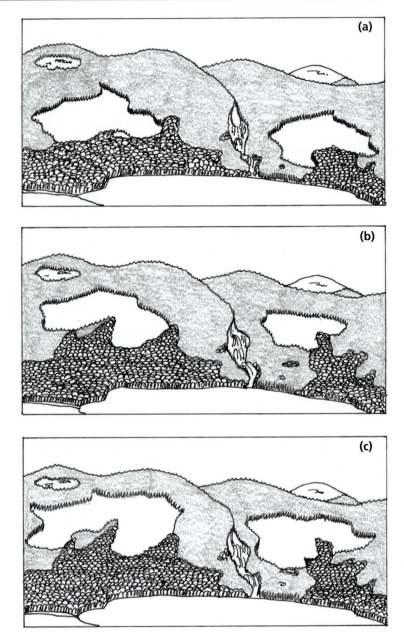

In British Columbia, the Ministry of Forests has, for many years, incorporated a quantitative evaluation into its visual landscape quality system. This is referred to as "percent alteration" and sets a standard of how much of a visual landscape unit can be altered by, usually, logging, at any one time. This can be deemed to reflect a measure of scale (see Chapter 3). Another measure is the "rule of thirds" which can also be used to measure scale or proportion, and which works satisfactorily where large created openings have proved to be acceptable to the public, as in Great Britain.

It is possible to develop similar tables for each resource objective, using a variety of attributes, so that an overall score for each option is obtained after which all values are compared. The

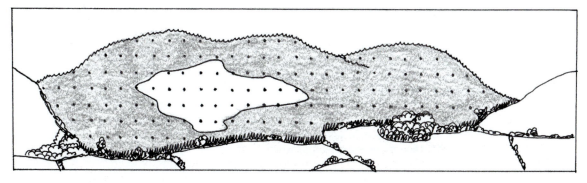

Figure 7.3
This example shows how "percent alteration" to the landscape can be measured using a dot grid over the sketch of the cutblock on a landscape unit. In this case 23.6% of the visible area will be altered by logging.

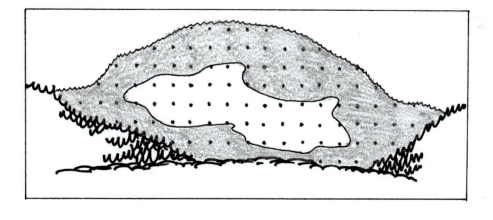

Figure 7.4
This sketch uses a dot grid in order to calculate if the "rule of thirds" has been met as a measure of landscape proportion. In this example the area to be felled is almost exactly 1/3 of the visible area.

design team can then decide which option is best overall, or they can look at ways in which some aspects of one or more systems can be improved to raise the overall standard. This focuses the attention on the areas of the design that are most difficult to resolve.

The use of any scoring system should be carried out with the agreement of the full design team and the stakeholders, and there should be a satisfactory basis for each grade. On a scale of A–E, it can be helpful to decide at which point an acceptable level is reached. For example, where scores relate to compliance with a set of guidelines or certification standard the following classes could be adopted:

A = Considerably exceeds guidelines/certification standard
B = Exceeds guidelines/certification standard
C = Meets guidelines/certification standard
D = Near miss – with extra work on the design it could achieve C
E = Fails to meet guidelines/certification standard.

In some circumstances, such as where environmental sensitivity is greater than average, it might be expected that higher standards of compliance than a C would be needed (Chapter 3 provides an example of a method for assessing landscape sensitivity).

Quantitative analysis of objectives

Some resource managers and forest interest groups do not trust qualitative measurements. "Show me the numbers," they say. In the example of water quality described above, this can be done fairly simply. The area most highly susceptible to erosion can be measured, and then the proportion of that area that is "protected" under a given alternative. The one with the greater amount of protected area can be expected to be better at meeting the water quality objective.

Earlier in this chapter a difference was noted between the outputs of the forest design and criteria related to the changing characteristics of the forest as the design is progressively implemented. In the next section some examples of both will be demonstrated and discussed. Timber production is one major output where an evaluation of quantities over time is important. The ability of the forest to continue to supply habitat requirements, or to improve in its ability to do so over time, is a very important aspect related to forest landscape structure and composition.

Evaluating timber production in the design

Evaluation of timber production can use a variety of approaches. What these all have in common is a spatially explicit database, usually in a geographic information system (GIS), where the forest is divided into compartments and sub-compartments or some other units. For each unit, information is collected on tree species, composition, standing volume of timber, growth potential (expressed as site index, yield class or some similar measure), age (if known), stocking density, average tree size and so on. Growth models may be available for each species, predicting volume production on different site types over time.

A design that divides the forest into harvest blocks, protection zones and inaccessible areas can be overlaid on a stand inventory map. The timber volume not available for harvest can be subtracted from the total, and the available volume assessed in terms of its flow over time, calculated by a combination of area, operational phase, silvicultural system and rate of continued stand growth. Clearly, different levels of protection will increase or reduce available timber, while alternative silvicultural systems will yield timber at varying rates over varying time periods.

As well as the basic volumes of timber, it might be important to evaluate the breakdown by species (as these may go to different markets), size assortments (which might relate to different product types and have different values) or quality classes, all likely to be important for marketing purposes.

While the phasing of the operations may have been anticipated within the design (see Chapter 5), it is also useful to try to optimise phasing to yield the most volume and increase efficiency of road construction, length of haulage to the market, etc. A number of computer program-based tools have been developed to try to optimise wood production over time. These use decision rules to calibrate the program, which can then evaluate a large number of options until an optimum solution is reached. Rules to be incorporated might include the selection of the oldest stands first, a need to maintain a minimum time period between certain types of logging on adjacent coupes to allow for regeneration to be sufficiently well established and the percentage of basal area to be removed per phase.

The evaluation of timber flow, especially by species or quality, may be more important in diverse natural forests compared with simpler plantations. In the case of the latter, volumes per hectare may be much more consistent across the forest so that coupe timing options may produce fewer differences in overall wood flow over time. Systems based on clearcutting are also simpler to calculate than more complex, selective systems.

Evaluating ecological functioning in the design

Quantitative testing or evaluation of the landscape characteristics produced by a forest design can be done by measuring key landscape attributes. This is relatively easy through use of GIS, which can quickly calculate areas, edge lengths or other landscape features. An emerging aspect of landscape ecology is the development of "landscape metrics", which are attributes that can be measured spatially. Landscape metrics are helpful but are highly sensitive to the scale and resolution of mapped information, and to the way in which patches are delineated (that is, which categories are used).

As discussed in Chapter 2, landscape patches, matrices and corridors are the key building blocks of the forest considered from a landscape ecological point of view. The types of patches, their size and their distribution over time and space can be organised by the design, and implemented by forest managers. Landscape metrics are a means of gathering and analysing information about patches, whether present in the existing forest or projected through the life of the design that can help to evaluate ecological functioning. Quantitative aspects of landscape patches which may be worth measuring include:

* The area in each patch type, existing and proposed
* The percentage of the total landscape in each patch type at any given time
* The total number of patches of each type
* The mean patch size
* The frequency of distribution of various patch sizes
* The amount of each patch in "interior" habitat
* Percentage of the landscape in interior habitats
* The total number of interior habitats.

Qualitative aspects could also be evaluated using GIS, such as:

* The length of patch edges relative to their area (also known as edge density)
* The degree of contrast between adjacent patches
* Characterisation of patch shapes (i.e. amount of sinuosity).

It might also be appropriate to measure the relationship between patches:

* Clusters of similar patch types close together
* Small patches that may be "stepping stones" connecting larger ones
* Patches that support larger habitat networks
* Patches located at key intersections or nodes.

HABSCAPES is a computer modelling program developed in the 1990s by the US Forest Service. It is designed to measure habitat availability for many species at the scale of tens or hundreds of thousands of hectares. It does this through a five-step process, as follows:

1 First define the relationship of local species to specific patch types.
2 Group species into guilds that have common habitat requirements.
3 Classify the habitat for each guild in terms of patch structure and composition.
4 Convert available vegetation data into a classification of patch types that reflect habitat values.
5 Evaluate the suitability of patches for local species guilds using landscape metrics (size, shape, clusters, etc.).

HABSCAPES allows forest design teams to check for what their plan will or will not achieve for most vertebrate species. For example, in a design option, one guild may rely on the continued presence of old growth or mature forest patches totalling at least 300 acres (120 ha) in area. One of the management goals is therefore always to have at least 10 patches of an average area of 300 acres available in the planning area at any one time. Using GIS, it is relatively easy to test the design for this attribute.

Step 1 has created a picture of the habitat requirements of the various species, drawn from available literature or information collected during the landscape ecological analysis (see Chapters

2 and 5). Where data is lacking, professional judgement of local wildlife biologists should be relied upon.

Step 2 might group a number of species into an old growth forest large-patch guild. These are the species that are reliant upon or favour this type of forest, while other species may be more adaptable to a wider range of forest conditions or smaller patches.

Step 3 classifies the available habitat. This is normally done by using existing vegetation classifications that come closest to fitting the desired habitat characteristics. In many cases, vegetation cover maps are derived from forest stand maps that classify the forest on the basis of timber type and size. Biologists often use size as a surrogate for forest age and overall structure. In other cases the mapping may have been done with habitat rather than timber in mind. Often, a range of vegetation classifications may be grouped together as being similar enough to be treated as one habitat type.

Each patch that appears to fit the required habitat type can be analysed for its suitability for the guild. Some patches may be too small in size, too linear a shape or otherwise deemed unsuitable. Those patches that meet the requirements are mapped and delineated. This process is repeated for each phase of the design. This analysis will identify any periods of time when the available habitat drops below the threshold of guild viability. The design might then need to be revised and re-evaluated.

The forest design in this example could incorporate a permanent network of at least ten old growth patches, protected over time, or a rotating network where some old growth disappears, to be replaced by mature forest that develops into old growth during the plan period, or ideally, some combination of the two, since "permanent old growth" may be an oxymoron (except for a few rare, well-protected places). Riparian areas and sensitive soils or slopes are often designated as long-term, or permanent, mature forest elements that anchor the more dynamic, shifting parts of the design.

Figure 7.5 An example of the output of the HABSCAPES program. The key shows habitat areas suitable for the species guild selected for analysis, which is species with a large home range that prefer late successional forest and also need a mosaic.

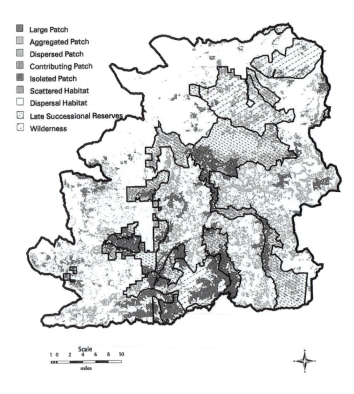

Large Patch
Aggregated Patch
Dispersed Patch
Contributing Patch
Isolated Patch
Scattered Habitat
Dispersal Habitat
Late Successional Reserves
Wilderness

Scale
1 0 2 4 6 8 10
miles

The above example is quite straightforward, because a species guild dependent on one habitat type was chosen. Where guilds require combinations of different types, especially ones that are relatively ephemeral in the landscape such as early successional forest, the value of the HABSCAPES program becomes much more apparent. It overcomes the near impossibility of testing a forest manually to see how suitable it is for a range of species, since the task is simply too complex. This is especially true where spatial relationships between habitat types are important (such as where a species rests in one type but hunts over another).

HABSCAPES and similar programs can also be used to quantify the available old growth habitat across a time continuum if they are combined with forest growth modelling programs. However, HABSCAPES requires the structure of each phase to be worked out in advance, because it does not grow and develop the forest forward over time. For this task, more sophisticated modelling is required (see below).

The examples and discussion above has demonstrated some of the complexities of evaluating the objectives of the forest design. Other objectives, such as protection of archaeology and cultural heritage, recreation provision or protection of soil, may be easier to evaluate.

Since archaeological and cultural heritage sites tend to occupy discrete areas, their identification, delineation and protection are more straightforward. There may be legal requirements for protection zones and buffer areas that can be easily marked on maps and on the ground. Following this the protection might involve more detailed, site-specific management plans at a finer scale of resolution than the landscape.

Testing the forest design for practicality

The next step, and a most important one as far as the successful implementation of the design is concerned, is to test it for practicality. Is it capable of being implemented with the currently available technology and resources? How can we know whether the coupe/cutblock shapes shown on the map can actually be created on the ground? While the only true test of this is the one of time, the authors are keenly aware that a design has essentially no value unless it can be implemented. Furthermore, it is critical that the local management team believes that the plan is practical, even if not ideal or as cost effective as it might be. There are usually three or four key elements of a forest design that have to be evaluated on a practical level.

The first problem is whether the unit shapes can be surveyed and laid out on the ground with any degree of accuracy. This depends on being able to find key points when in among perhaps very dense stands on rough, steep terrain. Complex shapes, though they may reflect topography and look great from viewpoints, may be either impossible to lay out beforehand or be extremely expensive to do so. The advent of Global Positioning Systems or GPSs has made this sort of task much easier, although in some areas accurate readings are hard to get because of dense canopy.

The second issue is that of accessibility by forestry machinery, such as harvest equipment. If the analysis phase of the design process has been thorough and an engineering and road construction analysis or "total chance plan" has been undertaken, no design units should be in inaccessible areas. However, even if a coupe is generally accessible from a road and/or landing, certain portions of it may be difficult, especially if the local topography is complex and cable logging systems are being used. Thus, experts on road construction and harvest planning should look at the design in some detail in order to check that each coupe/cutblock is accessible. Those planned for the first phase at least should be checked on the ground.

The third issue is the practicality of the proposed silvicultural system. Clearcutting, or other simple systems, are relatively straightforward, whereas selection systems, especially with a cable on steep, highly visible terrain, may pose more challenges. Not only does the logged timber have to be extracted but there should be limited damage to the site and remaining trees (site-level design and operations are discussed later in this chapter). It may not be possible to evaluate this aspect fully until the operational planning is underway.

The fourth issue, especially where commercial timber production is a key objective, is whether the harvest operation can be undertaken profitably. While it may be imperative that volumes of timber flow regularly to the mill, few land managers are willing or able to carry out operations at a loss unless these can be financed as investments that yield greater profits later, as might be the case with some thinning operations.

Often all these issues arise together, since accessibility to machinery is often linked to the cost of harvest, volume of timber removed per unit of road construction and so on. The following example illustrates the complexity of the issue, taken from a typical scenario in the US Forest Service in Oregon.

In this example, as part of a particular forest design, much of the harvest is to come from thinning Douglas fir out of stands where the objective is to restore them back to open Ponderosa pine forest. What are the requirements?

- There has to be direct road or track access for equipment to the stands in question, or landing areas within a certain distance if helicopter logging is being considered.
- The costs of the logging cannot exceed the market value of the timber, unless the project is being subsidised as part of an ecological restoration project.
- The foresters must be able to do the appropriate post-logging treatment, such as prescribed fire, seeding or planting, in a timely fashion.

To answer these questions, the designer consults with those who have expertise on logging systems, costs and markets, as well as those who specialise in restoration work. At this stage, the designer calls a meeting of the local experts and those managers who expect to be responsible for implementing at least the first phase of the plan. The designer presents the design and invites comments from the meeting. The logging manager might say: "Yes, we could do it with cable logging from the existing ridge-top road network, but only if we take xxx board feet of at least 14″ diameter trees per acre at each entry, because that is the minimum we need to make this operation pay . . . and only if we do not have to absorb the clean-up and fire costs!" The silviculturalist might respond: "We need to do a post-logging fire treatment and we need suitable fire breaks. The ridge-top roads are ideal but some of the curvy shapes downslope might be a problem unless we start the burn there and burn upslope . . . yes, I guess we could do it. But what the map says and what is actually there can be two different things. There may be some steeper slopes or rocks that prevent us from getting down to the bottom of the cutblock." The designer replies by assuring the forester that if such problems are discovered then the design can be revised (this issue is considered further later in this chapter).

The key is to determine whether the plan is operational at a level that meets the minimum needs of the landowner or manager. A private forest owner evaluating a forest design will have more rigorous requirements for economic return than a state or national forest agency, and may choose to conduct a more detailed analysis. Detailed evaluation of each coupe or cutblock on a large design is time consuming, and it is generally only important to concentrate on the early phases that are likely to be implemented before the plan is expected to be revised, after which case the design may change radically anyway.

Evaluating the design over time

It is not enough to evaluate a forest design as if it is going to be implemented completely at once. Implementation will take a long time, perhaps several decades or even centuries to complete. Therefore, we have to be sure that as it develops and as the original forest pattern becomes overlain with the new design, it continues to meet the objectives. Forest managers only have limited control over the forest and unforeseen events may occur to invalidate the design at any point. The appearance of the forest may look satisfactory once the design has been finally completed, but it may look less satisfactory when only half completed, unless we take the time to test it. (Figure 7.6 in the colour section illustrates a time series, showing the development of a design over several decades.)

Evaluating a forest design over time requires us to be able to visualise the design as it gradually unfolds. Generally, this is done by creating a series of snapshots or time sequences. Ironically, the first snapshot is not what the forest looks like today, but rather what we want it to look like at some distant future point, perhaps 40, 80, 100 or more years hence. If the forest in question has a 300-year life span, then our design ought to start by visualising the pattern and structure that we expect to have in about 300 years' time, assuming it has been followed fully during that period! Clearly, it is unlikely that future generations of forest managers are going to be slavishly following our design for generation after generation. They will probably have new knowledge, new skills, different needs and possibly be working in a different social and political framework. Consequently, a long-term forest design concept is more a direction than a destination. We are setting a trajectory that we know will be adjusted from time to time. If we do our job right, then the adjustments may not be especially radical.

Long-range designs need to be brought back in time to the present in order to be well grounded. Generally, we set the first benchmark for the design at 5–10 years after the start and continue to take snapshots at appropriate time intervals. This fits the planning time frame for most resource managers and enables initiation of the first step to be visualised reasonably accurately.

In order to carry out an evaluation over time it is necessary to have a complete sequence of phases, where each management unit, coupe or cutblock has been ascribed to a phase. This assumes that we can do this without checking the implications of one stage of implementation before deciding on the next, which obviously causes complications. Even the tools described above for modelling habitat values assume that the future landscape structure can be predicted before it can be tested to see how much habitat it provides. Thus, there is a clear relationship between developing the phasing and testing it over time. It needs to be an iterative process, where it is possible to track back a step or two if the trajectory towards the vision suddenly veers away from where it should be going. This is why it is important to concentrate on direction more than on destination.

One way to start is to overlay the present forest pattern over the long-range vision, and then sketch out alternative ways to begin getting from here to there. Then the results of the first intervention are added and each remaining patch in the landscape is "grown" over the chosen time period by estimating the natural growth and development anticipated to occur. There are sophisticated computer programs that can help to do this, although they are mainly research tools at present. If the landscape in question is very large and complex, then computer analysis can be quite helpful. For testing visual appearance, computer graphic packages linked to GIS can be used to generate perspectives showing the developing forest from key viewpoints (see Chapter 6).

This first-phase simulation can be a useful reality check on the long-range plan. It may turn out that, while over the long term our plan provides lots of timber and plenty of habitats, the present condition is so far off or already over-harvested that it may not help us reach the timber targets we need to keep the local mill operating in the short term.

Assuming the first-phase implementation plan has proved adequate, the next step is to create a model that shows what the forest will look like after a further time period. Fifty or one hundred years may be a good choice for a very long-range plan, while ten to twenty years is better for a shorter one. While it is unlikely that the planners, designers and managers will still be working after 50 years (or would even remember that they had once done a plan for the area), at least some of the younger participants in the planning effort may be around long enough to see the results. In other words, this time-scale reflects the human life span and therefore has some meaning to us. It is also a long enough period in the life of the forest to see real change in pattern and structure as successional processes have time to operate and some disturbance cycles are quite likely to come round.

A 50-year view can often produce surprising results. For example, in one project, a forest design for the Collowash Watershed on Mt Hood National Forest in Oregon, it was discovered that at about 40 years into the plan, the forest would enter a long period where timber outputs from thinning would far exceed the first 10-year production from limited clearcutting. This was because a lot of land area that had been logged in the 1950s to 1970s would be well into its production potential from commercial thinning in 40 years' time. Another surprise was that some of the old growth that had been saved would probably be lost to windstorms, fires or other natural disturbances, yet replacement old growth would still be many years away. This indicated that the risk to old growth-dependent species would increase, even though timber harvest had been significantly reduced since this area was placed in an old growth reserve (see case study in Chapter 8).

At times, these shorter-term visualisations and analyses may lead us to adjust the long-range concept. In other cases they may support the original vision. Using phases or time steps is convenient but hardly reflects the continual changes that take place. Computer models allow forest managers to plot the trajectory of each element of the design continuously over many years, estimating economic outputs, visual quality, habitat suitability and any other indicators together.

While a design may be implemented over several decades, it can be useful to continue to model it well beyond that time, in order to ensure that problems do not arise with the later phases that would otherwise go unnoticed. It is also possible in some of the more sophisticated computer models to introduce natural disturbances at intervals related to their expected cycles but timed using stochastic models. This enables managers to see how resilient the design is likely to be if such events take place, and to develop a management strategy to cope with them over the life of the design.

Testing the environmental impact of a forest design

Another method for evaluating alternative forest design concepts is to subject them to an environmental impact analysis or assessment (EIA). This can be a costly and cumbersome procedure, but is recommended in cases where the land manager or important stakeholders desire to have a legally binding plan for the area in question. In some jurisdictions an EIA may be required by forestry authorities if the plan is significant in scale or impact, especially if there is major road construction, for example.

Many developed countries have environmental impact procedures. The National Environmental Policy Act (NEPA) was legislated in the USA in 1969. This law was in response to problems created by numerous federally initiated or funded projects that caused substantial, but often unanticipated environmental harm. The core intent of the law was to improve the decision making that goes into federal projects in order to improve the relationship between development and nature. NEPA tries to accomplish this in several ways. Firstly, it requires a project proponent to state clearly what actually comprises the proposal and its purpose. In the case of a forest design project, the statement might be: "The purpose of this project is to draft a long-range concept for managing the pattern, structure and transportation system for a 50,000-acre area". This "purpose and need" statement, as it is known, may go on to describe or identify some of the main issues known about the area. In some cases, the purpose and need can be quite detailed and specific. They could call for conserving at least 15% of an area in old growth habitat, or for locating areas that can be clearcut harvested. Alternatively they can be vague, stating only that a plan will be developed aimed at balancing some set of objectives.

The second important component of NEPA is the setting of minimum requirements for public notification and involvement. Public involvement in forest design is inherently important (see Chapter 4). Bureaucrats are generally not trusted to act in the public's best interest in the absence of citizen oversight.

A third component of NEPA is for the project proponent to consider alternatives to the proposal, including simply not doing it. This particular part of NEPA has often been misunderstood. This provision does not force a proponent to develop multiple detailed alternatives just for the sake of it (the US Forest Service is rather famous for this interpretation). It only requires that the proponent consider alternatives that actually meet the stated purpose and need.

Thus, alternative forest designs can and should be developed if and only if this furthers the purpose of the project in the first place. In the authors' experience, in most cases there are viable alternative designs that ought to be sketched out and evaluated. But in some cases the objectives are so precise, and the landscape so constrained, that these may involve only very minor modifications around a core design.

However, what does it mean to consider the "do nothing" alternative? For years, public land managers in the USA never took the do nothing option seriously. Why would anyone spend money planning and evaluating only to choose to leave the forest alone? The point of environmental impact analysis is to achieve a greater public good, which might mean that to do nothing, at least for a time, may be the best policy. Choosing the do nothing option is and should be a rare occurrence, since environmental assessments are often cumbersome and expensive (which is one reason why most forest designs have not been subject to the NEPA process).

In Europe, there are also environmental impact assessment regulations that apply to forestry. In the UK, EIA regulations have been set to reflect a European Union directive. An EIA in forestry is typically required if there is a major road construction element or if there is significant deforestation as part of the plan (often needed in plantation forests of exotic species where trees were originally planted on what are now recognised as important open-ground habitats).

Implementation of a forest ecosystem design

Once a forest ecosystem design has been approved, it is presumably ready to be implemented. Generally, implementation is through standard forestry activities such as timber harvesting, silviculture, road construction or special ecosystem conservation projects, restoration, protection measures and recreational developments.

The main, significant aspects of implementation are to convert the landscape-scale plans into site-based projects and to apply them accurately on the ground. Three elements are involved with this:

1   Each management unit or cutblock defined by the forest ecosystem design has to be converted from a generalised description of what to do into a site-specific implementation plan based on the local conditions.
2   The implementation plan has to be set out on the ground to some defined tolerance in terms of the boundaries.
3   The implementation plan has to be carried out as accurately as possible using the available techniques within the assumptions about feasibility used when the forest ecosystem design was developed.

If the design was evaluated properly there should be few nasty surprises waiting for managers but, depending on the location and resources available, the degree of site knowledge may be higher or lower. For example, in parts of British Columbia that are not yet accessible by road and where the terrain is difficult, it is likely that no one has been on very much of the ground. Site knowledge is limited to what was gained from examination of aerial photographs, with little chance of seeing much beneath the canopy. Thus, when foresters visit the area to develop the site-level implementation plan, the forest composition and structure may be somewhat different from what was assumed, the terrain may be more difficult for road building and the choice of harvesting equipment may have to be changed. At the other extreme, in a plantation forest the level of site knowledge may be to a high degree of detail so that it is much easier to move from landscape level to site level.

Developing the implementation plan

The first task when developing a site-level implementation plan is to carry out a more detailed survey of the forest and the terrain, especially looking at aspects that are important for the particular unit. If the forest is to be clearcut, the description of forest structure and composition is not as important as it would be where continuous cover silviculture is to be used. Road and harvest access must be evaluated and surveyed in some detail and a harvest plan developed which may include roads, tracks, stream crossing points, no-go areas for wheeled machines, landings, oil storage areas and parts of the site needing special protection for ecological or archaeological reasons. A silvicultural prescription should be developed, and where significant differences have been found on site as compared with the inventories used for the landscape-level plan, reference back to the design team should be made because of the knock-on effect this is likely to have.

The implementation plan should be developed using maps at a suitable scale and written reports supported by photographs, which thus record the baseline situation against which monitoring will take place (see below).

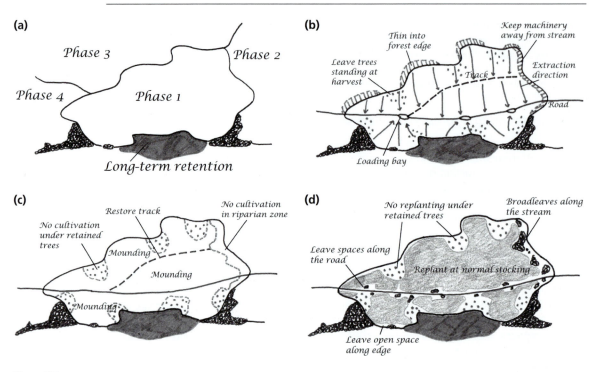

Figure 7.7
This series of diagrams shows different aspects of site-level planning, within the boundary of a felling coupe design: a) shows the coupe boundary set within a pattern of other coupes and long-term retentions; b) shows the detailed aspects of harvesting layout and planning – roads, access tracks, stacking areas, protected areas and zones for using different equipment; c) looks at the post-harvest treatment, including cultivation; d) presents the replanting plan in some detail, especially the treatment of small-scale areas within the coupe boundary, which could not be illustrated at the general design plan scale.

Setting out the plan on the ground

The boundary of a management unit or coupe may be determined as part of the initial development of the implementation plan or there may be revisions as a result of knowledge obtained at site survey at a later date. In difficult terrain, the surveying of a boundary may present challenges, but it is important that such boundaries are as accurate as possible. Often the designer has spent a lot of time ensuring that the outline of a coupe or management unit reflects topography and fits into the landscape well. This may result in an organic and quite complex shape compared with the rather rectilinear cutblock boundaries used extensively in the past. Foresters may find this a challenge and be tempted to simplify shapes. This temptation should be resisted.

Boundaries can be laid out in a number of ways, from using basic survey methods to sophisticated GPS satellite systems. The most basic is to survey around the boundary looking for key points defined from aerial survey and to interpolate the points in between. Some measurements may need to be taken from some key points that are not on the boundary, and some compass bearings and foot-pace measures used in places with fewer features. This is the least accurate approach but can be adjusted as harvest proceeds and the coupe takes shape. This rudimentary method has been used with surprising accuracy in Britain before the advent of more accurate methods. The most difficult units are the first ones. As a plan proceeds, units start to make use of the boundaries of adjacent ones, thus reducing the length of survey.

Where a road already exists, this can be used as a base line and offsets taken from it at compass bearings to a series of points around the perimeter of the unit. This is easier in flat terrain but still works in steep ground as long as some reasonably accurate distance measures can be made despite having to clamber over fallen logs or go around obstacles. If the forest is a plantation where the trees are in straight rows, the offsets can all be along these rows making setting out much easier.

A third method, which has the most accuracy and is also the easiest to carry out, is to use the Global Positioning System, or GPS. This system uses a network of satellites to give accurate locations nearly anywhere in the world to within a few metres (more than adequate for forestry purposes). The unit boundaries can be identified as a series of points generated from a GIS map of the plan and the surveyor visits these points in turn (directed by the GPS from one to another), marking the boundary with flagging as they proceed. Other features observed during survey can also be located to help develop the implementation plan. Since handheld GPS units are now readily available there should be no need to use anything else, except that in very dense forests it can still be difficult to get a satisfactory signal through the canopy.

Implementation

The last task is to implement the prescription using whatever techniques and systems are appropriate. Implementation is carried out by operational staff or contractors, including equipment operators, loggers or fire technicians. It is important that these people understand the intent of the plan, often a motivation for them to carry it out as well as possible within the limitations of contracts, health, safety and the capability of their tools and skills. Good supervision is essential, so it is worth explaining the full background of the plan to the field supervisors. Ideally the front-line supervisors were brought into the design team when the initial design was being evaluated.

However, do not underestimate the pride of good workers in a job well done. Since these people often work for long periods unsupervised, if they are able to spot opportunities to make the implementation even more successful they should be encouraged to do so.

Historically, foresters and loggers were understandably more concerned about what was leaving the forest, the timber, than the state of the forest left behind. This is no longer the case, and therefore it is vital that operational staff become at least as concerned about what they leave behind as about the products. In fact in some cases, the objectives of the job are to produce landscape or habitat, with timber production a by-product.

Figure 7.8
A felling coupe/harvest unit being implemented, in this case logging in an area of "new forestry" where minimal site disturbance is required. Good workmanship on the ground is essential for high-quality results. Source: Trout Mountain Forestry.

Figure 7.9
A site in Nova Scotia which has just been logged. As well as the accurate setting out, the harvest machinery operators had received training on how to spot special features, such as wetlands or important wildlife trees, so as to protect them during the job.

Mistakes made during implementation are difficult to correct – if trees have been felled that should not have been, they cannot be re-erected and may take 100 years to restore. Therefore accurate specification, good layout, unambiguous marking and good supervision are all essential if the project is to meet its goals.

Monitoring and adaptive management

Before project implementation has begun it is important to make sure that a well-designed monitoring programme is put in place. Since, as we have explained earlier, it is at present impossible to be absolutely certain that any forest ecosystem design will deliver all the benefits it is supposed to, despite the best evaluation and modelling techniques, monitoring must be used to ensure that if problems develop they are able to be dealt with as soon as possible. Without monitoring we cannot learn, and without learning we cannot adapt our practices. The essence of monitoring is to understand how well our design is working by collecting new data to bridge any gaps in our existing knowledge.

Monitoring can be defined, according to Reed Noss, as "the periodic measurement or observation of a process or object". Some monitoring is nothing more than rather pointless data gathering. It is often poorly funded at the outset, and the first thing to be dropped when budgets are squeezed. It is imperative, therefore, that monitoring is built into the project from the outset and mechanisms put in place to carry it out properly.

Monitoring is an activity that has been common in natural resource management for many decades. Foresters have long monitored the growth and development of plantations, often keeping detailed records that can now be used to develop a better understanding of plantation development and some of the problems associated with it. Wildlife managers have monitored game species, also keeping detailed records of bag numbers and the health of animals. Landscape architects have been monitoring changes to forest scenery for several decades. This means that the process is not new, it just needs to be extended, integrated and the right things monitored at the right intervals. The trick is to develop and apply more comprehensive ecosystem monitoring strategies in a cost-effective way.

Reed Noss makes the case for monitoring as part of a continuous feedback loop that allows forest managers to adapt plans and practices. He calls monitoring the "cornerstone of adaptive management" (adaptive management is discussed in more detail below). Noss identifies three basic types of monitoring, all of which can be applied to forest design projects:

1   Implementation monitoring
2   Effectiveness monitoring
3   Validation monitoring.

The first of these, implementation, is usually simple, straightforward and is the type probably most familiar to forest managers. It answers the question: "Did we do what we said we would?" Once we have a forest design, and perhaps a ten-year schedule of projects as part of the first phase (timber harvest, thinning, road building or removal, prescribed burns and tree planting, for example), we should periodically review whether the actions set forth were actually carried out. This is a way of ensuring that agreements made with clients and communities were kept and rules and regulations adhered to. If the forest manager was given funding to plant three miles/five kilometres of riparian woodland, then they want to know the money was spent for the right purpose. When the monitoring involves the rules and regulations or certification standards then it can also be described as compliance monitoring.

Effectiveness monitoring is a little more challenging. It asks the question: "Did the project actually work as intended?" For example, a forest design called for thinning and under-burning a mixed fir and pine patch with the intent of restoring an open pine woodland in order to improve habitat for western bluebirds. Questions may include whether the woodland conditions desired have in fact been created, and whether the bluebird is now thriving or not. Measures for these questions may include tree growth, changed composition, regeneration of understorey vegetation and the presence of bluebirds. Effectiveness monitoring must often take place at a time that differs from implementation monitoring. For example, understorey regeneration might not be expected to develop until several years after the initial work to prepare for it has been completed. The plan might have been implemented correctly and regulations complied with, but if the effectiveness was poor, then the plan has still failed.

Validation monitoring adds a new level of complexity. In essence, it asks the question: "Were the results a consequence of the actions taken, and were assumptions and models used in developing the plan correct?" Validation monitoring is a form of field research carried out at the same time as the plan is being implemented. For example, let us suppose that we had assumed that if we successfully restored the open pine woodland described in the example above, then western bluebirds would take up residence within a certain period of time. The project is complete (implementation monitoring) and we end up with the forest structure we had hoped for (effectiveness monitoring), but after ten years we do not have any bluebirds. The problem could lie at a different ecological scale. For example, it may be that the nearest bluebird populations are so far away that dispersal to the new habitat is a problem. It could be that the birds' winter habitat in Central America has been so depleted that there just are not enough bluebirds left to fill the available niches in the north. It may also be the case that what we assumed about bluebird habitat was incorrect. Perhaps there is some other element of the ecosystem that is still missing, for example an insect or seed source that is critical to the bluebird, and is still missing from our local patch.

To avoid the trap of monitoring becoming a pointless exercise in data gathering, or having unexplained results lead us astray, it is important that monitoring programmes are designed properly. If at the outset we choose the wrong indicators for any objective, measure the wrong elements or are working at the wrong scale, then we may well get the wrong message in the end. To be effective, a monitoring programme must have a clear vision of what it is intended to achieve, there must be a true commitment to implementation on the part of local managers and the information gained must be incorporated into management planning.

When setting up a monitoring programme it is a good idea to relate the factors to monitor against the objectives and the criteria developed to know if the objectives are being met. The tabular method of setting out objectives described in Chapter 5 can be used as a base, and by adding columns to it the method of monitoring, the frequency of data collection and the ongoing achievement or failure to achieve objectives can be recorded. In this way the monitoring programme is properly founded in the original aims of the plan and is transparent, so that if external auditing of the design is undertaken, perhaps as part of a certification protocol, it is clear and honest in its assessment. There could be a column for each of the types of monitoring and a comments section where the implications and remedial actions, if any, are also discussed and recorded. Another benefit of this type of record keeping is that as managers come and go there is a longitudinal record to help them understand what point the project has reached and where it is going next.

If the plan is implemented in phases that are spaced at relatively short time intervals, such as five years in the case of a plantation redesign in Britain, the main monitoring periods could coincide but be one year earlier, so that the plan can be reviewed and revised in time for the next phase of implementation. This is not to say that some aspects do not need more frequent or even continuous monitoring. This may be for operational reasons, such as where water quality in important streams is continually monitored to check the effect of harvesting operations on silt production, but this may not be relevant to the plan as a whole except where the next phase is being considered.

Where implementation phases are at longer intervals, such as ten or more years, monitoring may need to be carried out more frequently and if it takes the ten-year period to complete the first phase, then a more or less continuous system of monitoring may need to be developed. In a sophisticated system the implementation monitoring should be carried out soon after the activity has been completed; the effectiveness monitoring at a suitable time interval afterwards, depending on the subject area and the expected time for the result to manifest itself, and the evaluation monitoring at a later date still.

The records kept for the monitoring process should be full and comprehensive. The tabular forms are a good basic starting point, if possible related to a GIS record of progress so that the activities and the monitoring can be referenced to the areas involved. Aerial photographs and site photographs should also be taken and, where special requirements, such as basal area percentages or tree density per unit area specifications were made, on-site sample measurements to check implementation should also be made. For timber harvest outputs, records of production should be noted, and for landscape quality, photographs from viewpoints used in the design should be used to compare the original design with the results on the ground.

It may be necessary to engage a multi-disciplinary team to carry out each element of monitoring, because of the need to examine a wide range of factors involved in the original design and the expertise used to develop it. This increases the expense but is more likely to produce truly valuable results. Over time the process should become more efficient as the monitors become more familiar with the plan and develop a "nose" for where they should be concentrating their efforts.

Adaptive management

Adaptive management, again according to Reed Noss, is a "continuous process of action based upon planning, monitoring, evaluation, and adjustment." It is an evolution of the monitoring process which seeks to introduce a more active feedback of the findings. What we have to admit is that much of the implementation of forest ecosystem design is a grand experiment or a type of action research. The phases of the plan are implemented but as we have already admitted, we have imperfect knowledge and rely on expert opinion and various assumptions about what we expect to happen. Since it is impossible for reasons of time and money to set up huge experimental programmes fully replicated and conducted with scientific rigour, we must look to other ways to increase the knowledge base. We can do this by treating each element of implementation as an experiment and a chance to learn.

Figure 7.10
Cycle of adaptive
management.

The approach towards adaptive management is to plan each element of the implementation of a design with a detailed monitoring programme which has research content. For example, in the pine forest restoration project described above, instead of carrying out the validation monitoring as a separate exercise, without coming to conclusions and recommendations for management in the case of the expected results not being fulfilled, the exercise is set up so that a research question about the use of the woodlands by forest bluebirds and an experimental design are included from the outset. Then, if the expected results fail to appear and the conclusions from the researchers suggest that the silvicultural pre-scription should be changed, the plan can be revised and the implementation adjusted before too much damage has been done. It should be noted that it may be easier to remove additional trees rather than to grow new ones, so if the canopy density that

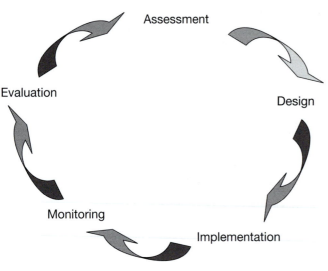

Figure 7.11
An example of an area where adaptive management is being carried out following the initial harvest. In this case the rather rectangular layout is in order to test the impact of a selection harvest and it is needed for experimental purposes. Cispus Adaptive Management Area, Gifford Pinchot National Forest, Washington, USA. Source: Tom Savage, US Forest Service.

bluebirds require was an unknown, then the initial entry should err on the light rather than the heavy side. This is sometimes referred to as the "precautionary" approach to forest management.

Adaptive management is particularly important in the area of forest ecology, due to the huge complexity and dearth of knowledge, but it applies elsewhere too, such as in innovative approaches to road construction, harvesting methods, planting techniques, successional processes, the visual appearance resulting from different silviculture and so forth.

Conclusions

Creating a design is one thing, implementing it is another. There are many stages to go through and because of the long time-scale over which plans are implemented there are many chances for things to go wrong and to change. If a plan has been prepared with a lot of community participation and it is implemented badly because of careless operational management a lot of the goodwill generated from the public can be wasted. Thus it is very important to take a draft design and to evaluate it properly to ensure that is the best that can be produced in the circumstances, is as practical as possible, is cost effective and meets the regulatory requirements and certification standards. It must then be implemented as well as possible and any problems sorted out before it is too late. It must then be monitored, with adequate resources made available, and the results of the monitoring fed back into revisions. Only then will the plan be adaptable, resilient to change and likely to be implemented properly over the long term.

Part III
Forest design application

Chapter 8

Forest design and ecosystem restoration

Introduction

In this chapter we explore one of the most challenging aspects of designing sustainable forest landscapes: ecosystem restoration. The international Society for Ecological Restoration (SER) defines restoration as "the process of assisting the recovery of an ecosystem that has been degraded, damaged, or destroyed."

Nearly every country has experienced severe deforestation, alteration through fire suppression or fragmentation of natural forests. In deforested areas, forest managers have often practised reforestation, sometimes to intensively managed plantations. In other places the land has become degraded, so that natural regeneration is slow to take place or is so unsuccessful that intervention to restore it is warranted.

Sometimes the problem is not the removal of forest but its degradation, leading to severe habitat loss. Degradation of natural forests in the western USA and Canada has occurred as a consequence of three elements: fragmentation through dispersed clearcutting, construction of dense road networks and suppression of natural or anthropogenic fire. Ecosystem restoration frequently involves a detective story, where evidence about the natural forest has to be pieced together before any plans, design or implementation can take place.

In the case of a Scottish glen, where the forest disappeared millennia ago, and where human activity, changes in the climate and increases in the number of grazing animals such as sheep and deer have occurred, there are real questions about what is meant by the term "natural", and thus what exactly is the nature of the forest ecosystem to be restored. Many rare or endangered faunal species may rely on the now open semi-natural landscape, so that large-scale restoration may be inappropriate. Cultural landscape values associated with land-use history and human remains may also limit what can be practically achieved. Finally, there may be problems of native tree seed sources for natural regeneration. All these issues complicate the situation, but must be faced honestly if meaningful results are to be achieved.

In the Pacific Northwest of the USA, the history of logging in the national forests has led to extensive fragmentation. The forest has changed from a pattern of very large areas of late successional forest to one of patches of different ages that bear little relationship to the pattern likely to have been present historically. It can be extremely difficult for planners, designers and managers to see beyond this fragmented pattern and unearth the natural character of the forest. In addition, there are severe difficulties and long time frames involved in changing the now established forest structure back into anything resembling its natural state.

At the outset it is worth stressing that any form of ecosystem restoration is unlikely to have as its objective a "pure" restoration to some completely natural state untouched by human activity.

Figure 8.1
An example of a typical bare Scottish mountainside, where restoration of native forest is a possible option. These landscapes are not natural, although they do possess certain natural elements. What kind of natural forest can be restored in these areas?

Figure 8.2
A fragmented forest landscape in Washington in the USA, where cutting has created a very patchy result and while the continuity of the forest canopy is technically present the narrow strips between cuts mean that much interior habitat has been lost. This fragmentation is at a much finer scale than the natural variation of ages or sizes of stands brought about by natural processes. It is also all at the same scale, with no variability from very small to very large. Source: Tom Savage, US Forest Service.

Restoration as an end in itself can be expensive to implement and of limited value except where it is necessary to expand or secure the habitat of rare and endangered species. It is more often the case that restoration is designed to meet a range of goals, nature protection and conservation being among many. Others may include scenery restoration, creation of wilderness as a recreation or nature tourism asset, and possibly an economic goal of producing future forest timber or non-timber products within a framework of sustainable development. There is no doubt that ecosystem restoration can make people feel good and gain a sense of paying something back in order to replace some of the resources that have been depleted. Ecosystem restoration is frequently associated with community participation, and can also provide important opportunities for public education and local employment. Planting trees or improving habitat can give people a feeling of connection with nature, and doing so in the context of ecosystem restoration can increase this feeling several-fold.

While recognising the multifaceted aspects lying behind the reasons for undertaking forest restoration, the practical methods of carrying this out, especially at a larger, landscape scale, rely on a good knowledge of landscape ecology and silviculture. The means of restoration may be as simple as erecting a fence to keep out grazing animals, or merely ceasing to carry out agricultural practices such as cultivation, so that natural regeneration of the forest from local seed sources starts immediately. These are passive, as opposed to active, restoration strategies. More often there may be inadequate regeneration, overcrowded young plantations or an excess of roads. These issues tend to require employment of a more active helping hand. In some cases, highly fragmented landscapes may need to be redesigned before restoration proceeds.

In this chapter we are going to concentrate on two areas where the authors have direct experience, although other examples may also be cited. These areas are the reforestation of deforested landscapes in Scotland and the restoration of natural landscape patterns in fragmented areas of the Pacific Northwest of the USA. These examples illustrate very well the issues and difficulties faced in such projects.

What do we mean by the term natural?

Whatever the landscape setting, one of the fundamental questions that must be asked in terms of the character of the forest to be recreated is: What exactly do we mean by the term natural? It is useful to consider three kinds of natural forest that can be described for any landscape: past, present and future natural.

Past natural

Past natural means the composition and character of the forest that existed in the past, perhaps at the last phase of evolution of the landscape before it was significantly altered by human culture. This definition can include some historic human engagement or impact. This period may have existed hundreds or thousands of years ago in the case of Scotland, Iceland, Denmark or parts of France, or only a few decades or hundreds of years ago in the case of the USA, Canada, New Zealand or Chile. Clearly, the further back in time this past natural condition existed, the harder it is to deduce what it was actually like, since it is highly improbable that much direct evidence exists. Pollen analysis from plant remains preserved in lake sediments may help to establish the proportions of different species over long time periods. Historical survey records may supply information. The earliest maps may convey something of the location and extent. Paintings may show something of the character, and within the last 150 or so years there may even be photographs.

However interesting it may be to reconstruct the nature of the past natural forest, unless it is in relatively recent times that the forest was cleared the exercise may be of academic interest only because it is probably not feasible to restore it entirely. Beyond providing a picture of what an ideally restored forest would look like, historic information may be useful for comparison purposes, in order to show how the modern condition differs from the past. There may be some key missing ecological elements that if brought back would restore forest health.

Present natural

Present natural means the character of the forest that would be expected to be on the site under current conditions. Here it is important to distinguish between what the landscape would be like if there had been a continuous presence of forest until the present time, and what a restored forest would be like following greater or lesser periods of deforestation. The latter scenario is of particular interest and will be discussed in detail later in the chapter. For all the reasons described in the previous section, the present natural forest following a long period of deforestation could be significantly different from that resulting from continuous evolution and therefore fulfil somewhat different ecosystem functions. It is important to emphasise here that restoration of ecosystem functioning may be different from restoring the ecosystem structure to its natural state.

Future natural

Future natural refers to the character of the forest as it may evolve or develop at some future time. The concept recognises the fact that there are longer-term changes in the wider environment, especially the prospects of climate change, that are likely to affect many aspects of the character of a forest ecosystem. Some of the changes may occur so quickly that the forest cannot respond at a sufficiently fast pace. A warmer, wetter or windier climate may quickly affect the growth conditions of some immobile species, such as certain plants, while very mobile species, such as some insect pests, may migrate and affect new areas with catastrophic consequences. The concept of the "future natural" forest should be borne in mind as an analytical tool. But there are so many uncertainties, particularly climate change, that it is not practical to take every possible scenario into account.

The process of forest restoration design

A restoration project might seem to be mainly driven by ecological principles, allowing little scope for design. Restoration could be approached as a purely technical exercise, with little room for subjective judgement. But any situation where an existing landscape is to be changed into something new, particularly at a large scale, can benefit from conscious design. Forest ecosystem restoration is not a precise science where the results can be accurately predicted; nor is it a deterministic process, since natural processes operate within various ranges of probabilities. This leaves a lot of flexibility to those undertaking a restoration project, and also means that the kind of landscape to be created can be designed to meet a set of objectives that are embedded within a social context. Few restoration projects are able to leave nature to take its own course, with no human intervention. It is important for planners, designers, managers and communities to create a vision for what they want to see achieved, set within an economic and social context as well as an ecological one. After all, such projects cost money and will have an effect on local communities. Therefore, forest restoration projects should follow the same basic steps of the design process set out in Chapter 5.

As in all forest design projects, objectives must be established that describe what is to be achieved. Ecological objectives may be considered as the most important ones, but there may also be a significant social dimension, whether to provide recreation and nature tourism opportunities, to beautify the landscape, to give employment or educational opportunities, to reduce fire risk or to help build community capacity. High-quality forest products may also be produced, possibly as secondary outputs from forest management, but with a particular quality, which may give them a special niche-market appeal. Water management may also be enhanced through restoration, landslide risk reduced and hunting values increased.

The scope for community participation in forest restoration projects can be very high. Where the main goal is restoration rather than economic exploitation, certain community stakeholders may have a greater interest than others. In some cases there may be many who resist restoration since they are happy with the current situation, or who fear the prospect of significant landscape change, even if will be slow to appear, and its effect largely benign. It may also be the case that communities initiate a restoration project themselves, and that they have optimistic and ambitious goals for the landscape (see Chapter 4).

The inventory stage of a forest restoration project may need to focus on parameters that contribute to the development of an understanding of what the present natural forest state is likely to consist of. This will involve a certain amount of detective work. It is important not to overlook non-ecological layers of information, especially cultural heritage aspects, which may be important in constraining or shaping the extent of the restoration project.

At the analysis stage the landscape ecological approach to help reconstruct the appropriate landscape structure is likely to be a crucial issue. Successional processes are also likely to be important to understand, especially when restoring deforested landscapes by both natural regeneration and planting. Modelling to test the development of the landscape over time may be a valuable aid to this. The role of disturbance could be expected to be important, though potentially very difficult to understand and predict, especially in deforested landscapes where there is no remaining evidence of the historical disturbance patterns. In this case, modelling may be the only route available to accomplish this.

An opportunities and constraints analysis will be important in order to set the limits of what can be achieved through restoration, since it is unlikely that a completely unconstrained approach will be possible. In fragmented forest landscapes there may be serious practical difficulties associated with changing the landscape structure, while in deforested areas values associated with open-ground habitats may be locally more important than the potential benefits of a restored forest.

Landscape character analysis is important in both of the above scenarios, because the end result is likely to be significant visual landscape change. In the restoration of deforested landscapes there may be significant issues associated with the loss of open scenery considered to be natural by many people. If it is not possible to reforest all of the area because of various constraints, it will be necessary to design the extent of the restored forest, to fit into the scene around the boundaries of the prescribed location, so that the final design appears to fit naturally within its surrounding context.

The vision for the design concept is likely to be mainly driven by a desired future condition that emerges from the ecological analysis, but related to the character of the landscape, especially where scenic values are high and where there is a strong spirit of place. A design strategy identifying zones of different forest types within the landscape, each with associated species composition, structural characteristics and degrees of connectivity, for example, is a very useful product before focusing on the development of the more detailed design.

Where natural processes are to be the main means of implementing the plan, the development of the plan beyond the concept or strategy stage is probably unnecessary, but where major surgery of a fragmented forest is needed a full design will be vital. The same applies to the design for planting a forest on a deforested area, where complete specifications for planting mixtures defined by possibly small areas will be needed.

The stage of monitoring and revision is vital, and restoration projects are appropriate candidates for ecosystem management procedures, learning from the success or failure of the restoration as it develops phase by phase (see Chapter 7). Of particular importance is the success of natural regeneration and the need to either manipulate the early results if the desired structure or balance of species does not occur, or to plant where it fails altogether. This to some degree depends on the time-scale over which substantive results are expected, as well as the philosophical approach to restoration and the appropriate level of human intervention or manipulation of nature.

The next two major sections examine approaches to forest restoration in two examples: re-establishing a natural forest ecosystem in deforested landscapes and restoring fragmented forests.

Restoration of deforested landscapes

Where extensive areas of forest have completely disappeared, as they have in northwest European countries, New Zealand, Chile and elsewhere, it is a very challenging task to try to restore significant tracts of forest ecosystem back to a "natural" state. In many deforested countries the landscape is now occupied by intensive agriculture, urbanisation and other land uses that do not allow forest restoration at a landscape scale. It may, however, be relatively easy to recreate small patches of new native woodland and to try to link them together by hedgerows or narrow strips of woodland.

In some countries and regions, such as Latvia and the other Baltic States, the Maritime Provinces of Canada and the New England States, there is a large amount of surplus agricultural land within a natural forest matrix. Such land, once abandoned, starts to become forest very readily, owing to the absence of sufficient densities of grazing or browsing animals to prevent regeneration. There are often abundant native seed sources, and it is easy to see how such regenerating areas develop through various natural successional stages towards mature forest. The challenge in Latvia (and to an extent in other areas) is more about how to ensure that valuable open traditional landscapes of cultural value can be retained and managed. Traditional forest hay meadows are often important habitat types within the landscape mosaic. In these cases, a design for forest restoration could identify any number of important open areas that would be retained, while at the same time exploring options for re-establishing extensive areas of forest that include interior habitat. The design of forest restoration in this case is more about controlling how much and where it occurs, and maintaining the range of other values in the landscape. See Chapter 2 for an example of the changing landscape of Latvia.

Figure 8.3 Colonisation of abandoned fields in Latvia.

Scotland, especially the Highlands, presents a very different picture. The extensive valleys, or glens, were largely deforested by a combination of clearance for

Figure 8.4
An example of a remnant fragment of old pine forest in Scotland, with some regeneration as a result of excluding grazing animals. The main ground vegetation is Calluna vulgaris.

agriculture, the prevention of regeneration by grazing animals (cattle and sheep) and later, the intensive commercial exploitation of remaining natural forests. This process took many centuries or even millennia. The climate changed over the period, becoming cooler and wetter, encouraging the accumulation of peat which, in turn, changed the vegetation cover, soil type and drainage characteristics. The resulting landscape is open, dominated by low-growing shrubs such as heathers and bilberry (*Calluna vulgaris, Erica* species, *Vaccinium* species), grasses (*Deschampsia flexuosa, Nardus stricta, Molinea caerulea* among a range of species), bracken (*Pteridium aquilinum*) and extensive bogs dominated by sphagnum mosses. The native woodland would generally consist of Scots pine (*Pinus sylvestris*) in mixture with downy birch (*Betula pubescans*), sessile oak, (*Quercus petraea*), juniper (*Juniperus communis*), aspen (*Populus tremuloides*) and occasionally ash (*Fraxinus excelsior*) and alder (*Alnus glutinosa*).

The open moorlands remain deforested because of grazing pressure from high populations of sheep (the Cheviot breed) introduced mainly in the eighteenth and nineteenth centuries, or red deer (*Cervas elaphus*) managed for sporting purposes. These practices mean that natural regeneration is not only impractical, but that there are very few sources of seed available should the sheep or deer populations be reduced below the threshold densities for regeneration to commence.

In landscapes deforested for millennia and used or managed by human populations during the years since, there are likely to have been significant changes to soil as well as to vegetation, through selective grazing, cultivation, erosion, effects on drainage and nutrient status. These effects may have been intended to make the soil better for agriculture by raising the nutrient status and lowering soil moisture content, but over time may have impoverished it by allowing nutrients to be lost and moisture content to increase. This implies that the kind of forest currently able to grow on such areas may not be identical to that which was originally present (see below).

The long history of human intervention also means that human remains are likely to have accumulated, albeit more densely in some places than others, such as lower valley slopes and valley floors compared with hill summits. These remains may have significant historical or archaeological value, and may constrain forest restoration in selected areas. Equally, the long-term development of the open moorland vegetation may have become a valued habitat type in its own right, notwithstanding its anthropogenic origin, while certain fauna may also have adapted to it, including rare and endangered species such as some raptors. It may also be valuable economically, for grouse shooting, and be a valued scenic asset for tourism and recreation. This may also set practical limits on what is possible.

This example provides a good testing ground for approaches to forest restoration because it presents an enormous challenge and thus, if it can be successful in Scotland, it should be possible elsewhere.

What are the main elements that must be considered in such a restoration project where so little is known about the past natural forest, let alone the present natural? Recent research undertaken by various agencies in Scotland, such as the Forestry Commission, Scottish Natural Heritage and the Macaulay Institute, has identified three main elements that need to be developed and then integrated together in order to form the complete picture. They are:

1   The character of the spatial mosaic of forest types across a landscape
2   The processes and time-scales of the colonisation and ecological succession
3   The nature and role of natural disturbance processes in the future development and management of the developing restored forest landscape.

Together these give a four-dimensional picture of the forest landscape. The spatial mosaic pattern provides the two-dimensional layout of the forest, which can be resolved to as coarse or fine a detail as needed. The colonisation and successional processes provide the third dimension of height and vertical structure as the forest develops and part of the fourth dimension, the temporal changes and development of the landscape. Natural disturbance is one further aspect of temporal change, which also contributes to the character of the two-dimensional spatial mosaic pattern and to the third dimension of vertical height and structure. All three of these elements must be considered together and integrated. In practice, all of this is even more complicated than it sounds.

Developing an understanding of the spatial mosaic pattern

While a landscape may be expected to develop a number of general types of forest, it is valuable to know whether these are likely to be simple and homogeneous in distribution or complex and heterogeneous. The pattern of different types will become the landscape structure of the eventual forest and, as demonstrated in Chapter 2, will affect ecological functioning.

The eventual pattern of the forest mosaic will depend on several factors. At the inventory stage a number of layers of information are needed. These were listed in Chapter 5 and are more specifically summarised here. The general climatic regime of the area will determine, first of all, the potential range of forest types that could occur. Climatic factors include:

*   The degree of continentality or maritime influence of the landscape, which primarily affects temperature range, continental climates having a wider or more extreme range than oceanic climates. A way of measuring this is the Conrad Index.
*   The accumulated temperature, which can be expressed as day degrees above 5°C, the lowest temperature for plant growth. This determines the length of the growing season and therefore affects what plants are able to grow in the landscape.
*   Wetness, expressed as soil moisture deficit, is the degree of drought in summer, rather than the degree of waterlogging in winter. This is used, together with key soil properties, to determine the soil moisture regime (see below).
*   Windiness, expressed as the average wind speeds experienced across the landscape, which vary according to the degree of topographic shelter and elevation. Wind has both an inhibiting effect on tree growth and causes tree stem deformation.
*   Soil moisture regime expresses the availability of water for plant growth and also its excess, which may inhibit growth. This can range from very wet to very dry. It is a combination of available moisture from precipitation combined with the capacity of the soil to retain water. Thus a very dry site could be one where there is a high summer soil moisture deficit because of low rainfall, low water retention capacity or both. For example, a sandy soil in an area of low rainfall has a high soil moisture deficit and may be unable to support tree growth or only trees suitable to those conditions. Conversely, low soil moisture deficit (or excess wetness) occurs in areas with high precipitation and moisture-retentive soils. For example, a peaty soil in an area of high rainfall is likely to be waterlogged for much or all of the year and to be unable to support tree growth.
*   Soil nutrient regime is expressed as the availability of the major nutrients such as nitrogen, phosphorous, potassium, calcium and magnesium. Very poor soils tend to be acid (low pH) and lacking in these nutrients. Very rich soils are neutral pH and have the nutrients freely available. There are also carbonate soils with a high pH and calcareous composition. Some trees and plants can cope with low pH and low nutrient conditions, while others need richer sites.

The factors listed above can be used to create a predictive decision support system such as an Ecosystem Site Classification (ESC) developed by researchers at the Forestry Commission. If the data on all the factors is readily available at a fine resolution it is relatively straightforward to produce a map via GIS to show different site types, which can be related to a range of present natural forest types. Unfortunately, such comprehensive information is rarely available. An alternative is to infer the soil moisture and nutrient regimes by a study of the existing vegetation supported by some calibration samples collected on site.

If deforested landscapes have a natural or semi-natural pattern of vegetation, it is possible to observe a relationship between vegetation type and soil type. This use of vegetation indicator is widespread and has a long history in forestry. In Finland the forest is divided into classes of quality or growth potential on the basis of vegetation association. Systems initially developed in Central Europe have been borrowed and applied elsewhere, such as in the Biogeoclimatic Ecosystem Classification used in British Columbia. What is special about their use in deforested landscapes is that there is no obvious link between the vegetation of the forest floor and the trees, because there are no trees!

Research has been able to determine the means by which vegetation correlates to soil and present natural forest types. In the UK this is a system known as the National Vegetation Classification (NVC), which relates forest (or woodland) type to what are known as "optimal precursor" vegetation communities. Using soil samples taken from a landscape it is possible to develop more localised relationships which can then be applied across the landscape to produce an initial map of potential forest types, broken down further into, for example, elevation classes.

Such maps may reflect the maximum resolution of the spatial forest mosaic, or they may still be rather generalised and simplistic. This depends on the degree of resolution of soil or vegetation maps, which may be quite coarse-grained for areas with low-value land-use potential, compared with high-quality agricultural land. In the case of former farmland to be restored to forest, the use of optimal precursor vegetation is not possible, but the quality of soil maps may be much better. Soil nutrient regimes may have been altered by fertiliser application, although over time many of these may become leached from the soil. Agricultural drainage may also have affected soil moisture regimes and their effect may continue until they start to fail. Thus, restoration of deforested landscapes used for agriculture presents its own set of challenges.

In order to work out the finest spatial scales of mosaic heterogeneity it is possible to use micro-topography as a guide. Within the general characteristics of soil moisture regime, for example, there are likely to be localised variations due to soil type changes related to topography. An example could be the relationship of convex landforms to better drained, more acidic (podzolic) conditions, owing to the presence of coarser glacially deposited materials, and the relationship of concavities to wetter, acid but gleyed (anaerobic) conditions. It might be expected that different vegetation would occur on the convex topography to the concave, which might also relate to different present natural forest types. In agricultural areas this use of micro-relief can be useful because of the absence of vegetation differences. A practical means of assessing these relationships is to use a digital elevation model of the topography, resolved to as fine a scale as possible, and to analyse it within a GIS to detect the pattern of concave and convex landforms, perhaps to a set of steps from strongly convex to strongly concave. An ortho-rectified aerial photograph on which vegetation differences are visible can be classified and overlaid on the topographic analysis and correlation sought. Field-testing and calibration is also feasible. It is a short step to developing a predictive model for the detailed forest mosaic for a landscape once the input data has been established.

Developing an understanding of the processes and timescales of colonisation and ecological succession

It is one thing to be able to describe the present natural forest that would be found in a landscape, and quite another to achieve it on the ground. Moreover, as the forest develops the ecosystem changes together with its functions and habitat values. In places with little or no available seed source, or only limited species, it may be necessary to plant. In this case the forest mosaic pattern can be created by planting the correct mixtures in the relevant places according to the principles described earlier. This leads to an even-aged forest, although differentiated growth rates and development in different areas may produce a more varied structure. Even when planting, it may not be appropriate to plant the component species of the late successional forest if these are different from the early successional or pioneer species. In these circumstances it may be better to plant pioneers and then either let Nature take its course or else introduce later successional species by planting as the conditions of the developing forest permit. This means that an understanding of successional stages and their expected duration is needed as much for a planted forest as for one created through natural regeneration.

In many places where there is already extensive forest of the same type on the same sites as the restoration project, it will be relatively straightforward to develop a model of colonisation and succession. Indeed, it may be possible to observe it, such as the situation in Latvia described earlier, where abandoned fields readily revert back to forest. Elsewhere it may be less clear and a deductive process combining observations and samples from a range of similar areas together with expert knowledge may be necessary. Depending on the predicted degree of heterogeneity of the forest mosaic, it might be necessary to develop several succession models to apply in different places. In landscapes where frequent catastrophic disturbance is also a feature it will not be possible to develop complete succession models without an understanding of the role of disturbance within the succession system.

Of course, colonisation of open ground does not occur simultaneously across the entire area. If the seed sources are unevenly distributed and are limited to the periphery of the area to be restored, the rate of colonisation spreading from these sources has to be taken into account. Some tree seed is light and spreads far and wide (e.g. poplar or birch) while dispersal of others is limited and dependent on animals and birds which themselves need a forest to use. Thus, a second element of the colonisation model is the rate at which the landscape is filled up with forest. In some circumstances there may be barriers, so that while part of the landscape should theoretically become completely forested, in practice parts are unlikely to be colonised and remain bare owing to the absence of seed sources.

This information about colonisation, succession rates and pathways is also important because of the habitat function each element supplies over time. It is therefore possible to relate each successional stage to its use by different animals and birds using the virtual forest of the spreadsheet to determine the relationship between flows and structures demonstrated in Chapter 2. When the colonisation and successional stages are related to the spatial mosaic it is possible to obtain a good idea of how the natural forest will function and to use GIS tools or programs, such as those briefly presented in Chapter 6 or HABSCAPES described in Chapter 7, to evaluate the proposed design, given certain assumptions about its evolution.

Using decision support tools in GIS, for example by applying rates of colonisation, a series of maps showing how the landscape could change and develop over time can be produced. This can be tested by changing the assumptions used in the model. The effects of land-use decisions affecting part of the landscape can also be tested, for example in areas where there are different landowners or some parts of the semi-natural moorland or mountain vegetation have been set aside for protection.

Developing an understanding of the nature and role of natural disturbance processes

Of all the elements of landscape restoration, the role of natural disturbance is the most difficult to determine. If the area has been devoid of forest for millennia and there has been a degree of climate change, evidence contained within pollen deposits or carbon left by ancient forest fires may be no use in predicting current disturbance agents. If there are natural forests in similar landscapes elsewhere, then it is probably safe enough to base the predicted disturbance patterns for the design area on this, at least until sufficient local data has been accumulated.

One of the disturbance agents that can be modelled and predicted is that of wind. Britain is a particularly windy country and its forests are prone to wind damage by events of different scale and intensity at frequent intervals. There is also data available on the strength of wind needed to blow over different sizes of different species of trees on different soils and in different degrees of exposure and shelter. Thus, using a probabilistic model on a landform where there has been a calculation of the degree of topographic shelter and for which there is knowledge of the local wind climate obtained from anemometer readings, a map of anticipated disturbance can be produced.

As we know from the discussion about disturbance in Chapter 2, it is difficult to predict the precise frequency and intensity with which disturbances are likely to occur. Thus it is necessary to introduce some stochasticity into modelling of the likely effects, based on the average expected return cycle for each major disturbance type. Additionally, disturbances are often interrelated. Thus an insect outbreak, one form of disturbance, leads to trees dying on their feet, which results in windthrow, followed by intense fire as fuels dry out.

The result of the interaction of the above elements is often a complex picture that needs to be untangled before designing restoration of a natural forest landscape. The spatial mosaic pattern is the most predictable element, the pattern of colonisation and succession open to a degree of uncertainty and the expected disturbance regime the most unpredictable of all. Thus it is very difficult to obtain a clear picture of what is likely to happen without the benefit of a computer-based modelling program that looks at the landscape-level patterns and processes. These are currently under-developed but hold great promise. Some were introduced in Chapter 6 and one will be briefly demonstrated as part of the following case study.

Case study: Restoration of native pine forest at Glen Affric in Scotland

Glen Affric is a remote valley in the Scottish Highlands north of Loch Ness. It was once a part of the great forest of Caledon, consisting of native Scots pine and birch. Part of the area has been reforested since the 1950s by means of fencing out the red deer so as to allow natural

Figure 8.5
The location of Glen Affric in Scotland.

Figure 8.6
A view of part of Glen Affric showing a remnant pine stand surrounded by areas where regeneration is to take place.

regeneration to occur. This was possible because there were remnants of the original forest left after various periods of logging, the most recent of which took place during the Second World War. This period of some 50 years since regeneration commenced has enabled foresters to study the processes of colonisation and disturbance. Since the early 1990s, the scale of restoration within the Glen has been significantly increased and from 1994 onwards a programme of applied research was carried out.

Developing an understanding of the spatial mosaic

The first element of this research was to construct a map of potential forest types across the landscape study area (bounded by mountain ridges and other natural boundaries). This was prepared by applying the ESC described above. In the absence of soil maps, data from the Land Cover for Scotland 1988 (LCS88) data was used to provide proxies for soil type and thence to soil moisture and soil nutrient regimes. The use of the land cover map was supported by a series of sample transects taken in various locations to examine the relationship between soil type, especially peat depth, and vegetation in relation to moisture and nutrients. The resulting map contained a number of distinct forest types determined by elevation, soil characteristics and peat depth.

LCS88 classes were amalgamated, taking into account their likely soil moisture and nutrient regimes, to produce 13 classes. The next stage was to use the information from the transects referred to above to refine the characterisation of the soils from the LCS in the different climatic zones. Not only were there apparent differences in soil types and peat depths between climatic zones, but it was possible to subdivide the largest zone (Upper Forest) into two sub-zones based on peat depth. Thus, in this zone, each LCS88 class was allocated to either a shallow peat or deep peat type of soil. Soil moisture and nutrient regimes were assigned to the other vegetation types based on the indicator values of the principal species on their respective transect points. A map of the four zones and the two sub-zones, the "potential vegetation zones", was then developed.

It was not possible to produce a map showing the mosaic at finer resolution due to the limitations of available soil data, but it was possible to identify the elements of the mosaic in the different zones. The potential for a higher resolution mosaic was tested in a pilot study and found to be predictable but could not be taken further.

Woodland suitability ranges were used to decide which NVC woodland communities and sub-communities could occur in each of the potential vegetation zones. Only three woodland communities were considered likely to be important in Glen Affric, the W18 pinewood, the W17 oak-birch with *Dicranum majus* and the W4 downy birch with purple moor-grass (*Molinia caerulea*). In addition, the W7 alder woodland (*Alnus glutinosa – Fraxinus excelsior-Lysimachia nemorum*) would be found on alluvial and flushed streamsides.

Figure 8.7
A map generated by the use of ecosystem site classification and the Land Cover for Scotland 1988 data, showing predicted broad forest types. Source: Forestry Commission. © Crown copyright material is reproduced with the permission of the Controller of HMSO and Queen's Printer for Scotland.

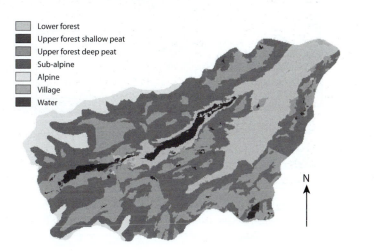

Lower forest
Upper forest shallow peat
Upper forest deep peat
Sub-alpine
Alpine
Village
Water

N

No woodland growth is possible in the Alpine zone. Within the Sub-alpine zone the climate, although very severe, will permit scrubby growth of downy birch and pine in patches of relatively well-drained soil. Such growth will develop into "krummholz", where the clumps of trees are semi-prostrate and have a wind-shaped profile. In the Upper Forest zones conditions are climatically suitable only for the pinewoods and the downy birch woods. In the Upper Forest zone, however, the deep peat sub-zone is considered too wet for appreciable amounts of woodland colonisation. Only in the Lower Forest zone are conditions suitable for all three communities.

Developing a succession model

By knowing the main forest (or woodland) type as predicted by the ESC and related to the national vegetation classification, it was possible to develop a model of succession in terms of composition, structure and duration, with feedback loops where disturbance would have the effect of reverting succession to an earlier state. This concentrated on the pine forest elements which are seen as being the more significant proportion of the landscape and the more valuable from a habitat point of view, especially when they reach old growth condition.

The development of the model for natural disturbance focused on wind as the most significant type to be found in British forests. There is no record of the occurrence of wind damage in Glen Affric, and the forest area has changed so much, and in character to such a degree, that a history

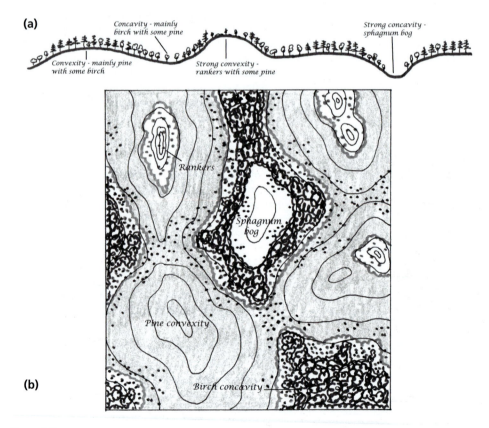

Figure 8.8
Designing the restoration of the forest mosaic using topography, soils and vegetation indicators at Glen Affric: a) a section of the terrain in Glen Affric, Scotland, showing the micro-topography which determines soil and site types; b) a design for an area using the micro-topography as a guide.

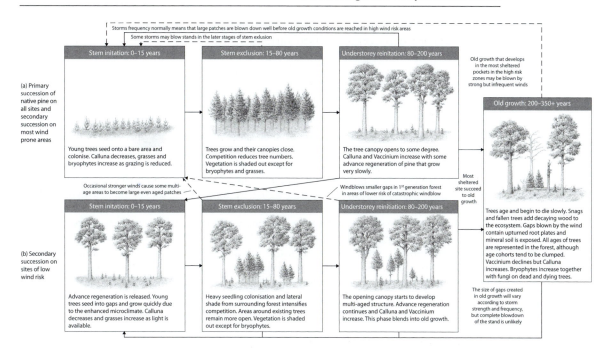

Figure 8.9
This series of sketches shows the anticipated succession and disturbance pathways for pine forest at Glen Affric. Source: Forestry Commission. © Crown copyright material is reproduced with the permission of the Controller of HMSO and Queen's Printer for Scotland.

Figure 8.10
An example of the effect of a catastrophic wind storm on a British forest. This has resulted in large-scale disturbance to the landscape.

would be difficult to interpret. An alternative approach is to use existing models to consider the likelihood of forest disturbance, and then interpret the information in the context of disturbance types and their spatial pattern.

Developing a disturbance model

A model to assess the risk of wind damage to coniferous forests has been developed by the Forestry Commission. The Forest*GALES* model calculates the threshold wind speed required to break or to overturn a typical tree, based on mechanical characteristics of the stem form and the species. The likelihood of the threshold wind speed being exceeded is then assessed for location in the country and topography, using a conversion of mean wind speed to extreme wind speed. Mean wind speed is estimated using the Detailed Aspect Method of Scoring (DAMS) system which combines the influence of regional location, elevation, topographic shelter, funnelling and aspect on site windiness.

The vulnerability of Scots pine and birch was assessed by calculating the threshold wind speed required to overturn, or snap, the average-sized tree at various stages in the stand growth, which can be interpreted as catastrophic damage to the stand. The likelihood of these threshold wind speeds being exceeded was assessed for a number of topographic positions represented by a range of DAMS values. Peats provided a more stable substrate than other soils (gleys and freely draining soils), and on these the threshold wind speed for breakage approached the values for overturning. Changes in DAMS score, within the range present in Glen Affric, were very influential in determining the probability of damage.

Five disturbance classes were interpreted from the frequency (and implicit scale) of likely disturbance. The highest DAMS scores represent a zone (class 1) where wind exposure and lack of warmth make tree growth impossible. Below this is a zone (class 2) where growth of woody plants is possible, but will be limited to dwarf or other highly adapted forms; these are unlikely to be overturned by strong winds as the canopy of the stand is streamlined. In the zone of good tree growth (classes 3–5), there is a zone (class 3) where catastrophic disturbance is likely to occur very regularly and lead to a mosaic of even-aged patches of several hectares in size. In the most sheltered zone (class 5), catastrophic disturbance is extremely rare, and there is adequate time in the intervening periods for an uneven structure to result. In between these two zones is an intermediate zone (class 4), where a mix of even-aged patch and diverse remnant may emerge – depending upon the precise frequency of the strong winds. The spatial pattern would reflect

Figure 8.11 A map showing predicted wind disturbance zones for Glen Affric. The zones can be translated into possible silvicultural systems for managing the developing forest. Source: Forestry Commission. © Crown copyright material is reproduced with the permission of the Controller of HMSO and Queen's Printer for Scotland.

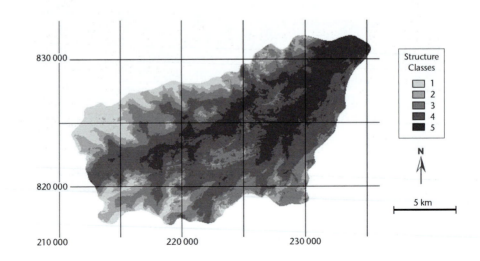

the mosaic of varying site conditions, the particular features of damaging winds (e.g. landform aspect versus wind direction) and the initial conditions.

Once the three elements of the landscape restoration are in place it is possible to develop a dynamic model of the developing and changing landscape, examining the rates of colonisation, the development of the forest through the various successional stages and to insert a stochastic element of wind disturbance. Such modelling can then be linked to the habitat values of the different forest stands. An attempt at such modelling was made for the Glen Affric project, the first such approach to be developed in Britain.

Simulation of forest landscape dynamics in Glen Affric

The Glen Affric Landscape Dynamics Reconstruction (GALDR) model, described in some detail in Chapter 6, was used in this part of the landscape reconstruction project. It produced a series of images at different time steps, showing the predicted expansion of the forest over time using the decision rules for all the different factors such as colonisation rates, succession stages, wind effects, site types, herbivory and so on. It can be seen that a significant extension of the forest would occur over a 100-year period.

Redesign of fragmented forests

There is a whole class of forest landscape where previous harvest practices have led to a great deal of fragmentation and/or permeation. Fragmentation is usually defined as the breaking up of large habitat blocks into smaller ones. Permeation is the incursion of roads, powerlines or other corridors into otherwise intact habitats. The former occurs where the main timber harvest method is by clearcutting blocks of a limited size (to meet regulations), dispersed about the forest so as to prevent large clearcuts and limit local impacts. In the USA, this technique is referred to as "staggered setting". The result is frequently an accumulation of rectilinear cutblocks over time, with different successional stages in each block, and the breaking up of mature or late successional forest matrix into small "habitat islands". Wildlife populations are faced with four interrelated challenges to habitat in highly fragmented forests: direct loss, subdivision, isolation and edge effects.

Direct loss of habitat is self-explanatory. For species dependent on mature and old growth forest, after logging there is less of this available. Subdivision results in smaller patches, which may be less suitable than large ones. Isolation occurs when remnant patches get so far apart that wildlife are unable to "aggregate" them. For example, if a species needs 100 hectares of mature forest and there are four 25-hectare blocks in an area, this would be inadequate if the blocks were too far apart for the species to travel between them. Edge effects pertain to how much of a habitat patch is "perimeter" versus "interior". At the edge of a patch, light, wind, temperature and predators reduce the value of the habitat for species that require more protected conditions.

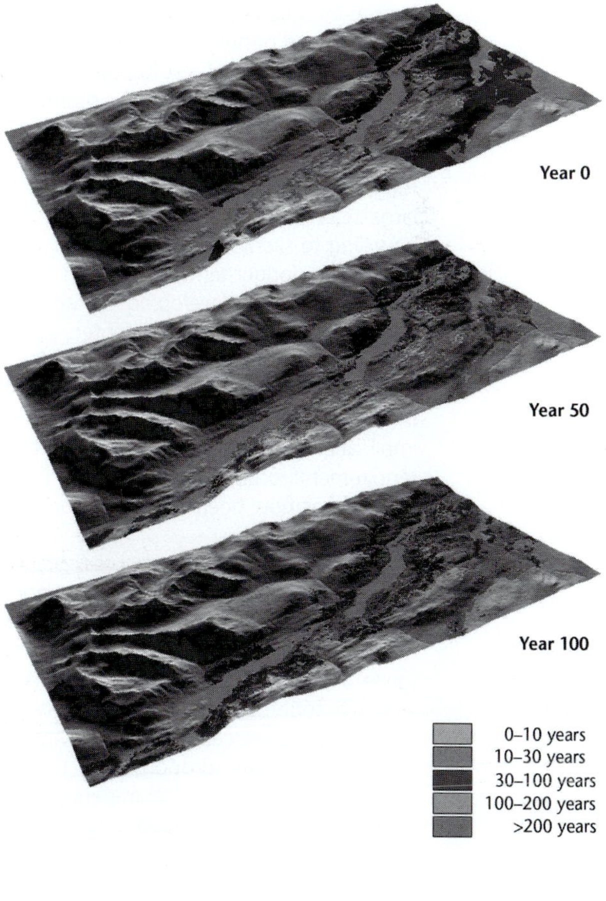

Figure 8.12 An example of the output of the Glen Affric Landscape Dynamics Reconstruction model showing the expected expansion over time. Source: Forestry Commission. © Crown copyright material is reproduced with the permission of the Controller of HMSO and Queen's Printer for Scotland.

Year 0

Year 50

Year 100

0–10 years
10–30 years
30–100 years
100–200 years
>200 years

Figure 8.13
A fragmented forest on Vancouver Island, British Columbia, Canada. This example has been caused by the over-large scale of the patches logged over a short period. The cutblocks are not representative of the natural scale of disturbance, which here would be very fine.

Figure 8.14
This view shows a similar forest with no logging, consisting of a very uniform canopy. Nootka Island, off the coast of Vancouver Island, British Columbia, Canada.

A key problem facing forest managers in fragmented areas is that very few remaining mature stands are of a size or in a pattern where there is sufficient interior mature or late successional habitat to satisfy the requirements of certain species. In some models, logging 50% of an area in a dispersed pattern reduces interior habitat to zero. Moreover, since it takes many decades for young stands to reach ecological maturity, the established pattern that has developed will persist for very long periods. One of the problems when faced with such a situation is that the fragmented pattern is so strongly defined that it is difficult to see past it to envisage the pattern that should be there, given new objectives related to sustainability of all native species. The only way to break the pattern in the short term may be to log remnant mature blocks in ways that reflect natural large-scale disturbance patterns. Managers, their ecological staff or conservation groups may be unwilling to sacrifice further areas of late successional forest in the short run in order to achieve medium- or longer-term aims. It was this type of situation faced by the authors and our colleague Nancy Diaz when developing the application of applied landscape ecology to "Leoland" at Mt Hood National Forest (see Chapters 2 and 5).

This problem is not unique to the Pacific Northwest of the USA, but it is exemplified by it. In some ways it reflects the desire of managers to organise and control the forest and to assemble it into a mathematical model of yield regulation. Without a full understanding of the ecological implications, however, such drastic restructuring of the forest could not hope to meet the requirements of sustainable forestry beyond timber production. While the practice of dispersed logging has now been recognised and reduced in some regions, the problems it created remain a daunting legacy.

The steps of the design process presented in Chapter 5 include setting objectives. In cases of highly fragmented forests, the restoration of ecological functioning may be particularly significant, while timber production may drop down the scale. At the inventory stage a good deal of detective work should be spent to find out what the forest was like before the fragmentation began (generally prior to the Second World War). Old aerial or fire lookout photographs are especially valuable, together with early inventory maps. These records reveal the scale of the natural forest mosaic. In some cases this information may date from after the period of successful fire suppression, so the proportion of the forest in an undisturbed state may be over-represented. Thus some earlier historical material from explorer accounts, traditional ecological knowledge or pollen analysis may be needed to get a better picture of the natural range of variability.

Landscape ecological analysis

The ecological analysis steps laid out in Chapter 5 are useful when faced with redesign of frag-mented forests. Through this process, the project team can identify present ecological functioning, and note the problems and liabilities presented by the current landscape structure and pattern. In particular, the focus should be on gaining an understanding of the relationships between key flows and the landscape mosaic. Habitat modelling through GIS-based metrics can be a great help in objectively determining whether the present landscape provides adequate habitat for species of concern.

An evaluation of disturbance and succession dynamics should be used to build a picture of the kind of forest landscape mosaic that would probably have existed if logging had not taken place. This pattern can then be compared to the existing, fragmented condition to see where there are major discrepancies occurring between the two. By overlaying existing and "natural" mosaics, opportunities for reconciliation may emerge. The key is to help the team (and community) to see beyond the existing, dysfunctional pattern to other possibilities.

It is also very important to use the analysis stage to gain an understanding of landscape context. How does the area being designed fit into the larger pattern? If it is part of a critical system of

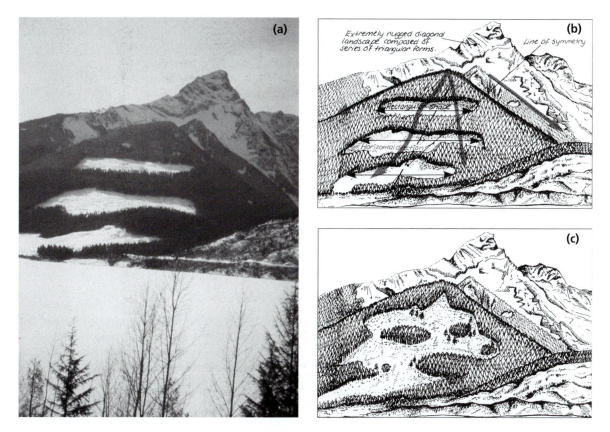

Figure 8.15
This example shows a) a poorly designed and fragmented forest on a hillside near Revelstoke in British Columbia, b) an analysis of the problem and c) an option to redesign it and restore a more connected forest over the longer term, by expanding the area, breaking through the horizontal banding and creating a more organic shape. Although larger in scale this reduces fragmentation. Source: British Columbia Ministry of Forests.

mature forest habitats, then this may constrain opportunities to log remnant patches in order to establish a more natural disturbance pattern. If it is not part of a regional network, then this may open opportunities.

Landscape character analysis

Once the desired future ecological condition has been developed, the landscape character analysis should be prepared. If the area being planned is highly visible, restoration to an acceptable aesthetic state may be desirable. The analysis of landform and the identification of the worst visual eyesores should enable the designer to focus their efforts on correcting the most blatant mistakes. Often, the main problems include the shape, scale and proportion of cutblocks, their position and balance, their conflict with visual forces and landform and their strong contrast with their surroundings. Solving these problems may suggest fairly bold and radical approaches which would need to be assessed against their ecological impact.

Constraints and opportunities

This analysis is particularly important, given the existing state of the forest. There is likely to be an existing road network, although some of this may have been de-activated or it may present liabilities due to poor construction and maintenance. Age structure of adjacent stands may present serious constraints to redesign, riparian zones may not be intact, and the lack of merchantable timber left in the forest may present problems as to how to finance restoration. There may, however, be opportunities through silviculture, such as thinning, to improve stand structure and to help advance stand succession. The road system may help keep costs down and permit very small-scale harvest, perhaps to get rid of particularly intrusive shapes, to take place economically.

A thorough brainstorming by all the team members is likely to be the best way to ensure that all the constraints and opportunities are identified.

Concept

This stage is crucial, because it is so important to be able to see beyond the existing landscape into the future potential. The concept should, once again, be presented as a series of zones or elements. For example, the desirable pattern and extent of riparian zones, old growth or connected habitat should be located in general terms, even if it does not yet exist. This allows silvicultural prescriptions to be developed with long-term aims for these areas.

It is vital that a creative element be introduced to the concept stage, notwithstanding the significance of the various constraints. These tend to be most pressing during the early stages of implementation, whereas the concept must be long term. The potential ecological patterns together with the landform analysis will drive the concept design.

Sketch design

This is one of the most difficult stages, because of the tendency to get caught up in detail. The concept of cutblocks should be replaced with design units or management units, discrete areas where a particular stand structure is desired at some defined point in time. These and the restored riparian and other areas should be designed as far as possible using landform, although it is inevitable that some sub-optimal elements will be forced on the designer due to some of the severe constraints, such as where a badly shaped old growth patch lies next to early successional forest. Some of the units may be able to be restocked sooner than others, more easily or less expensively, so a phasing plan is still necessary. Each phase then becomes more of a restoration project that needs to be planned carefully at a detailed level. As the restoration proceeds there will undoubtedly be a need for revision, because progress and experience will probably change the original anticipated design.

Three case studies are presented in this section, all in the USA. The first, the Cispus AMA in Gifford Pinchot National Forest in Washington State is an Adaptive Management Area, a special type of land designation, while the second, the Collowash watershed, is a more standard forest planning project in part of the Mt Hood National Forest in Oregon. The third, the Little Applegate watershed, is also in an AMA and also had considerable public involvement (see Chapter 4).

Case study: Landscape design for the Cispus AMA

The 1994 Northwest Forest Plan initiated by President Clinton established ten Adaptive Management Areas (AMAs) on national forests in Washington, Oregon and northern California (see Chapter 7). All of these are near communities that were greatly impacted as a result of reductions in timber harvest in order to conserve old growth conifer forests crucial to the survival of the Northern Spotted Owl and numerous other species.

Figure 8.16 Map showing the location of the Cispus Adaptive Management Area.

The purpose of the AMAs is to develop and test new management approaches that attempt to integrate ecological, economic, social and community objectives. Each AMA was directed to develop a management plan based on a shared vision of the forest landscape. In the Cispus AMA, located near Mt Rainier in Washington State, this vision is to integrate commodity production with conservation of mature forest ecosystems, aquatic and riparian functions and recreational opportunities, while being responsive to a continued role for people in the landscape. Learning through experiment and adaptation is a key objective for all AMAs, as the name implies.

Today, about 5,000 people live adjacent to the Cispus AMA, mainly in the towns of Packwood and Randle to the north. The timber industry has been the backbone of the local economy for most of the last half century. Area residents traditionally collect firewood and other forest products, including mushrooms, berries, transplants for landscaping and greenery for floral arrangements. A combination of mechanisation, consolidation and harvest reduction led to a chronically high unemployment rate and generally depressed condition over the past several decades. Opportunities for diversification are limited by the relative remoteness of the area and the lack of infrastructure suitable for modern industries. There has been some increase in tourism- and retirement-related economic development due to the proximity of the area to Mt St Helens National Volcanic Monument and Mt Rainier National Park. The project was carried out by Forest Service planners with a lot of public involvement using many of the tools described in Chapter 4.

Physical description of the Cispus AMA

The Cispus AMA is located within the Gifford Pinchot National Forest, west of the Cascade Mountain crest. It is about 143,000 acres (59,500 ha) in size, and is approximately 12 miles east to west and 20 miles north to south. Most of the Cispus AMA is heavily forested, although in higher elevations a combination of climate, soil and past wildfires have combined to produce open lands dominated by grasses and shrubs.

Major streams include the Cispus River and the north fork of the Cispus River. There are several other streams and lakes in the AMA. Prominent ridgelines flank the broad, glaciated valleys of major streams. Elevations range from 300 metres in the Cowlitz River Valley to nearly 2,000 metres on the highest ridges. Part of the southeastern boundary of the AMA borders onto the Yakama Indian Reservation.

Vegetation within the area has been shaped by numerous disturbances. Eruptions of nearby Mt St Helens have altered the landscape dramatically at least three times over the last 4,000 years. Wildfires have played an important role in the evolution of vegetation pattern, composition and structure. Prior to 1920, reforestation following natural disturbances resulted from natural succession. The dominant forest structure tended towards large, fairly continuous patches of one age class, with relatively little tree species diversity. Douglas fir, which thrives after disturbance, was widespread over much of the area. Some stands survived multiple fires and grew into more complex forests that included shade-tolerant conifers and hardwoods, thus forming a mix of tree species and ages commonly associated with late successional stands in the western Cascades.

Figure 8.17
Photo of the character of the Cispus AMA looking to Tower butte. Source: Tom Savage, US Forest Service.

Figure 8.18
Another view of the Cispus AMA showing scrubby summits to the hills as the result of harsh weather, grazing and fire. Hamilton Northwest. Source: Tom Savage, US Forest Service.

Many of these older, complex stands were concentrated in riparian zones and wet areas, particularly at lower elevations. In higher elevations, open areas were maintained in grass and shrub stages by wildlife grazing, as well as by burning of berry patches and meadows by Indians over many hundreds of years.

Large fires swept over much of the Cispus AMA in 1918, with some of the burned areas planted by the Forest Service beginning in 1920, and continuing into the 1930s. Modern logging and forest management began in the 1940s. By 1993, a total of about 19,000 acres (14% of the area) had been harvested, primarily as clearcuts. In addition, about 3,000 acres of commercial thinning in younger stands had been done over this period. All the clearcuts have been reforested, either by natural seeding or hand planting.

A combination of timber harvest and wildfire suppression has led to a vegetation pattern and structure that differs quite a bit from that of the past. Much of the timber harvest has occurred in older stands within riparian areas that had survived previous fires. As a result, the location of remaining late successional stands is generally higher on the slopes. Dispersed clearcutting has resulted in many small (10 to 20 hectare) rectangular patches of early and mid-successional forest within remnant patches of late successional forest.

Table 8.1 Land-use zones at Cispus

Management allocation	Acres
Deer and elk winter range	6,382
Mountain goat summer range	352
Mountain goat winter range	7,864
General forest	68,900
Developed recreation	462
Roaded recreation	18,557
Unroaded recreation	17,879
Visual emphasis	9,637
Wild and scenic river	8,097
Special interest	130

Source: US Forest Service

Management direction

The Gifford Pinchot National Forest Land and Resource Management Plan (GP Forest Plan) was approved in 1990. It was amended in 1994 by the Northwest Forest Plan. As with all national forest plans, the end result is essentially a complex zoning strategy that subdivides the entire forest into individual management areas, each associated with a land-use emphasis. The lands within the Cispus AMA have been assigned to ten separate land-use zones.

In addition, there are two small parcels of privately owned land included within the boundaries of the AMA that are not under the jurisdiction of the Forest Service. The Northwest Forest Plan overlay included designation of riparian reserves as buffers to all water bodies and flowing streams.

Landscape design

Subsequent to formation of the AMA, the Forest Service spent about two years getting organised, conducting outreach, developing a watershed assessment and forming a broad-based stakeholder group to brainstorm management strategies. This process reached an impasse, after which the decision was made to experiment with a long-range design, generally following the process laid out in the *Forest Landscape Analysis and Design* publication.

The broad objective for the Cispus AMA was to integrate timber production with maintenance of late successional forests, healthy riparian zones and high-quality recreation. More detailed objectives to guide the landscape design were developed during the autumn of 1995. Several design themes emerged, as follows:

- Maintain late successional forest structure sufficient for dependent species
- Ensure that high-quality recreation areas are protected
- Maintain healthy riparian areas, to protect water quality and associated aquatic species
- Provide a predictable and ecologically sustainable timber yield
- Encourage experimentation and flexibility
- Provide opportunities for further interaction between communities and land managers.

Over a concentrated period of a few weeks, project landscape architects and an interdisciplinary, interagency team worked with ideas and suggestions generated in a public workshop setting to produce an initial concept design. This concept was then reviewed by a steering committee, which recommended further analysis, development of quantitative measures and inclusion of a wider range of design options. Subsequent results were presented at a public workshop in December 1996, which generated further recommendations and ideas. The project team incorporated these ideas into the design options, which were presented to the AMA Steering Committee in March 1997 for review. The committee recommended that the team consolidate similar options, re-frame management objectives in terms of ecosystem management, define learning objectives and analyse the consequences in terms of acres to be treated.

The steering committee also recommended participation by the Southwest Washington Provincial Advisory Committee (PAC), a broader team dealing with sub-regional issues, to better integrate the design into the Southwest Washington province. This 28-member committee, composed of representatives from various public, county, state and federal agencies, assists in the implementation of the Northwest Forest Plan. In July 1997, one design option was selected for the Cispus AMA by the Gifford Pinchot National Forest Supervisor, based in part on recommendations from the PAC.

Landscape design units

The selected landscape design includes the following landscape design units:

- old growth forest
- riparian reserves
- managed habitat
- habitat development
- managed mosaic
- natural mosaic and lodgepole pine.

Old growth

These are areas where the intent is to retain or develop older forest with large-diameter trees. The basic goal is to maximise the presence of stands 170 years old or more, with average diameters of 30 inches or greater if environmental conditions allow. Important structural elements include standing dead as well as living trees, coarse woody debris, downed logs in various states of decay, patches of shrubs and multiple canopy layers. Frequent canopy gaps and small openings provide structural and species diversity. Dominant trees include Douglas fir, western hemlock, western red cedar and grand fir. The overall goals are to maintain and restore old growth forest habitat and connectivity, and emphasise low-impact recreation. Managers will not schedule regeneration harvest in old growth design units, but may use thinning and selective harvests to restore or maintain structure.

Riparian reserves

Riparian reserves were designed around all permanently flowing and intermittent streams, lakes, ponds, wetlands and unstable or potentially unstable areas. This design unit includes the body of water itself, the inner gorges, all riparian vegetation, 100-year flood plains, mapped landslides and high-risk landslide areas (i.e. steep, concave headwalls). Widths of riparian reserves are based upon ecological and geomorphological factors associated with different water bodies. These widths are designed primarily to provide a high level of fish habitat and riparian protection, and thus take account of the interaction of vegetation and the water body. Generally, reserves parallel the stream network, but may also include nodes necessary for maintaining hydrological, geomorphological and ecological processes. Vegetation management is limited to restoration and road removal.

Managed habitat

These are broad areas characterised by a mix of young, middle-aged and older forest stands. Composition varies across a range of elevation bands. Small man-made and natural openings are fairly common, so the forest has a "patchy" appearance. The objective is to integrate timber harvest with the maintenance and restoration of late successional, riparian and aquatic habitats. Timber harvest is designed to maintain high canopy closure and structural diversity. High numbers of large green trees, snags and down woody debris (biological legacies) are retained within regeneration harvests. Road densities are to be reduced through decommissioning, with priority given to high landslide risks. Moderate volumes of timber production are expected from managed habitat areas. In essence, these areas are fairly aggressively managed old growth and mature forests.

Habitat development

Habitat development design units are similar to managed habitat ones. However, openings from timber harvests are generally smaller in size than in managed habitat. The intent is to mimic small, naturally occurring openings of only a few acres in size. There is increased emphasis on the restoration, maintenance and connectivity of late successional forests, including mountain goat winter range. A fairly active thinning regime is anticipated in order to develop structural features characteristic of older forest types. Silvicultural treatments include manipulation of the distribution and abundance of coarse woody debris, down logs and snags and the creation of cavities for dependent species. Commercial thinning is initiated at about 40–60 years of stand age, and continues at 20-year intervals until the stand reaches 140 years of age. Openings from regeneration harvests are small in size. Moderate volumes of timber are expected.

Managed mosaic

Managed mosaic areas are subject to more intense human activity, with vegetation alteration planned at large scales. These areas include an abundance of early and mid-successional stands, with fast-growing young trees. The patterns of stands with different tree ages are shaped to reflect the underlying landforms, with old growth islands in riparian or specially protected areas. The objective is to design timber harvests that mimic large-scale natural disturbances, thus avoiding the fragmentation that resulted from dispersed harvests. Areas subject to timber harvest will include retention of biological legacies, including large green trees, snags and down woody material. Stands of varying ages are maintained across the landscape in amounts and patterns likely to be found in a natural forest.

Forest Structure and Landform Compatibility

● Optimum Forest Structure   ◉ Possible Forest Structure

Landform	Old Growth Natural	Old Growth Managed	Continuous Forest	Perforated Forest	Mosaic Natural	Mosaic Managed	Unforested
Ridgelines > 4,000'					●	◉	◉
Steep or Unstable Slopes < 4,000'	◉	◉	◉	●	◉		
Mod. Steep Slopes < 4,000'	◉	◉	◉	◉		●	
Lodgepole Flats			◉		◉	◉	
Alluvial Fans	◉	◉		◉	◉		
Valley Bottoms and Large Streams	◉	◉		◉	◉ (Old Growth Dominant)		

Figure 8.19 A table showing how desired future forest structures relate to the landform types at the Cispus AMA.

Natural mosaic and lodgepole pine

Natural mosaic design units are located in the highest elevations of the AMA, and include high meadows, sub-alpine parkland and alpine lakes. Forests naturally thin out with increasing elevation, due to a short growing season and a deep, long-lasting snowpack. Scattered clumps of higher-elevation tree species, along with open heath and meadows, occur along upper slopes and ridges. These large-scale, natural openings are visually attractive, and offer long, sweeping vistas. Lodgepole pine design units occur in similar settings, and sometimes intermingle with natural mosaics, but are more dominated by closed canopy lodgepole pine and mountain hemlock forests. These tend to be concentrated in the southeast corner of the AMA. The objective is to maintain high-elevation forest habitats. There are no scheduled timber harvests or road-building activities. Primitive recreational sites are maintained. Fire is the prime management tool in both natural mosaic and lodgepole units. Prescribed burns are anticipated on an experimental basis. Natural fires will be either suppressed or allowed to burn within containment areas on a case-by-case basis that relies on careful monitoring.

Cispus design summary

The selected landscape design is expected to result in about 76% of the total AMA being in a late successional forests condition at any given time (although this number will not be reached for many years into the future). Gradually, old growth habitat will be allowed or encouraged to develop in old growth and other design units where it is currently lacking. As old growth forest develops, it will become a connected matrix, linking the north fork and mainstream Cispus Rivers to the Woods Creek Late Seral Reserve to the west, and east to the Goat Rocks Wilderness area. Cispus area managers believe that one consequence of developing a high percentage of late successional forests may be an increased risk of stand-replacing fires. This risk could be reduced if thinning treatments are in part designed to reduce fuel ladders. Reduction of road densities and the restriction of ground-disturbing activities within riparian areas are expected to result in a gradual decrease in sediment delivery to streams. Water quality, channel conditions, riparian conditions and flood plain, meadows and wetland conditions will recover in part through restoration, and in part at a natural rate.

Figure 8.20 Map showing the long-term management plan for the Cispus AMA. Source: US Forest Service.

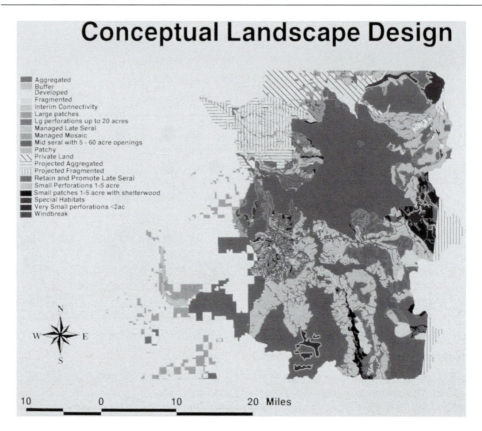

Figure 8.21 Sketch illustrating forest management ideas for part of the Cispus Adaptive Management Area in Washington State. Illustration by Tom Savage, US Forest Service.

Probable Sale Quantity (PSQ) is an estimate of the quantity of timber (in thousands of board feet) that can or might be sold from a given land area. The plan anticipates various levels of harvest, from low impact/high retention to high impact/low retention. Pre-commercial thinning is anticipated for stands that are 10 to 20 years of age. Commercial thinning may be used twice or more during the life span of a given stand.

Long-term PSQ represents the potential harvest once the landscape reaches the vegetation pattern and structure illustrated in the design. Short-term PSQ is the expected harvest level from now until the time that the design is realised (50–100 years).

220

Table 8.2 Areas available for harvest at Cispus AMA

Total acres available for scheduled harvests	37,000
Average rotation length (years)	116–143
Acres light retention regeneration harvest per decade	800–1,800
Acres medium retention regeneration harvest per decade	100–200
Acres high retention regeneration harvest per decade	50–100
Acres regeneration harvest per decade (subtotal)	950–2,100
Acres pre-commercial thinning per decade	800–1,900
Acres commercial thinning per decade	6,100–8,500
Estimated short-term annual PSQ (mbf)	7,000–9,000
Estimated long-term annual PSQ (mbf)	12,200

Source: US Forest Service

The range of values in the above analysis anticipates a 50-year moratorium on the harvest of stands that are 170 years or greater in age, until 80% of the old growth design areas have attained old growth characteristics.

The landscape design for the Cispus AMA demonstrates the potential for using design to resolve very complex and controversial issues surrounding continued management of natural forests that have been impacted by dispersed clearcut logging over many decades. While the sustainable harvest level anticipated is low compared with recent historic outputs, it is a reflection of the high levels of ecological protection now governing national forests in the Pacific Northwest.

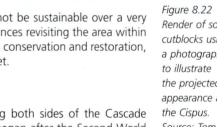

Figure 8.22
Render of some
cutblocks using
a photograph
to illustrate
the projected
appearance at
the Cispus.
Source: Tom
Savage, US
Forest Service.

It is fair to argue that the high level of old growth retention will not be sustainable over a very long time frame, given the likelihood of large-scale natural disturbances revisiting the area within the life of the plan. However, given the present focus on old growth conservation and restoration, the managers had little choice other than to strive for a high target.

Case study: The Collowash/Hot Springs watershed

Mt Hood National Forest covers over 400,000 hectares straddling both sides of the Cascade Mountains in northwest Oregon. Large-scale commercial logging began after the Second World War, with the heaviest cutting taking place in the robust old growth Douglas fir forests of the Clackamas River Basin. Generally, logging started at the lower end of the watershed, near the town of Estacada, and proceeded upriver as roads were extended. All of the Clackamas Basin lies squarely within the range of the Northern Spotted Owl (*Stryx occidentalis*). The following description of a landscape design for the Collowash and Hot Springs Fork watersheds is summarised from an unpublished manuscript of the Mt Hood National Forest.

The 97,000 acre (40,400 ha) Collawash and Hot Springs Fork watershed lies at the southern end of the Clackamas Basin (see map). Typical of other watersheds of the western Cascade Mountains, the topography is rugged and steep, with elevations ranging from 1,480 feet at the mouth to nearly 6,000 feet within the Bull of the Woods Wilderness. Climate is temperate, with heavy rain and snow through late autumn, winter and spring, followed by warm, dry summers. The winter snowpack starts at about 3,000-foot elevation.

Little is known about early landscape history of the Collowash area. There is archaeological evidence of Indian use, including numerous fishing sites. High ridges may have been burned off to facilitate hunting, travel and huckleberry gathering. Wet meadows may also have been

Figure 8.23
Map showing
the location of
the Collowash/
Hot Springs
project area in
Oregon.

Figure 8.24
A view of the
existing
fragmented
landscape in the
Collowash area.
Source: US
Forest Service.

occasionally burned. A general map of early twentieth-century forest conditions was pieced together through ecological detective work that included study of 1940s-era aerial photography, panoramic photos taken from fire lookouts in the 1930s and early ranger reports. From this, it was determined that the vast majority of the area was mature or old growth conifer forest. Much of this forest was 200–300 years of age, which indicates that large-scale stand replacement fires had taken place in the early eighteenth century. This is consistent with estimates regarding fire regimes for the area, which tend to range from 100 to 430 years for stand replacement burns.

Higher elevations and north-facing slopes tend towards lower fire frequency and larger burned areas, up to tens of thousands of acres at a time, leading to fairly large, uniform conifer patches. In lower elevations and on south-facing slopes, burns tended to be more frequent (25–150 years) and smaller in scale. The overall landscape condition would be much more patchy and fine-grained.

The Collowash area was first opened to logging in the 1950s. Eighty-seven miles of road were built in the 1950s, 144 miles in the 1960s, 79 miles in the 1970s and 49 miles in the 1980s. By the end of this period, concern over the fate of the Northern Spotted Owl and many other species of wildlife dependent on mature and old growth forest resulted in a political crisis that led to a virtual shutting down of further logging and road building.

The Northwest Forest Plan, developed under the Clinton Administration in the early 1990s, sought to put an end to the dispute by setting aside large "reserves" of mature forest. "Watershed Analysis" was a new tool that required local forest managers to conduct a thorough study of conditions to set the stage for everything from further commercial logging to restoration forestry. The Collowash-Hot Springs Watershed Assessment, completed in 1995, included a conceptual-level forest design as a way to visualise future forest patterns and prioritise projects.

Much of the area was included within "late seral" reserves. This meant that the management goal had shifted from emphasis on timber harvest to one of protection and restoration. But these reserves are not pristine old growth forest. They tend to have high road density and a high degree of fragmentation, with numerous 40–60 acre patches of young to mid-seral regeneration punched into remnant mature and old growth stands. In some areas, the proportion is about 50/50 between young and old forest. The Northwest Forest Plan envisions silvicultural operations within stands younger than 80 years, aimed at moving them towards mature conditions by reducing density to speed growth and/or provide space for understorey regeneration. In all, about 70% of the entire watershed is contained within late-seral and riparian reserves (which have similar goals). The remaining 30% is intended to continue to be managed for commercial forestry. The pattern and structure of these areas can range from uneven aged, multi-storey to large-scale clearcutting with plantations.

However, there are further complications. The Collowash-Hot Springs watershed is believed to have the most unstable terrain on the entire Mt Hood National Forest. This limits road construction and large-scale clearcutting due to potential stream impacts. Much of the area is also highly susceptible to catastrophic windthrow, which tends to be exacerbated at the downwind edges of clearcuts. Since the old growth reserves of the Collowash are expected to serve as a "core area" for wildlife, a high level of reserve edge protection is warranted. Aesthetically, the fragmented clearcut pattern is far below current forest plan standards.

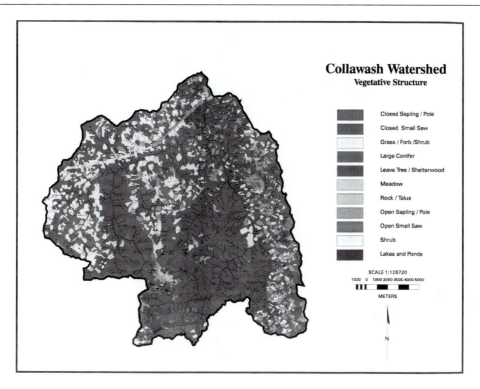

*Figure 8.25
A map showing
the existing
vegetation types
and the
consequent
degree of
fragmentation of
the Collowash/
Hot Springs area
of Mt Hood
National Forest,
Oregon.
Source: US
Forest Service.*

And finally, regeneration along high ridgelines is problematic, particularly in the southeast portion of the watershed.

These characteristics were recorded onto an opportunities and constraints map. Design objectives were essentially drawn from the forest plan, supplemented by team findings and informal consultations with interested citizens. In summary, the objectives were:

- Restoration forestry within young stands in reserve areas
- Small patch forestry within remaining commercial areas to minimise landslide and windthrow risks
- Improvement to aesthetics by softening or rounding rectangular openings where possible
- Decommissioning of roads no longer required for recreation access, fire management or silviculture
- Reintroduction of fire within the wilderness area to re-establish historic huckleberry openings.

From these objectives, and mindful of the landscape opportunities and constraints, a long-range concept plan identified two large areas where contiguous old growth forest structure would be restored through active and passive means. The ridges and slopes that bordered these areas would be managed as "perforated" patches, meaning that a 0.5–2.0 ha clearcuts would be dispersed within young to mature forest stands. This pattern would mimic small-scale windthrow disturbances, while preventing triggering of large landslides or catastrophic wind damage.

The long-range concept has a 200-year time frame, and does not provide much guidance for short-term projects. Consequently, the next step was to develop a short-term concept to guide the next 10 to 20 years of forest practice. This concept anticipated small-scale timber harvest within young plantations and at the edges of existing clearcuts. Timber outputs would be modest and secondary to the goal of "re-shaping" the landscape pattern to a more natural mosaic. Fire would be reintroduced along high ridgelines, and redundant roads would be decommissioned.

*Figure 8.26
The target
landscape
structure for the
Collowash/Hot
Springs project
showing a much
more connected
forest matrix and
different
structural types
related to soils
and landform.
Source: US
Forest Service.*

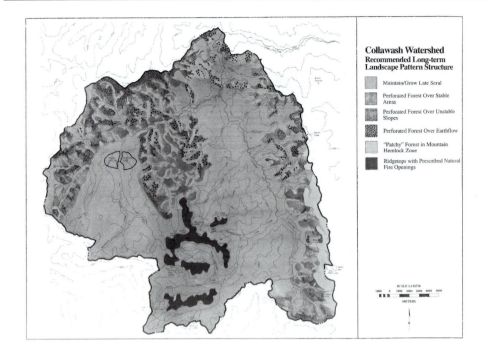

Collawash Watershed
**Recommended Long-term
Landscape Pattern Structure**

- Maintain/Grow Late Seral
- Perforated Forest Over Stable Areas
- Perforated Forest Over Unstable Slopes
- Perforated Forest Over Earthflow
- "Patchy" Forest in Mountain Hemlock Zone
- Ridgetops with Prescribed Natural Fire Openings

SCALE 1:126738

METERS

RECOMMENDED LANDSCAPE PATTERN

PATTERN TYPE	LANDSCAPE OBJECTIVE	INCLUDES:	VEGETATION MANAGEMENT	PATTERN ILLUSTRATION
Retain and Promote Late Seral Forest ■ Existing late seral ▨ Existing early seral ▨ Existing mid seral	Aquatic Habitat Protection Late Seral Connectivity	• Late Seral Reserves • Riparian Reserves • Owl Activity Centers • Rare and unique habitats	• Thin plantations to produce large trees • Accelerate regeneration of conifers • Reduce risk from fire insects and disease • Create snags and coarse woody debris • Plant unstable areas • Release young conifers from overtopping hardwoods in Riparian Reserves • Reforest hardwood stands with conifers in Riparian Reserves	
Interim Retention of Late Seral Forest ▨	Interim late seral connectivity	• Recommended blocks of late seral habitat from Watershed Analysis • District Connectivity Net		
Retain Forested Conditions with Small Perforations to maintain root strength ■	Aquatic Habitat Protection Landform Stability Timber Production	• 50% slopes on unstable landforms	• Thin but maintain root tensile strength • Create small regeneration openings	
Aggregate Patches on Stable Ridges ▨ Primarily early seral ■ Primarily mid seral	Timber Production Create Early Seral Habitat Increase Patch Size Decrease Patch Frequency	• Matrix allocation • Mosaic habitat along Fish Creek divide	• Retain 15% of live trees and snags singly and in patches in timber sales • Mimic natural fire caused openings • Create early seral habitat	

*Figure 8.27
Some of the silvicultural models used in the Collowash plan.*

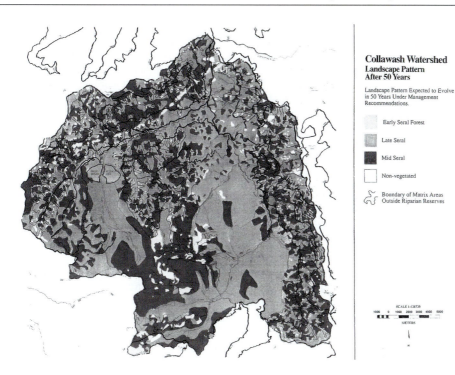

Collawash Watershed
Landscape Pattern
After 50 Years

Landscape Pattern Expected to Evolve
in 50 Years Under Management
Recommendations.

Early Seral Forest

Late Seral

Mid Seral

Non-vegetated

Boundary of Matrix Areas
Outside Riparian Reserves

SCALE 1:126730

1000   0   1000   2000   3000   4000   5000

METERS

*Figure 8.28
A map showing
the anticipated
structure after
the first 50 years
of plan
implementation
for the
Collowash/Hot
Springs project.
It demonstrates
progress in
reducing the
fragmentation
and increasing
the proportion of
stands in older
age classes.
Source: US
Forest Service.*

Finally, in order to provide a gauge of sustainability, a simple conceptual model of how the landscape would develop after 50 years was created. Interestingly, this model found that a very large portion of the land would be in a "mid-seral", or 50–80-year-aged forest structure. At this stage, future managers would have an option to generate fairly substantial amounts of timber strictly from thinning, with no further need to establish patchy openings. Essentially the forest matrix will have been woven back together, although it will be many more years before mature and old growth forest reaches the proportion that existed in the early twentieth century. Furthermore, this model did not account for potential stochastic disturbances that probably will exert their own influence on forest pattern and structure. Nevertheless, the combination of these three time frames has provided forest managers and the public with a much clearer picture of how the landscape ecology of the Collowash could take shape under the new "ecosystem management" paradigm.

*Figure 8.29
Map showing
the location of
the Little
Applegate
design project in
Oregon.*

Case study: The Little Applegate watershed

Background

The Little Applegate Forest Design process was initiated in the autumn of 1996 by the Rogue River National Forest. All of the federal lands within the 500,000 acre (208,300 ha) Applegate watershed were incorporated into an AMA as part of the Northwest Forest Plan of 1993 (see above). This plan calls for development and testing of progressive forest management practices that provide a broad range of values. Experimentation in planning, project implementation, community relations and ecological monitoring are encouraged.

A technical analysis of the Little Applegate watershed, completed in 1995, had stopped short of providing clear ideas or "vision" with regards

CANADA

Washington

Oregon

Idaho

Little Applegate Watershed

California   Nevada

to the future management of the landscape pattern and structure. Forest managers wanted to have a clearer picture of how the landscape should look and function across the entire watershed. They wanted this picture to cover the federal and private lands if possible. They felt that such a picture could be used to guide short-term projects and help establish priorities regarding silvicultural treatments, restoration projects and fire management. They also wanted to use the design process to help build better relations with the local community living in and around the Little Applegate watershed, as well as with other area stakeholders.

Environmental context

The Little Applegate watershed is located south and a little west of Medford, Oregon, in Jackson County. Overall it comprises over 72,000 acres (30,000 ha), of which about two-thirds are in federal ownership (US Forest Service and Bureau of Land Management), the remaining one-third private. The private lands include large timber industry holdings in the mid- to upper watershed, and ranches, small farms and forested homesteads in the lower end. The pattern of land ownership follows a township, range and section line grid that has been superimposed onto an irregular mountain topography.

The Little Applegate River is part of the Applegate watershed. It joins the Applegate River near the unincorporated town of Ruch. The Applegate in turn joins the Rogue River near Grants Pass and flows some 75 river miles to the Pacific Ocean.

The landscape of the Little Applegate watershed is quite diverse, ranging from a low elevation of about 1,466 feet at the river mouth up to 7,418 feet at the Siskiyou Mountain crest. The topography is rugged, and includes high mountain glacial features, steep slopes, dense networks of streams and gentle valley floors. Vegetation communities include farm fields, pastures, bunchgrass prairies, oak woodlands, chaparral-type brush fields, pine–oak woodlands, mixed conifer forests and high mountain sub-alpine parklands.

The diversity of landform and vegetation leads in turn to a great diversity of wildlife habitat, including as many as 272 species of native vertebrates. Some of the more well known of these include chinook and coho salmon, steelhead, trout, spotted, flammulated and great grey owl, black-backed and white-headed woodpecker, northern goshawk, red tree vole, western pond turtle, Siskiyou mountain salamander and several important bat species. A number of these are either listed, or are candidates for listing under the federal Endangered Species Act.

Figure 8.30
A view of the landscape at Little Applegate, showing bald summits to the hills and forested slopes, reflecting the climatic conditions and disturbance history. Source: US Forest Service.

Figure 8.31
Another view of landscape at Little Applegate, showing the valley bottom landscape. Source: US Forest Service.

Design issues and objectives

An *ad hoc* citizen and agency task force met over a two-day period in early December 1996 to develop a list of broad issues and objectives to guide the watershed design process. The issues and objectives were drawn from three main sources. Firstly, a watershed analysis completed by the Forest Service and Bureau of Land Management in 1995. Secondly, the document "Words into Action" by Kenneth Priester of the Rogue Institute for Ecology and Economy. This was a comprehensive study of local values and concerns. Thirdly, information came from the direct experience of those on the task force. It is important to note that a key assumption of all those on the task force was that the landscape objectives had to be consistent with existing land-use policy direction for federal and private lands. In other words, even though some members of the task force did not support all existing policies, they agreed to try to work within the existing policy framework and see where this might lead, rather than to spend time debating policies that this group could not easily change.

The watershed analysis had focused on technical issues related to fire hazard and risk, water quality and quantity, wildlife habitat and vegetation. Recommendations from this analysis were restated and incorporated into the design issues and objectives. The "Words into Action" document provided a social values framework for the entire Applegate watershed. Essentially it documented what the local citizens wanted, and how they felt about the landscape and natural resources of the area. The members of the *ad hoc* team were some of the most knowledgeable about the ecology and social dynamics of the immediate area.

The issues and objectives developed at the December meeting were presented to local citizens and federal land managers, and were agreed to by all present. These objectives incorporated restoration, but also included economic and other considerations. There were restoration objectives for nearly every landscape community within the watershed. These were refined at a two-day workshop several weeks later, and are described below.

Landscape pattern and structure

It was agreed by all that a key issue was the large-scale changes in landscape pattern and structure that had occurred over the past 150 years as a result of displacement of the Indian people and subsequent settlement by Euro-Americans. These changes had in effect resulted in:

- Less acreage, and more fragmentation of mid- to high-elevation old growth conifer forest
- Loss of large, old open-grown pine and oak stands due to fire suppression
- Loss of bunchgrass prairies to brush encroachment (also due to fire suppression)
- Damage to streams and riparian woodlands as a consequence of road building, agriculture and logging
- Severe degradation of high alpine meadows as a result of historic sheep grazing
- Overall crowded and hazardous forest conditions due to an abundance of small-diameter conifer trees, particularly Douglas and grand fir.

Initially, the *ad hoc* task force identified 18 distinct landscape patch types that would form the basis for the design. They recognised that these only represented broad "headline" descriptions of landscape types, since the Siskiyou Mountain region is noted for its almost infinite variability of vegetation composition. Consequently, within each patch headline there would be a lot of variety of "fine print". For example, "old growth forest" patches would probably have small openings, groups of hardwoods and other variations related to micro-site conditions or disturbances. Nevertheless, they developed a list of patch types that corresponded to broad composition and structure variations that are common to the area. These are described as follows:

Pastures and fields

These are the patchwork of cultivated areas that generally are found in the lower valley, along the main streams. Although they are "introduced" patch types that are primarily composed of non-native plants, they are very important culturally, aesthetically and economically. The team felt that while pastures and fields ought to continue as part of the landscape, some portions should give way to riparian forest along streams. Also, improved management of weedy species is needed in some fields.

Brush mosaic

This patch type refers to the shrub-dominated vegetation found along the dry, south-facing slopes primarily at lower elevations. These are fire-dependent plant communities that include manzanita, *Ceanothus*, bunchgrasses and herbaceous species. They have high value for some wildlife species in early- to mid-seral stages. As brush patches age, they become unpalatable and quite flammable, posing risks to nearby homes and forests. The Bureau of Land Management has begun active management of brush patches by using a combination of hand clearing and fire. The team felt that this type of management should be continued and adopted by private landowners as well where possible.

Perennial grasslands

These are native bunchgrass openings found at very dry sites throughout the watershed, but mostly on south-facing slopes in the lower mountains. They are important as a landscape diversity element. They also help "absorb" fire, and are important aesthetically. The team believed that some prairie could be restored from present brushy areas, perhaps building out from existing grass patches.

Riparian woodlands

These forests are mixed conifer and deciduous species, including: willows (*Salix* spp.), cottonwood (*Populaus trichocarpa*), bigleaf maple (*Acer macrophyllum*), Douglas fir (*Pseudotsuga menzisii*), sugar pine (*Pinus lambertiana*) and Ponderosa pine (*Pinus ponderosa*). Riparian woodlands are found along stream terraces in the lower watershed, and along main streams reaching up the main tributaries. This patch type has lost ground to farm fields, housing and roads over the years. It is now the focus of restoration work due to the State of Oregon's salmon recovery plan and overall concerns regarding water quality. The team agreed that it is a very important priority to re-establish riparian forest wherever possible. A good strategy might be to establish "stepping stones" between existing forest patches.

Sub-alpine forest mosaic

This patch type is found in the upper elevations, primarily associated with the Siskiyou Mountain crest and glaciated headwater basins. Much of this area is characterised by forest and meadow stringers whose location and distribution are influenced by seasonal snow accumulations, soil conditions and past land uses, particularly burning and grazing. Some portion of the open character will recover to forest over time as grazing pressures are reduced. The team agreed that this recovery process should be encouraged, but that natural fires should be allowed to burn in order to maintain the mosaic character over time.

Pine–oak savanna

This is open-canopy woodland found mostly in the lower elevations, but can reach fairly high on south-facing ridges. It is a patch type that is much diminished from its former range, due to the absence of periodic ground fires. Many areas of former savanna have given way to brush, or in some cases to young Douglas fir forest. The team agreed that the extent of savannas should be increased, mostly at the expense of brush lands and dense forests on dry sites.

Old growth conifer forest
This is the characteristic mature forest found in the middle elevations of the watershed, primarily in the "white fir" vegetation zone. It contains large, old Douglas fir and sometimes Ponderosa pine in the overstorey, with white fir (*Abies grandis*) in the understorey. It can also include some hardwoods. There are numerous snags and downed logs, providing important habitat for several sensitive or endangered species, including spotted owls. Maintenance of a minimum amount of this patch type is a requirement of the Northwest Forest Plan that governs federal lands in the watershed. The team felt that this patch type should be located to be as contiguous as possible, and should be concentrated in the middle elevation portions of the watershed.

Dark, dense forest
This generally describes forests that are primarily coniferous, have a single canopy layer, with fairly small stems that are very close together. This patch type is presently widespread in the watershed. For the most part, dark dense forests have resulted from a combination of logging, turn-of-the-century burning by gold miners and decades of fire suppression since that time. They are felt to be vulnerable to disease and fire, and are of marginal value for many wildlife species. While this patch type has probably always been a part of the landscape, and has some important values, its increased extent poses a threat to the overall health of the watershed, in that it is subject to hot, crown fires that are difficult to control. As an illustration, over 10,000 acres in and around the Little Applegate watershed burned quite severely in 2000. The team agreed that this patch type should not be designed into the overall pattern, but should be viewed as a transitional stage towards other, more desirable patches.

Serpentine/Jeffery pine
This is a patch type that is specific to serpentine soil. It is an open, pine-dominated plant community with a number of important endemic plants. It is not a patch type that can be "designed" into the landscape, in that it can only occur where very specific conditions allow. Where these conditions are found no other patch type can grow.

Industrial forest
This patch type is found on the large areas of private timberland in the middle to upper part of the watershed. Typically, it is composed of young, small conifer trees, mostly white fir, with a dense network of skid roads. While clearcutting is generally not practised on industrial forest land in this watershed (due to regeneration problems), these areas are managed intensively by removing merchantable-sized trees at frequent intervals. The team felt that this patch type is generally detrimental to most important values, and should be replaced by one or more of the following four options.

Below are descriptions of forest patch types that are not widespread in the watershed at the present time, but represent important elements that can be created and maintained on commercial forest land. They can be managed at a profit by private landowners or public managers, while also benefiting native wildlife and protecting water quality.

Multi-layer canopy forest
This is similar to old growth, but lacking in snags, down wood and very old trees. It is primarily coniferous, but contains an important hardwood component. It is an intensely "managed" forest that is maintained through frequent harvest of individual trees. While it can provide habitat, scenic and watershed values similar to old growth forests, it requires a relatively high density of roads to maintain. These roads can cause other watershed problems. If located in the "white fir" zone, this patch type may gradually lose Douglas fir and pine as components. It would thus be most appropriate in the dry Douglas fir or Shasta red fir (*Abies magnifica*) zones.

Two-layer forest
This patch type refers to an intensely managed forest where both an upper and lower canopy are retained. As the upper layer is thinned and opened, the lower layer develops and gradually replaces the one above. Some private woodland owners within the watershed are already maintaining this forest patch type. It is primarily composed of Douglas fir and Ponderosa pine, but can include hardwoods and other conifer species. It allows for a high rate of harvest, but can be expensive operationally. The best areas for this type are below 3,800 feet, in the dry Douglas fir zone.

Perforated forest
This is a forest that is mostly closed canopy, but has frequent small openings (half to two acres) that allow for regeneration and diversity. In silvicultural terms, it translates to "group selection" forestry. It works very well for many wildlife species, including spotted owls, if the amount of small openings do not exceed 20% of the total forest area, and if the forest in between the openings has complex structure, including snags and down logs. It is fairly efficient operationally, while also protective of water quality and scenery. The best locations are in both Douglas fir zones, and in the Shasta red fir zone. This type can be used as a transition towards multi-layer patch types.

Patchy forest
This patch type is characterised by clearcuts of 5–25 acres within a forest matrix. It can be valuable for edge-dependent forest species, particularly deer and many raptors. It works best in forest types that are moist enough where regeneration is not a problem, particularly the white fir zone in the mid-elevations of the watershed. It should not be designed where erosion hazard is high (i.e. shallow granitic soils on very steep slopes). This patch type does allow efficient, economic timber harvest and regeneration of tree species not tolerant of shade (i.e. Douglas fir and pine).

Matrix

A key challenge in designing restoration of the Little Applegate watershed was determining what the "matrix" should be. As described in previous chapters, a matrix is the most dominant, connected landscape patch type, and tends to control the key landscape flows. Because of the high level of landscape diversity of this area, no single patch type dominates the Little Applegate watershed. However, a broader heading of "mature mixed forest" (including all mostly conifer communities over 80 years of age) indicates that there is a matrix, mostly in the mid- to upper elevations of the watershed. The team felt this matrix could be improved by reducing fragmentation of mature forest, restoring dark dense patches to other conditions and restoring riparian forest.

Landscape flows

There are thousands of landscape flows that could be considered in a landscape as large and complex as the Little Applegate watershed. These could range from grass pollen and the tiniest insects up to the largest carnivores. The design task force chose to focus on 16 important "landscape flows" that should be considered in the design. These were selected due to ecological, social or economic importance, based on present policies and local knowledge about the area. Some of these flows should be enhanced by the design, others discouraged. These are as follows:

Fish
Fall chinook, coho salmon, summer steelhead, winter steelhead, trout and several other fish species inhabit some 50 miles of streams in the watershed. A number of these are declining or "at risk". The decline of fish in the Little Applegate watershed is a result of many factors, including some that are manifested far downstream, such as ocean condition fluctuations and over-harvesting. Nevertheless, there are a number of important factors that do influence the ability of the watershed to provide habitat for fish, particularly high erosion rates, hydrological changes, water withdrawals and diversions and loss of streamside vegetation.

Water

Water is, of course, the flow that most affects the status of fish. An important characteristic of water flows in this watershed, as is the case throughout much of southern Oregon, is that flows are highly variable with the season. Winter and spring flows are very high, and summer flows very low. This results from the fact that only 6% of the annual precipitation (on average) occurs from July through to October. Late spring and summer snowpack is essential in maintaining summer flows. Water withdrawals for irrigation in the lower valley, and diversions through Wagner Gap, are also critical factors.

Sediment

This flow has an important influence on the flow of fish and other species. Fine sediments from granitic soils in MacDonald Basin and elsewhere have been identified as important factors that have damaged spawning areas, raised water temperatures and generally degraded water quality in the lower watershed. While some erosion is entirely natural in any mountainous landscape, the pattern and structure of vegetation and roads has created conditions that have increased sedimentation over natural rates.

Deer

Deer, while not an endangered or sensitive species, are important both culturally and as an ecosystem element that provides food for predators. The Little Applegate watershed provides important winter range habitat in the lower elevations, and summer range in the sub-alpine areas along the Siskiyou crest and upper basins.

Amphibians

This is an important but highly variable group of species that make use of a wide range of landscape conditions in the watershed. They are very sensitive to water quality, riparian conditions and the presence or absence of complex structure in upland forest areas.

Human use/recreation

There is a "flow" of humans in the watershed strongly tied to road and trail corridors leading to favoured or special sites. This flow is influenced by streams, ridgelines, vegetation type and other landscape features.

Human use/economic

This flow relates to the access and removal of products from the landscape. The most important landscape product in the Little Applegate watershed has been timber over most of the twentieth century. In the nineteenth century it was gold, as well as grass for grazing animals. More recent trends have seen the increasing importance of a wide range of "wild craft" products, including floral greens, mushrooms and berries.

Spotted owls

This flow is strongly related to the presence and distribution of mature and old growth forest patches. The spotted owl is an "indicator" of the health of the entire old growth forest system, and as such should be retained and enhanced within the watershed, particularly on federal lands.

Great grey owls

This is a sensitive species that is tied to a wide range of forest and landscape conditions. Its overall presence is favoured by the maintenance of a very diverse landscape pattern, from old growth to sub-alpine to pastures and fields.

Weeds

This includes a wide group of introduced annuals and perennials (most notably star thistle) that become established in disturbed areas. Weed occurrence can threaten the health and productivity of several local plant communities, including perennial grasslands, pastures and fields, pine–oak savannas and the sub-alpine mosaic. This is a flow that should be discouraged, if possible, through the design.

Cattle and other livestock
This is an important economic and cultural flow, but one that is in decline for many reasons. It is a flow that is felt to have a negative effect on several natural patch types, particularly sub-alpine mosaic, pine–oak savanna, riparian forests and perennial grasslands.

Fire
This is a critical flow in that it is strongly influenced by landscape pattern and structure, and in turn has the power to shape this pattern and structure. Historically, the Little Applegate watershed experienced frequent but low-intensity fires, many of which were set by Indian people. Settlement by Euro-Americans and removal of Indians resulted in a period of very intense, land-clearing fires that augmented grazing and mining. Fire control over the past 50–90 years, combined with logging and forest recovery, has resulted in a landscape structure that is now very prone to high-intensity fires. High-intensity fires may risk habitat for many species, water quality, homes and property.

Snow
This is a "flow" that is very important in augmenting summer stream levels. A landscape structure and pattern that captures and retains snow for long periods would release water slowly over the spring and summer.

Snag users
This "flow" refers to a wide-ranging group of species that make use of snags, or standing dead trees, for either nesting or feeding. It includes bats, birds, mammals and insects.

Northern goshawk
This flow makes use of a wide range of forest patch types in the middle to upper elevations of the watershed. It particularly benefits from forest patches that include small to moderate openings that facilitate food sources.

Designing restoration of the Little Applegate Watershed

There were four overriding restoration issues that needed to be resolved in the design, as follows:

1. Restoring an old growth forest network
It was clear that the design would have to develop a workable, logical network of areas where old growth conifer forest would be retained and restored. This network would have to be designed to facilitate the important flows described earlier. The design attempts to organise the old growth into a connected east-west band that linked prominent gaps in the ridgelines. This old growth band was focused on Forest Service administered lands, but incorporated upland riparian and unstable slopes on private holdings. The result of this approach would be a fairly "organic" shaped old growth pattern that follows landforms rather than strictly adhering to property boundaries. One implication of this approach might be the need for land trades in strategic places between the federal government and private timber interests.

2. Fire management network
The threat of stand-replacing wildfire that can destroy homes and set back watershed restoration is very real. The design needed strategically to restore patch types that could block, or absorb, fire rather than magnify it. The result was to advocate restoration of large areas of "pine–oak" savanna along south-facing ridges in the central part of the watershed. These restored savannas

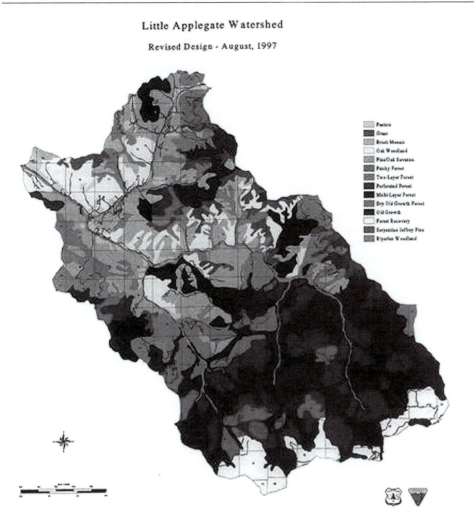

Little Applegate Watershed

Revised Design - August, 1997

Figure 8.32
The revised design for the Little Applegate watershed following consultation and discussion, depicting the desired future vegetation types in the landscape. Source: US Forest Service.

(located mostly on BLM administered lands) would "absorb" the effects of fires that might have their origins in the upper watershed (from lightning or slash burning) or in the lower watershed (from homes and farms). Thus, if successful they might prevent crown fires from sweeping across the entire watershed, no matter what the direction or origin. These savannas would also improve habitat for many species, and would return a good portion of the landscape towards its historic, "pre-European settlement" condition.

3. Stream restoration
Due to overlapping issues related to diminished water quality and the listing of several fish as threatened or endangered, it was critical to design a landscape pattern and structure that offered the best chance to improve water quality and quantity in the streams. This included augmentation of snowpack retention, reduction in fine sediment erosion and increased shading. Wide riparian

buffers in the lower watershed would be restored at the expense of existing pasture and fields. The ultimate extent of riparian planting would, of course, be determined by the private property owners in that area. In addition, redundant roads in the middle to upper watershed, particularly those on granitic soils, would be decommissioned to reduce sedimentation.

4. Integrating restoration with economic land use
The design recognised the importance of maintaining economic opportunities in the watershed. These include farming, ranching, timber removal and other forest products, as well as aesthetic quality. The Little Applegate watershed is not a wilderness landscape nor is it a park. It is and has been a "working" landscape for thousands of years. Present land-use policies on private and federal lands support this concept. The challenge to the designers was to incorporate economic uses as tools for shaping a pattern and structure that would result in an ecologically functioning landscape. One where the salmon spawn, the owls survive and fire is a friend rather than a foe. Economic uses were integrated with restoration objectives primarily by retaining significant areas of open, agricultural land in the lower watershed and by re-tooling forest practices in the middle watershed to reduce erosion and stitch the forest matrix back together. Time will tell whether or not this strategy proves to be successful.

Conclusions

This chapter has shown that ecosystem restoration is an important activity in many areas but that it also poses great challenges. Merely taking a fragmented or deforested landscape and planting native trees across it, while preventing soil erosion and restoring forest cover, may be insufficient. If the complete forest ecosystem is to be recreated it takes a good deal of detective work to determine the spatial pattern and structure of the landscape and the kind of ecosystem processes at work there. There is also the question of what is meant by the term "natural" and how knowledge of a historical landscape pattern affects the understanding of what would be natural today or in the future. Luckily, there are many tools now available and in development to help foresters and ecologists to restore forest ecosystems. It is now easier to see past the current landscape patterns and to envisage the forest as it could be. This is not a deterministic process but is based on knowledge of dynamic and often stochastic processes, some more predictable than others. In this chapter we have demonstrated the essential elements of the process of ecosystem restoration design and illustrated it with a series of case studies with which the authors have been directly involved. Some of these also included significant elements of original research as well as the development of practical plans capable of being implemented on the ground. Clearly, the subject of ecosystem restoration continues to develop, especially in the realm of modelling these dynamic patterns and processes, so that better understanding of the field should endure and that better results are achieved over time.

Chapter 9
Design in managed natural forests

Introduction

This chapter explores how the design process can be applied to forests of natural origin that are subject to commercial forestry. Inevitably, management is likely to focus on timber harvest followed by regeneration, but will also include conservation objectives. In the majority of temperate or boreal coniferous forests timber harvest is normally carried out by regeneration cutting (increasingly with legacy retention) followed by either natural regrowth or planting. This chapter commences with a discussion of how the design process as described in previous chapters applies in this situation and then illustrates the application with the use of three case studies, one in British Columbia in Canada, one in Nova Scotia in Canada and one in Finland. Each generally followed the process described in Chapter 5, although each also applied it slightly differently. In two examples, those from Nova Scotia and Finland, a community participation process took place at the same time, following some of the methods outlined in Chapter 4. These examples are presented in some detail so that readers can see exactly how the process can be followed. They are also comprehensively illustrated with maps, diagrams and sketches.

Management in natural forests

Until recently, harvest plans were almost exclusively driven by a combination of economic and practical factors, including the means of access to stands with a timber volume that made them economically worth harvesting, the fibre needs of local mills, the type of logging equipment and expertise locally available and the ability to construct access roads. Early harvesting in natural stands of North America used the practice of "high grading", where only the most valuable trees were cut. Some years later a second or third pass might follow as other species became valuable, or as the highest-quality trees were exhausted. In later years "progressive clearcutting" was the preferred logging method, characterised by all the merchantable timber in an area being systematically logged, probably by starting at the valley bottom and working gradually up the slope. High grading led to a gradual transformation of forests towards hardwoods and stunted or deformed conifers, yet substantial forest habitat remained. Progressive cutting led to vast areas of cleared forest with few, if any, remnant patches left standing. Assuming that regeneration was successful, such areas would eventually revert to an expanse of even-aged forest of almost plantation-like quality.

The practice of regulated clearcutting was first developed in Germany in the fifteenth century. It has since become the most widely employed form of silviculture in temperate forests, particularly in North America. When natural forests are clearcut, they are usually replaced with plantations, although this is not always the case. Always controversial, large-scale clearcutting became much less ecologically, socially and politically acceptable in most temperate forests by the late 1980s. Harvest plans began to be more tightly regulated, usually through the practice of dispersed patch clearfelling. The size of clearcut patches was limited in some jurisdictions, for example British Columbia. Individual clearcuts are usually separated by intervening stands of remnant forest, often of a minimum width, and these remnants cannot be felled until the original patch has been regenerated and reached a certain stand structure (sometimes a height is stipulated, such as two metres). Other new regulatory measures introduced in this period included the retention of forest buffers along some or all watercourses, and along the shoreline of estuaries, lakes and wetlands. These regulations have attempted to achieve a better balance between the desire to maximise wood production and minimise costs, with an attempt to limit ecological and social impacts by reducing the rate of harvest across the landscape. This administrative control has been applied more rigorously in some places than in others. With the more recent advent of third-party certification, some of these controls are incorporated into certification standards, but there is now more scope and incentive to develop integrated forest planning (see Chapter 1).

Regulatory-based forestry systems have their limitations. An important one is that they are not flexible enough to be tailored to variability in different forest ecosystems, where the patterns of stands and the dynamics of the forest through the agencies of natural disturbance and succession vary widely. One of the main problems to emerge from limiting clearcut size is fragmentation, as the forest pattern becomes a patchwork of small-scale stands with little or no connectivity provided by retained late successional forest. Another problem occurs when the balance of stand types tilts in favour of early and mid-successional stages at the expense of late successional ones, especially old growth or "antique" forest. Other problems include severe visual impact from the unnatural appearance of a series of geometric shapes scattered across the landscape, a scene that is especially obvious during the winter months. Road construction can also be a problem as greater lengths of road are usually needed to gain access to dispersed harvest units, with all the attendant risks of erosion, siltation of water courses and long-term maintenance requirements.

The recent emergence of ecological forestry (see Chapter 2) has created the opportunity to apply an approach to harvesting natural forests that can be more closely tailored to the natural dynamics of forest ecosystems. The essential goal of ecological forestry is to "mimic" natural disturbance dynamics in terms of scale and frequency, and to use legacies as a "coarse-filter" conservation strategy. Ecological forestry seeks to reverse the trend towards "homogenisation", or simplification of forest structure, and move towards retention of natural levels of diversity. The basis for ecological forestry is to understand how the natural system functions, particularly the role of disturbance and succession. Most natural temperate forests are shifting mosaics of variable-sized patches that differ in composition, age and structure. Legacies are the "carryover" elements within a stand, and across a landscape. In the Pacific Northwest, Jerry Franklin and others developed "New Forestry", where clearcut prescriptions in dispersed patterns are modified to leave "legacies" such as old trees, snags

Figure 9.1 Clearcuts on a mountainside near Clearwater, British Columbia, Canada. Such practices are no longer acceptable, though their legacy in the landscape will persist for a considerable time.

and fallen dead wood (termed "coarse woody debris") after harvest, in order to provide carryover habitats for wildlife as well as to achieve other objectives. Foresters in British Columbia and across the boreal forest of Alberta and Saskatchewan have also begun to embrace this approach. In Sweden this idea has been taken up and promoted extensively to the small private forest owners as well as to the larger companies. However, such approaches often failed to address the landscape scale, particularly across multiple ownership boundaries.

The place of design in managed natural forests

"Managed natural forests" is a category that covers a wide range. Projects that the authors have worked on include plans for large areas where there has never been commercial timber harvest, where there are no roads and where the forest is completely natural except for the effects of fire suppression. There are also areas where some limited timber harvest took place, perhaps loggers selectively removing the best trees or certain species from easily accessible areas, often some 40–50 years ago. The composition of the forest may have changed to some degree, but often only in a minor way and the area retains a mainly natural structure. Other forests are of natural origin and contain native species but have been managed more or less intensively for several decades or even centuries. In some instances, such as parts of Scandinavia or Central and Eastern Europe, the proportions of species have been changed and the forest structures simplified into a pattern of compartments. Finally, there are the fragmented forests found in parts of the Pacific Northwest and Canada, where strongly contrasting structural patterns caused by past clearcut logging activities present serious ecological problems related to fragmentation and excessive amounts of roads. These may need extensive restoration such as that described in Chapter 8, or else they can be incorporated into the type of design described here.

Figure 9.2 Harvest unit carried out using "New Forestry" practices.

The challenges presented to planners, designers and managers are different in each of these cases. Where the forest is in pristine or close to pristine condition it is possible to start with a full ecological analysis and to use this to develop a spatial pattern with a range of ecologically driven silvicultural models. With a forest that has been managed for a long time the main opportunities are to change the direction of layout and management towards a more ecologically derived future condition without breaking the general continuity of management.

Figure 9.3 Area where the forest is yet to have any timber harvest planned or carried out.

Pristine natural forests

Pristine natural forests in temperate zones are globally rare, though they may seem locally still abundant. Large areas of northern Canada, Alaska and Siberia fit this description, as do smaller areas in the western USA. Where logging in these areas is being considered, it is incumbent upon managers to not repeat the mistakes of the past, but to start out with a search for an acceptable balance and approach that has its roots in our improved understanding of forest ecology.

We see three critical considerations that must apply to forest design in pristine areas. First is an analysis of the wider "conservation context". By this, we mean that before committing the forest in question to active management, the proponents must consider regional forest conservation issues. In most cases, there are existing studies that have already determined what these issues are. Canadian and American foresters in particular have, over the past few decades, done regional-scale analyses of remaining pristine areas. As a general rule, one key outcome of these analyses has been to designate certain remaining pristine forests as reserves, while others are "opened up" to commercial forestry. But even so, there are many conservation issues still unsettled. In the western USA for example, there is still much controversy over the fate of "roadless" areas. These are public forests that are pristine, yet

are not part of designated wilderness or other reserves. Environmental groups have lobbied for decades to prevent road building and commercial forestry in roadless areas. The Clinton Administration studied the issue, did extensive public outreach and implemented a rule that essentially forbids new road construction, although it does allow very limited logging. The Bush Administration has been looking for ways to relax or limit road building restrictions under these rules, and recently opened some 300,000 roadless acres of the Tongass National Forest in Alaska to logging. We must also point out that even where it may seem that the issue has been "settled", i.e. where some portion of pristine forest has been set aside while other areas have been opened to logging, a new controversy often re-erupts as soon as logging plans are developed. In other regions, such as Siberia, regional analyses may be lacking altogether, and the basic sorting out of areas to remain natural versus those that will be logged may not have been previously addressed.

A second key is to include suitable "set-asides" within the area to be logged to serve as biological reservoirs and hedges against the future. The extent and configuration of these is best sorted out within the design process itself, but in nearly every case it is critical that some area be reserved and left in its natural state. As a rule, these reserves will include riparian areas, rare habitats and inoperable places. Taken together, these are normally referred to as Forest Ecosystem Networks, or FENs. It is important not simply to identify inoperable areas and assume those will be sufficient to conserve all species. The key is to choose reserve areas that maximise the total number of species protected while minimising the amount of land set aside. To do this, managers usually look at a number of criteria, such as species diversity, representative habitats, level of previous disturbance, education value, uniqueness, system integrity and opportunity cost.

Thirdly, we feel strongly that areas of pristine natural forest newly subject to active management should use ecologically based forestry as the foundation for harvest plans. This may include large-scale fire emulation (where extensive cutting takes place to recreate the scale, severity and frequency of disturbance produced by a wildfire, for example), or may rely on selective cutting where stable multi-age natural forests are found. Roads should be minimal, and where possible temporary. The key is to understand the ecosystem and then to plan accordingly, rather than to impose pre-conceived management techniques that may not be suitable to the area in question.

Landscape ecological analysis

There is no case for initiating timber harvest in a pristine forest without a thorough analysis of the forest ecosystem, including physical, biological and social factors, using a process similar to that described in Chapter 2. This analysis must be done at an appropriate unit of scale, such as a watershed, regardless of how much of the area is ultimately likely to be subject to direct management. The analysis is often successfully subdivided by the main forest types found there, since these may be subject to different processes and use by wildlife. Other subdivisions based on landform, aspect changes or micro-climate variations may also be of use. The larger the study area the more useful it is to subdivide it into smaller, distinct units.

One of the disadvantages of working on a pristine area is often the lack of direct field knowledge available, especially about wildlife use, soils, disturbance history, potential for non-timber products and the true level of timber production potential. It may be necessary, therefore, to first undertake some specially targeted surveys in order to build up knowledge of particular landscape aspects. Where good-quality forest cover maps are available much can be gleaned from analysis of them in order to build up a picture of the ecosystem, even if the information they contain has not necessarily been collated for this purpose. By using GIS, it is possible to define age classes into successional stages, for example, and to produce maps which become the basis for landscape structure maps. Once such basic maps have been produced it is usually a good idea to visit some examples of them on the ground, to record vegetation structure details and to take some photographs. A reconnaissance trip by helicopter or small plane to look at the range of patch types present is also extremely helpful. Taking a series of oblique aerial photos can provide extremely useful reference material. This information is used to build up the description of each landscape element ready for analysis.

Actual wildlife use of remote, pristine forest areas is often not well known. The best source of information may be from local hunters, trappers or indigenous people. In a project one of the authors worked on in Alberta, Canada, a local trapper pointed out that wildlife was abundant in one valley, yet nearly absent in an adjacent one, even though the forest cover appeared to be identical according to the inventories. Without this, unless special surveys are undertaken, it is best to base analysis on experience and expert knowledge from similar places, perhaps neighbouring areas where the conditions are similar. Habitats can be classified using the same information from aerial photos or satellite imagery that is used to determine timber types and classes. This remote sensing information can and should be updated over time as field knowledge is improved. The use of an adaptive management approach, where initial assumptions are placed within a theoretical framework that is tested and adjusted over time, is an extremely useful practice in these circumstances (see Chapter 7).

The interaction of structures and flows (Chapter 2) can be estimated on the basis of available knowledge plus expert opinion, and from this the pattern and extent of use can usually be assessed with sufficient confidence to allow landscape-scale strategic planning to proceed.

We can assume that one of the reasons for developing a harvest plan is that there is sufficient merchantable-quality timber to make it worthwhile to exploit. This usually means that there is a significant proportion of late successional or mature forest present, often providing a large matrix element possibly consisting of old growth or antique forest. The presence of such areas may be valuable of itself if many surrounding landscape units have already been logged to greater or lesser degrees. It is very important to consider the wider landscape context in order to understand the role these forests play in the bigger picture. Is the area in question a core habitat? Is it a key part of a habitat network? How will these functions be conserved once logging begins?

In order to understand the key ecological dynamics of the forest several approaches can be adopted. From the forest cover map and aerial photographs it may be possible to determine a pattern of distinct age classes that reflect large-scale disturbance such as fire. From the age of the trees and ancillary evidence such as blackened stems, stumps and charred wood, a picture of the scale, frequency, distribution and intensity of fire can be built up. The same can be applied for windstorms and insect epidemics. Smaller-scale but widespread disturbance by fungi, snow break, some insects and minor fire and wind damage can also be estimated from aerial photographs and site sampling.

If there are oblique or aerial photographs taken several decades ago, it is possible to see how the forest has changed and the impact of any disturbances that have occurred. Once the basic picture has been built up, expert knowledge and reference to information such as ecological site classification handbooks for each basic ecosystem type can be used to piece together the anticipated successional pathways and probability of disturbance in different places, especially according to aspect.

One of the important outcomes of this analysis is to identify the places where catastrophic disturbance is least likely to occur. These are the places where stand-level dynamics occur and where old growth conditions are most likely to develop, usually in topographically sheltered locations. Such areas should be considered for retention, as a type of reserve, at least until replaced by other stands presently at earlier successional stages.

Other important features to identify are riparian corridors and places where a high degree of connectivity of late successional forest is desirable. Typically, riparian forests along larger streams do not burn as frequently or severely as upland areas, although they may be subject to replacement through flooding. Upland riparian areas generally burn at the same frequency and intensity as does the adjacent forest. Important non-forest landscape elements (whose functioning is to some extent dependent on the maintenance of forest around them) should be identified. These may include meadows, wetlands, outcrops and talus areas. All such areas will form part of the non-operational forest and may function as part of a forest ecosystem network.

A key output of analysis should be a series of silvicultural models to be applied in different parts of the forest according to ecosystem type, elevation, aspect, etc. These describe the way areas should be harvested and regenerated, together with the type of structures to be left behind or created through management. A number of these models may apply to each sub-zone to reflect the fact that several disturbance regimes may occur. Characteristic patterns of natural disturbance, such as a "log normal distribution", where the largest events occur the least frequently and vice versa, can be reflected in the distribution of different models in space and time. Larger-scale harvest, where a greater proportion of trees are removed in a pattern that mimics a fire, should be used more sparingly compared to models that create patterns of smaller openings in clusters, or where small groups or single trees are removed. Natural disturbance dynamics tend to show large numbers of small-scale events happening frequently, and small numbers of large-scale events happening infrequently. However, the total area of the large-scale events may be much greater than the sum of small-scale disturbances. These silvicultural models describe eco-logically desirable methods, but may be subject to change once other factors (such as social acceptability) enter the equation.

Therefore, in pristine natural forests the desired future condition is likely to be one of retaining ecological functioning that exists at present rather than trying to change it or direct its development elsewhere. In all cases a precautionary approach should be adopted and adaptive management used to build up knowledge as the plan is implemented.

Landscape character

While many pristine forests lie in remote areas far from public roads or settlements, and therefore visual sensitivity might be expected to be comparatively low, others lie on prominent slopes or along highways seen by large numbers of people on a daily basis. Often a key reason for their pristine nature is their visual sensitivity, meaning that forest managers have avoided logging these areas to forestall public controversy. Staff in the US Forest Service sometimes refer to this as "leaving an area for the night shift". However, because of the need to distribute logging activity widely around operating areas, some of the timber on these visible slopes is now required to keep local mills supplied. Understandably, local residents, tourism businesses and water users are often extremely worried about any changes to scenes that they have been used to for as long as anyone can remember. Thus, if management is to be responsive to local concerns, good visual design is essential. The process of landscape character analysis using a number of public viewpoints should be very comprehensively undertaken and used as the basis for the pattern of retained areas and potential coupes/management units/cutblocks. This could include consideration of aerial views if this is the method used by tourists to reach destination points within or beyond the area in question, as is the case in much of northern Canada.

The landscape ecological analysis may have produced information indicating that disturbance regimes of large extent are part of the usual dynamics of the forest ecosystem. It is often difficult, if not impossible, to persuade local conservationists to accept large-scale harvest of sensitive visual areas, even if ecological appropriateness based on disturbance dynamics can be demonstrated. Normally disturbance regimes show a range of natural variability. For example, it might be likely that due to forest type, landform and other factors, fires would tend to fall in a range of 10–10,000 hectares in size. It may be ecologically appropriate, and more socially acceptable, to design units in the range of 10–20 hectares instead of aiming for the 10,000-hectare theoretical maximum. Another option, where large-scale disturbance is seen as essential to be in step with natural processes, is to err on the side of leaving higher levels of legacies behind, particularly green trees and islands.

One of the interesting features about some pristine, visually prominent slopes is that they are often more green and visually intact than they may have been had natural processes been allowed to follow their normal course. Fire suppression, among other factors, may mean that the forest retains a uniform green of mature trees well beyond the age at which it would have been disturbed. To avoid abrupt visual disturbance, managers may be faced with a long period of harvesting smaller, dispersed units over an extended period. While more socially acceptable, this approach may be very expensive, operationally difficult, and may not be the best in terms of ecological dynamics.

In other cases pristine forests may not be what they seem. For example, in one case on the Mt Hood National Forest in Oregon, the Still Creek roadless area had burned heavily in the early 1900s, and was replanted with off-site genetic tree stock in straight rows running up and down the mountainside. An insect infestation in the late 1980s killed off many of the planted trees, yet logging was prevented by environmentalists who believed the area to be "pristine and untouched".

Another challenge is where fire suppression has gradually led to a formerly open forest becoming closed. In the western United States, there are many cases where what had been open oak or pine woodlands are now dense fir forests. The ecological goal may be to restore open conditions, which implies heavy logging with substantial visual impacts. The public may or may not be ready for this, and local wildlife, including sensitive species, may have adapted to closed forest conditions. In these cases, again it may be best to start small. Clear away from the drip lines of "granny" oaks and pines, and gradually work outwards, rather than initiate clearfelling of all the fir at once. In extreme cases, the old pines have been "high graded" out of the forest altogether, resulting in a 100% conversion to fir. Thus, selective logging will only perpetuate the fir, and will prevent pine reintroduction.

Constraints and opportunities

This is a particularly important planning step when developing designs for pristine areas. The strategic planning of wood supply from a plan area may be based on the availability of certain volumes over a particular period of time with minimal amounts meted out for regulatory reasons. This strategic plan may be based on rough-and-ready analysis scarcely supported by ground truthing. The economies of road construction and the use of less intense harvest methods may change the pattern of wood flow if not its total amount.

While harvest plans are no longer based on the simple implementation of an analysis of operable area, road construction and logging system, such technical information remains highly relevant. In an area that contains no roads, the potential access routes are critical. Limitations of different harvesting equipment and terrain problems also need to be identified. In obviously difficult areas, such as steep-sided valleys, it may be advisable to prepare a comprehensive engineering plan to identify the location of the road line landings, bridges and culverts, terrain to be avoided and the means of extracting (yarding) timber to the landings. The operability limits should be clearly identified and, if possible, the whole operable area divided into units based on the areas expected to be extracted (yarded) to each landing.

Terrain limitations may not completely preclude harvest but may limit it to selection systems due to the need to avoid large-scale removal of overstorey and the loss of root anchorage on potentially unstable slopes.

Preparing a design concept

In the example described above, the design concept is mainly generated by determining the desired future condition of the landscape arising out of the ecological analysis, adjusted in the light of the other analyses. The typical result is a map showing the forest divided into two main categories: the operational and the non-operational. The operational forest consists of those areas where logging of some kind can take place. It is subdivided into zones based on topography, ecosystem type or aspect, with the associated silvicultural models. The non-operational forest comprises those areas that are technically inoperable (not merchantable, uneconomic, inaccessible, etc.) overlaid with the areas that need to be retained for ecological, water protection or visual reasons. The resulting map often shows the operational forest split by the non-operational (especially by riparian corridors, etc.) into larger units (macro-units) which are then subdivided into actual cutblocks/coupes/silvicultural units at the sketch design stage.

To present the concept it is a good idea to describe in summary form the main attributes of each constituent zone, the main values associated with it, the objectives to be met and the type of management to be undertaken.

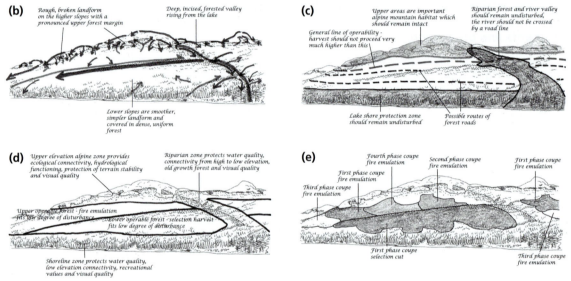

Figure 9.4
The main elements of design in managed natural landscapes: a) an example of a visually prominent area of forest visible from public roads. This is one of a number of viewpoints that might be used for analysis and design. Northern British Columbia, Canada; b) the landscape character analysis, focusing on the different types of landform and the simplicity of the forest cover; c) constraints and opportunities analysis, focusing on protection zones and operability; d) the design concept developed for the landscape based around the combination of two analyses; e) the sketch design for a pattern of coupes/cutblocks/management units, based on the concept and the analyses.

Developing a sketch design

The first task in the sketch design is to establish the boundaries of the larger non-operational forest areas, combining regulatory restrictions with response to natural landform, thus avoiding the creation of parallel lines along streamsides by altering the regulatory limit in as appropriate. The result should be that the macro-units have a well-defined shape that fits into the larger forest pattern and landform. The operable areas can then be subdivided into coupes or units that are managed as a whole. They should vary in size and shape in response to landform and disturbance dynamics and, as far as possible, incorporate one or more harvestable units to ensure practicality.

Once the units have been designed, the next step is to assign silvicultural models to them, based on guidance from the ecological analysis constrained by the operational limitations. Following this the phasing of activities is worked out. The central idea is to spread the rate of change of the forest over time, so that at any given time over the period of the plan and beyond, the balance of stand types and their distribution pattern accords with the requirements for ecological functioning, wood flow, water protection, visual quality and so forth. This process can be carried out manually, by trial and error on a map, or can be developed by computer using a designer support system based on a GIS (see Chapter 6). The situation is complicated when there are multiple interventions within different management units/coupes/cutblocks which are phased to start at different times. It is a good idea to develop a spreadsheet to organise the flow of timber and the changing proportion of different successional stages over time. From the spreadsheet, graphs and charts to monitor the flow can be produced.

To illustrate this important but complex process, the following example is taken from a project in British Columbia (more examples of design projects can be found later in this chapter). In this example, four different silvicultural management models were derived from the ecological analysis of the project. There was a degree of simplification to make it easier for managers to specify when working silvicultural prescriptions. The four models were to be used in different parts of the forest; some areas had mainly one type, other areas a greater mixture to reflect the pattern of disturbance. The timing of the phasings was based on the expected time for the stand initiation phase to succeed to the stem exclusion stage, probably around 10–20 years. Thus, if phase 1 started in 2005, phase 2 could start in 2025, but so could the second intervention in those phase 1 units which had more than one stage of activity, while some units could have phase 2 at year 10.

The silvicultural models were as follows:

1   *Fire emulation* This is a single-entry removal of 10% of the standing volume to result in a stand of even-aged forest containing some patches of trees in wet hollows or rocky knolls–"refuge patches" (or wildlife tree patches)–as well as individual standing trees (of thick-barked, fire-resistant species) and a number of dead snags. A fringe around the edge is to be thinned or selectively logged. A light burn of the slash introduces charred material and ash into the soil. Therefore, this model is completed at one go.
2   *Clustered patches* This is a two-stage entry where the unit is harvested in medium-sized patches removing around 40–45% of the volume, with 10–15% retained in variably sized clumps, single trees and snags. This reflects the medium scale of such as that caused by wind. The two removals take place 20 years apart, so that a stand of mixed-aged patches results. Thus this silvicultural model takes 40 years to complete.
3   *Group selection* This is a three-stage entry where small groups of trees are harvested, some 30% of the volume removed at each stage with 10% being left behind, as in the previous examples. This model reflects the dynamics of areas where fire is rare but insects or disease ensure a multilayered stand type develops. Each stage takes place 20 years after the previous one, so that the whole unit is completed over 60 years.
4   *Single tree selection* This is a five-stage entry where single trees are removed across the whole unit at intervals of ten years, with 10% left as legacies as before. This model applies to the most stable sites in mixed stands where individual trees die and are replaced by shade-tolerant species that may already be in the understorey as advance regeneration. This type of unit takes 50 years to complete.

Thus, with four phases, a timing of 20 years between each phase and a maximum duration of management for each phase, the entire plan could take up to 130 years to implement for the last phase of single tree selection. Clearly this has implications for wood flow and for issues such as forest health, because if a forest is already old (late successional), 80 years is a long time to wait to start a silvicultural prescription that can take up to 50 years to complete: the forest could have suffered many disturbances by then and the wood quality could have declined. Therefore, it is a good idea to start most of the single tree selection units early in the sequence and leave more of the fire-emulation or medium-sized patches until later on, if stand conditions will allow it. However, depending on stand type and site age, volume increment continues to increase, albeit slowly, so that calculations of future wood flows have to take it into account.

What can be seen from this is that even in the "operational forest" zone the design maintains a significant component of late successional forest until late in the sequence and that by the end

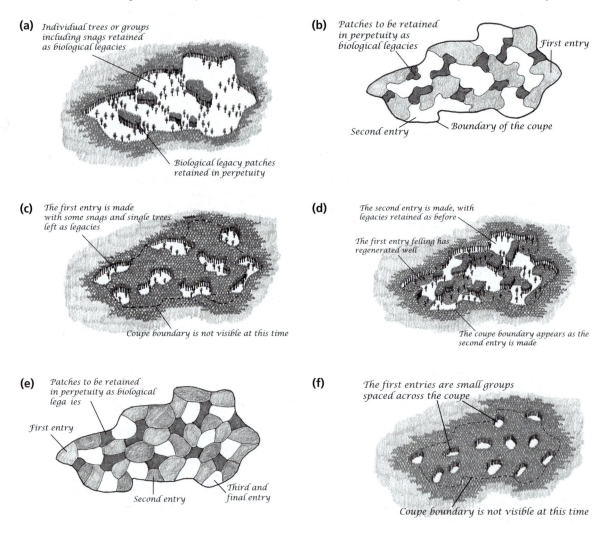

Figure 9.5

This series of sketches illustrates a number of silvicultural models based on natural disturbance types:

a) "fire emulation", where 90% of the standing volume is removed in one operation to result in an even-aged forest patch. The remaining 10% of the volume is used to provide legacies in the form of small patches, groups and single trees distributed in a pattern reflecting the likely areas that could survive a fire. b) the layout for the model based on "clustered patches", a two-stage entry system where the unit is harvested in medium-sized patches, removing 40–45% of the volume in each stage, the two stages being around 20 years apart. The remaining volume is used to create the pattern of legacies. The model for this type is wind damage. c) the appearance after the first entry. d) the appearance after the second entry. e) the layout for the model based on "group selection", where the timber is removed over three phases, 20 years apart, some 30% at a time, in of the plan units that were harvested earliest are well on their way to providing replacement late successional forest.

The sketch design is then evaluated using the methods explained in Chapter 7, and presented using the techniques described in Chapter 6. It is a good idea to use a range of colour-coded maps to present the somewhat complex combination of ecological zone, silvicultural model and phasing, with each individual unit being specified on a database in a GIS.

(g) *The second entries are also small groups and the first entries have regenerated*

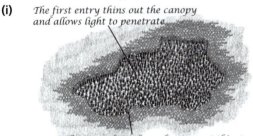

(h) *The final entry shows three age-classes of regeneration and some legacy clumps*

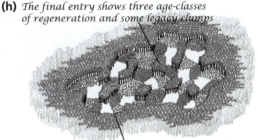

The coupe boundary shows up at this stage

(i) *The first entry thins out the canopy and allows light to penetrate*

The coupe boundary shows up at this stage as a result of the change in texture

(j) *At the second entry the canopy is thinned further to release the regeneration beneath*

(k) *At the third entry the overstorey is reduced further and a multi-aged forest develops in the coupe*

Figure 9.5 (continued)
small groups distributed across the unit, 10% being left as legacies as in the previous examples. This model is based on the small-scale changes that gradually regenerate a stand, such as insect attacks, fungi and wind. The result is a multi-layered stand. f) the appearance after the first entry. g) the appearance after the second entry. h) the appearance after the third entry. i) the appearance after the first entry of the model based on "single tree selection", where harvest takes place over five periods at intervals of ten years. Single trees are removed, being well distributed over the unit, and 10% of the volume is never harvested. This model applies to the most stable stands where individual trees naturally die and are replaced by shade-tolerant species that may already be in the understorey. j) and k) show the second and third entries respectively.

Figure 9.6
An example of an area that has always been forested, but where the forest has also been managed continuously for several hundred years. Germany.

Figure 9.7
This example, from southern Finland, shows a logged area with views towards a stand of Norway spruce. The proportion of spruce in the forest has been artificially increased over the last 100 or so years at the expense of pine and broadleaves, in order to increase timber yields. The area has been continuously forested but the ecology has been altered by management.

Design in previously managed natural forests

Given the long history of commercial forestry in most of the temperate world, previously managed but still somewhat natural forests probably represent the majority of cases confronting the forest designer. Some examples have been under management for many decades or even centuries, so that many of their natural features have long since been altered. The design process in these cases can be used to redirect management, if necessary, towards a more sustainable pattern.

In Chapter 1 some of the results of long-term planning in originally natural forests were noted: the layout of compartments, the emphasis on yield regulation and the intensification of management. This has resulted in forests that are significantly different from completely natural ones, which may have greater or lesser impacts on ecological, aesthetic or recreational values. As non-timber values increase in importance, the management of such forests may need to change. One of the problems facing managers is the tendency to define all prescriptions at the compartment or sub-compartment level, rather than to consider the forest as a whole. Such small-scale details at the stand level may be important for conserving biodiversity at one level, but miss important opportunities at the landscape level. Another issue facing many of these forests, especially in Scandinavia and Central Europe, is that the majority of forests are owned and managed by small owners, so that large-scale analyses present many practical problems.

The role of analysis is to evaluate the current state of the forest landscape and to see how well it functions ecologically. The role of design is then to define one or more target landscape patterns towards which the forest can be managed over time. Since many of these forests are in populated landscapes, aesthetic values may be important, especially of the internal landscape, since in places such as Scandinavia, where everyone has the right to wander around the forest (Every Man's Right), they are frequently popular for recreation of all types.

Since these forests have all been under some degree of management for many years there is usually a good deal of information available, at least about stands and their history. Less information may be available for ecological values, although because of the traditional interests in the outdoors in hunting, fishing and use of non-timber products, local knowledge is often available for the asking. The main problem is that all information tends to be to a small-scale resolution – the stand or ownership, rather than at a landscape scale, so that a degree of simplification is often necessary. This may worry many people at first, because it may seem as if valuable knowledge is being ignored. However, too much detail can soon become confusing and obscure the bigger picture being developed.

Owing to the many stakeholders involved in most of these forests, it is important to adopt some form of participatory planning, using some of the techniques described in Chapter 4. Where there are a number of landowners it is very important to find ways of incorporating their own values and objectives into the larger-scale plan. Their knowledge may also be the main source of information available and the designer needs to develop ways of recording this for analysis and design purposes. It may also be the case that GIS databases are not yet available for private forests.

Figure 9.8
A forested hillside in Norway. This may appear to be a single landscape but is composed of several small-scale ownerships, making landscape-scale management difficult.

Landscape ecological analysis

For natural forests owned and managed by large landowners such as timber companies, a state or provincial government, the "standard" design approach from Chapter 5 can be adapted. The forest, depending on its size, should be broken into landscape units and subject to the process already described. Stand data defined by age class should be converted into successional stages or structure classes and described. It may be appropriate to simplify the information so as to see the wide picture among all the detail. It is likely to be particularly important to identify landscape elements (biotopes) that are natural or otherwise especially ecologically viable. In Europe many of these may be classified under European Union regulation on biodiversity, or they may be identified in Certification Standards. Man-made elements will also feature in landscape structure maps, some of which may cause adverse effects to ecological functioning. The analysis of landscape structure will show what the forest is like now. It may be helpful to try to reconstruct what it might have been like in the past or in its natural state. This is likely to be easier when management has only been practised for a few decades compared with places that have been managed for centuries. In these circumstances it is probably irrelevant as such a state can probably never be recreated, cultural landscapes themselves have value and many biotopes or habitats created through human activity and traditional management have become ecologically valuable, although their future may be uncertain.

Wildlife use is likely to be affected to a greater or lesser degree by human activities and the human-altered landscape. It is important to reflect this fact in the choice of species to use in analysis. It may be particularly helpful to use the concepts of umbrella, keystone or sensitive species so as to focus more on those of specific interest for biodiversity than casting the net too widely (see Chapter 8).

Disturbance and succession may be difficult to understand in the analysis, especially in places with a long history of management. The pattern of forest stands currently present may be very different from that of a natural forest and disturbance agents may not be able to affect the forest to the same degree. Silvicultural practices may have changed the risks from disturbance – increasing some, such as insects, while decreasing others, such as fire. Successional pathways are also likely to be altered by management, especially accelerating processes in order to achieve more timber production, but also reducing rotation lengths and thus the proportion of the forest in later successional stages.

Potential desired future conditions for the landscape are therefore unlikely to include restoration to a completely natural state, but may choose a managed state that better ensures ecological functioning, especially with regard to the protection and enhancement of important species and habitats. Instead of trying slavishly to emulate patterns and structures created by the natural disturbance and succession, it may be more appropriate to use these as a reference point against which to evaluate the eventual design. One change to current management that should be included is the identification of stands that are to be retained beyond the normal timber rotation age in order to provide future old growth. The ecological analysis can be used to identify the locations where such stands might be expected to develop, in places least likely to be disturbed, and where gap-phase stand replacement processes are dominant.

An important component of the desired future condition is the protection of valuable biotopes important in their own right as a habitat for key species. These may have been affected by previous forest management, so that it may be necessary to establish buffer zones around them, restore key linkages or expand them. Stream corridors, wet areas, naturally open patches of grassland, heath or bog, wet woodland, lichen or bryophyte-rich patches or remnant ancient woodlands are examples of biotopes that normally should be given greater attention in the design.

Landscape character

The landscape character of managed natural forests may be influenced in several ways. Firstly, previous management activities may have introduced patterns that are visually intrusive. The layout of compartments in a geometric pattern may conflict with landform or produce awkward shapes of felling or silviculture. Sometimes, on rolling topography, angular shapes crossing the skyline can produce intrusive results such as notches or steps. Some "classical" silvicultural systems such as strip-shelterwood can produce unattractive results. Roads cutting across the contour on steep mountain sides may create visual scars and disrupt natural drainage patterns. The overall pattern of shapes may be chaotic and disunified. Furthermore, when the forest area in a landscape belongs to many different owners, the legal boundaries may be defined by different, contrasting management practices.

The second effect on landscape character is the contribution of human activities, particularly over long periods of time. In many European countries, New England and the Maritime Provinces of Canada, the landscape is often a mosaic of forest interspersed with farmland. It is difficult to separate a consideration of the forest from the rest of the landscape. The pattern of fields interlocking with forest, the elements of farmsteads and buildings and other features such as hedges, trees, tree clumps and the forest edge may all contribute to the overall character and affect activities and changes to the forest landscape.

Thirdly, the visual experience of many of these landscapes is from individual homes or local villages, as well as from roads and paths. These routes may connect settlements and run through fields for part of the time, affording short- to medium-scale views. Stretches of road within the forest give short-distance views, as do tracks and paths. The prevalence of internal views may mean that forest activities, such as felling, designed to reflect the scale of the landscape as seen from external viewpoints, may be perceived as too large and dramatic. Thus, a balance may have to be developed between

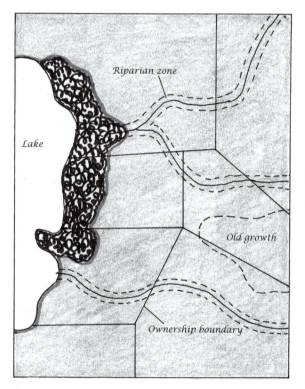

Figure 9.9
This diagram shows the complications of managing multi-ownership forests: the ownership unit boundaries cut across many of the features that need to be carefully protected and managed, such as the riparian zones.

external and internal views, depending on the sensitivity of the landscape to each, a balance that may also affect the ecology, depending on the desired silvicultural regime.

Constraints and opportunities

Since the type of forests discussed above have in many cases been managed for long periods, many of the key physical constraints (unstable slopes, wetlands) will have been recognised and worked around or already overcome. Relevant constraints are likely to centre on land ownership, particularly finding an approach to converting established small compartment patterns into a pattern that supports larger-scale ecosystem dynamics. Not only the pattern, but also the internal structure of the forest and silvicultural history may create difficulties for the implementation of a new design. A landscape-scale plan will require landowners to co-operate and to agree to implement a plan by interpreting its components within their own area, and in the right sequence.

Opportunities may include enabling landowners to achieve more objectives by co-operation than if they were trying to manage for all elements on small acreages by themselves. In Scandinavian countries, the marketing of timber has often been carried out by a commercial arm of the private forest owners' association. This association also provides management advice, draws up plans and helps ensure that legal requirements are met. Thus, there is an existing framework to develop a landscape-scale plan and to help in the reconciliation between landscape and ownership. Furthermore, non-timber values may be best developed co-operatively, including recreation and tourism infrastructure, marketing and biodiversity conservation.

Countries such as Sweden have developed excellent documentation to provide forest owners with advice on biodiversity and cultural heritage protection and management which go a long way to ensure that stand-level issues are well resolved. Landscape-level planning enables these to be set into a wider context and to ensure that connections across the landscape are made where needed, for example riparian zones or animal migration routes.

Preparing a design concept

The development of a design concept in managed natural forests is partly defined by the desired future ecological condition, but may be shaped much more by the visual landscape character, particularly when there are strong social constraints that reflect natural patterns and processes. It is often best if an initial concept is developed that initially ignores land ownership divisions, otherwise it may be too ecologically compromised, and important opportunities to consolidate habitat are missed. Thus, the initial concept should consist of broad zone maps showing where different management approaches should be carried out, together with the major components of protection and connectivity, both of ecological and cultural elements.

Figure 9.10 Location of the Bonanza Creek project area in British Columbia.

Sketch design

In cases where the landscape comprises only one land ownership, the sketch design can be prepared as described in Chapter 5. The proposed coupes or management units may reflect a compromise between long-established compartments and the attempt to respond to larger systems, particularly if there are substantial age, species or structural differences between what exists and what is desired.

Where the landscape is divided into many small-scale ownerships, the approach should be to obtain first an agreement from all or most of the landowners that they will follow the principles that shaped the concept, and which protect the major elements of biodiversity and cultural history that were identified. Secondly, an agreement to apply a set of design principles should be developed that describes how the concept will be implemented. This could

Figure 9.11
Photos of Bonanza Creek area: a) shows the landscape in a more distant view while b) shows the internal character.
Source: Slocan Forest Products.

include guidelines for coupe design, development of riparian corridors, forest edges and the treatment of skylines, roadsides and paths. When more detailed management plans for each ownership or working area are drawn up or revised, these principles would be applied. Thus, each owner would be able to exercise some freedom in managing their land, but within a framework of a whole landscape. In some cases, if a few owners are expected to carry greater conservation burdens, then some financial compensation from other owners or the government may be required, possibly through purchase of conservation easements.

The next sections present the case studies. The first one, Bonanza Creek in British Columbia, on public land managed by Slocan Forest Products, involved an area that had seen some limited small-scale logging but was otherwise fairly intact. The second project, Sutherlands Brook in Nova Scotia, on public land managed by StoraEnso, a timber company, had already been partly logged in the years before the project was initiated but had also been cut over in previous decades. However, it was not fragmented. The third project, Vuokatti in Finland, is in Europe and had been subject to all sorts of management practices including slash-and-burn agriculture as late as the nineteenth century. It is unique among the case studies, being in multiple private ownership.

Case study: Bonanza Face, British Columbia, Canada

This case study describes a project carried out by Slocan Forest Products, a company based in British Columbia, Canada. The project was carried out in 1995–1996 and the final design is being implemented at the time of writing. Simon Bell was a consultant to the multi-disciplinary project team and helped guide the members through the process, which follows fairly closely that described in Chapter 5. The purpose of the ecosystem design (the term adopted by the team) is to provide an ecologically, socially and economically sustainable landscape-level planning framework to guide operational forestry prescriptions for the Bonanza Face Planning Area (BFPA).

The Bonanza Face covers some 2,000 hectares of rugged, heavily forested and diverse terrain at the northern end of Slocan Lake in the Kootenay area of British Columbia, Canada. Located in the Selkirk Mountains, the general character is of steep rocky peaks; sharp, incised tributary creek valleys and flat, U-shaped major valleys. The climate is influenced by the mountains and prevailing westerly winds. It is characterised by hot, moist summers and mild, moist winters. Peak monthly precipitation occurs in the months of December and January, as snow and overall precipitation increases with elevation.

Elevations in the area range from around 600 m to over 2,000 m. The lower elevations lie within the interior cedar–hemlock (ICH) and the higher elevations within the Englemann spruce–sub-alpine fir (ESSF) biogeoclimatic zones.

Resource values and objectives

One of the first tasks of the design team was to identify the broad set of resource values and objectives in the Bonanza Face area, including any relevant issues and the opportunities and constraints regarding those resources. The following resource values were identified, although space does not permit a more thorough presentation of the values and other aspects:

- Biological diversity
- Timber-based forest products
- Wetlands, riparian areas and water
- Wildlife and fish
- Soil productivity and integrity
- Forest productivity and forest health
- Non-timber forest products
- Spiritual and cultural assets
- Recreation
- Aesthetic and scenic character.

Inventory and analysis

Data was assembled from existing inventory sources and from special surveys undertaken to collect specific information, especially ecological information, terrain stability, engineering and landscape character.

The ecological landscape

The analysis of the forest ecosystem was based on a number of sources of inventory information which provided a basis to categorise the landscape features into structures, flows and dynamics, and also served as a means to understanding the ecological processes within the planning area and between the planning area and its surroundings.

Forest landscape structures

Matrices
There are two biogeoclimatic zones represented in the project area: the Englemann spruce–sub-alpine fir (ESSF) and the interior cedar–hemlock (ICH) zones. The 1,400-metre elevation line was used as the approximate boundary between the two zones, which were chosen as the two matrices for the area.

Both matrices comprise mainly mature and mid-seral (successional) age forest. This is because of large fires, which burned most of the area around the turn of the twentieth century, the suppression of natural fires since that time and the lack of any other substantial disturbances. Thus, the forest cover is relatively homogeneous and there are few natural openings in the canopy. Nevertheless, the matrix has important stabilising effects on the ecology of the area, including important hydrological effects. The matrix also provides recreational opportunities for hunting, wildlife viewing, hiking and berry picking. It also provides habitat and shelter for many species of wildlife, particularly large mammals.

Patches
At the scale of resolution adopted for the project, the patches in the BFPA are considered to be contrasting areas over 1 ha in size. The patches are as follows:

- *Early successional forest* is represented by a 2 ha cutblock undergoing a transition from a shrub- and forb-dominated flora to an early successional deciduous and conifer tree-dominated flora, and a few smaller natural openings resulting from natural and human disturbances.
- *Old successional forest* The only identifiable area occurs as one 10 ha patch in the ESSF. It provides cover for wildlife and has moisture interception characteristics similar to the mid-successional forest. A number of small patches of less than 1 ha in size are scattered throughout the BFPA, typically in wet areas.

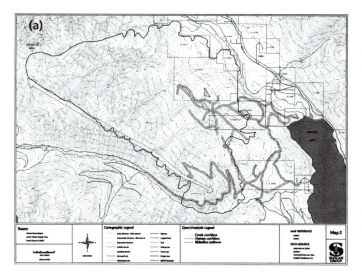

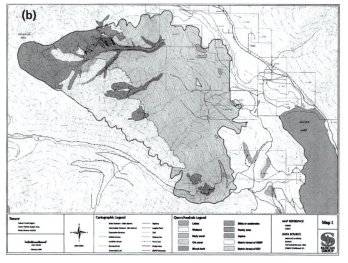

- *Rocky outcrops* are areas where exposed bedrock or colluvium is the most significant physical site attribute. They are normally dominated by hardy forbs and bryophytes. They are often important for birds and small mammals, used for nesting, roosting and hunting purposes.
- *Shrub and forb* sites are not productive in growing merchantable timber but support shrub and herb species and provide important habitat and diversity. Such patches are mainly alder, devil's club and white rhododendron, with occasional coniferous trees. In the ESSF they coincide with avalanche tracks.
- *Alpine* Although the alpine area is considered a patch in the context of the BFPA, it actually represents a small part of a more extensive alpine matrix to the north of the design area. This is characterised by rocky, rugged terrain, low-growing shrubs and forbs and a lack of trees. Due to the harsh climate, the flora and fauna are quite distinctive from those found in the rest of the BFPA.
- *Wetlands* The wetlands in the BFPA are unusual for the area and the marsh complexes near the mouth of Bonanza Creek are considered to be regionally significant. The wetlands provide habitat structure that is significantly different than the surrounding forest matrix, including higher snag densities and increased shrub cover.

Corridors and pathways

Corridors present within and surrounding the BFPA are of both natural and human origin.

*Figure 9.12
The ecological analysis: a) the map of patches and matrices at Bonanza Creek; b) the map of corridors at Bonanza Creek.
Source: Slocan Forest Products.*

The hydroriparian ecosystem
The hydroriparian ecosystem includes the streams, wetlands and adjacent terrestrial habitat. The BFPA contains seven natural stream corridors in well-defined valleys. They are particularly important as cross-elevational corridors. They are also the sites of avalanches, debris flows and mass soil movements.

Road and trails
There are many kilometres of roads and trails in the BFPA. These include a Forest Service road used for recreational, hunting and forestry purposes and old mining and forestry roads which are now overgrown, eroded or washed out.

High-elevation ridge
A distinct and continuous high-elevation ridge provides a connecting corridor between high alpine areas to the north and the open, south- and east-facing slopes above Shannon Creek to the south.

Avalanche tracks

These move moisture, sediment and coarse woody debris from higher to lower elevations. They often coincide with erosional gullies and stream channels, and provide corridors for stream flow, particularly in the spring and summer.

Landscape flows

The major flows within the BFPA include movements of ungulates, large mammals, people, fish and other wildlife, water, snow, sediments and other organic and inorganic materials.

Ungulates and large mammals

Migration patterns of ungulates and large mammals often focus on ridges and riparian areas, through saddles, along patch edges and within corridors. Most seasonal movements of ungulates and large mammals in the area are between high- and low-elevation areas, emphasising the need to maintain connectivity between valley bottoms and ridges. Ungulate winter range areas occur on the south-facing slopes above Shannon Creek and there are probable migration routes from this area up and down Bonanza Ridge and towards the middle of Bonanza Face.

People

Most human activities in the area are limited to the lower slopes – particularly related to the old mining and forestry road systems, and along the Shannon Creek Forest Service Road. These roads provide important access for activities such as hiking and skiing, snowmobiling and all-terrain vehicle use, hunting and trapping, as well as the harvesting of alternative forest products – such as berry and mushroom picking.

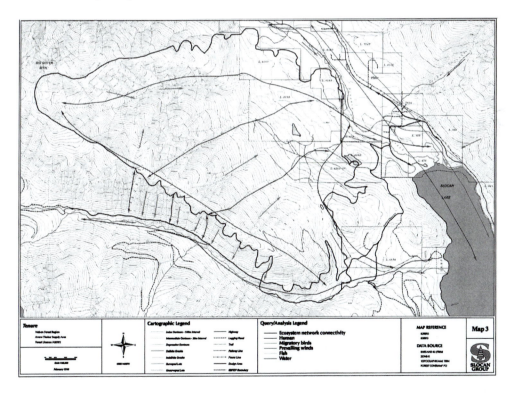

Figure 9.13
The map of flows at Bonanza Creek. Source: Slocan Forest Products.

Water

Water is a dominant flow. The upper and middle slopes of the Bonanza Face are the major source of water flowing into the tributaries of Bonanza Creek. Snow accumulated over the winter melts gradually, especially on northeast-facing slopes, and feeds the network of streams throughout the spring and summer. The forest tends to hold and slowly release water down the slopes, buffering maximum and minimum stream flows over the course of the year.

Soils and sediments

Soils and sediments can be transported in the form of debris flows, landslides and other mass movements. In the steep, well-developed gully system on the upper slopes of Bonanza Face and given the types of superficial materials present, these flows are important. The upper headwaters of the drainages are generally very steep with many active avalanche paths and gullies, typically the areas with the most soil erosion and transport with corresponding deposition in areas of decreasing stream gradient on the lower slopes.

Relationship between structures and flows

Once the landscape structures, patterns and flows are identified, it is necessary to determine how they interact with or relate to each other (see Table 9.1). Due to the dominating cover of the mature forest matrix in the BFPA, most landscape flows are influenced in some way by it. The movements and activities of animals and people are shaped by the stand characteristics and spatial extent of the mature forest, which extends, almost unbroken, from ridgetop to valley bottom. Water flows also interact in a particular way with the mature forest cover, in terms of flow rates, temperature regulation and filtration.

Despite the dominant nature of the mature forest matrix, the various landscape patches have important effects on flows. The wetlands, shrub and forb, and rocky outcrops are the most likely to influence landscape flows. The shape, size and orientation of patches and corridors are important determinants for the nature of flows occurring within, across or between them – such as water flows or bird migrations, as are the stand structural elements such as spatial heterogeneity, coarse woody debris and forest stand character.

Disturbance, recovery and succession

Disturbance

In the Kootenays, the predominant agent of change in natural forest ecosystems has historically been fire. Other disturbances, such as windthrow, stream bedload accumulation, debris torrents, insect or disease outbreaks and natural water diversions, also affect the area to a lesser extent. The historically occurring disturbances have been compared with the models provided by the Ministry of Forests in terms of Natural Disturbance Types (NDT) related to each BEC zone.

Fire

Before recent human intervention (>200 years ago) Bonanza Face, like much of the surrounding area, was characterised by natural fire disturbances of various sizes, ranging from small, low-intensity spot fires to very large, high-intensity infernos.

By far the majority of terrain in the BFPA lies on north- and east-facing aspects with some southern aspects and very few western aspects. South and west aspects tend to be drier and warmer in character, experience larger, higher-intensity fires and leave fewer surviving single trees and refugia patches. Conversely, north and east aspects tend to be cooler and wetter in character, experience smaller, lower-intensity fires and leave more surviving single trees and refugia patches.

In addition to the influence of aspect on fire disturbance, the general disturbance patterns of the ESSF and ICH zones are also distinguishable, in terms of fire periodicity and intensity, primarily due to climatic, topographic and tree species differences.

Table 9.1 Potential disturbances in the ESSF (NDT 1)

BEC	Fire	Pathogens	Insects	Human	Weather events
ESSF (all)	Periodicity: 200–350 years	Agent: *Armillaria*	Agent: bark beetles	Agent: logging and mining activity	Agent: avalanches, snow and ice, windthrow
	Size: 50–100 ha	Potential openings: single trees and small groups	Potential openings: openings vary with species composition, size and vigour	Potential openings: openings can be single trees and small groups, minor openings and larger openings with variable retention of trees	Potential openings: vegetation in avalanche tracks is maintained at a forb state. Snow and ice storms may change tree species composition
	Character: contain refugia patches and irregular shapes following landscape patterns		Openings can be single trees and groups, minor openings and catastrophic openings		
	Larger openings on S and W aspects than on N and E. Veteran and refugia patch distribution is a function of aspect			Disturbance can include road and trail construction, mine sites and cutblocks	Windthrow of single trees, small groups and minor-sized patches

Table 9.2 Potential disturbances in the ICH (NDT 2)

BEC	Fire	Pathogens	Insects	Human	Weather events
ICH (hygric sites)	Periodicity: 250 years Size: 1–10 ha Character: single tree lightning strikes and small irregular patches with unburned veterans	Agent: *armillaria* Potential openings: small groups and individuals	Agent: bark beetles Potential openings: individual trees to very large openings Higher probability of spruce beetle in valleys	Agent: logging and mining activity Potential openings: openings can be single trees and groups, minor openings and larger openings Disturbance can include road and trail construction, mine sites and selective or clearcut cutblocks	Agent: avalanches, snow and ice, wind Potential openings: single trees, small groups and minor-sized patches Vegetation in avalanche tracks is maintained at a forb state. Snow and ice storms may change tree species composition
ICH N and E aspect (mesic sites)	Periodicity: 200 years Size: up to 150 ha Character: finger-like openings with burned patches and single veteran trees Veterans have small to no surrounding refugia patches	As above	Agent: higher probability of mountain pine beetle as pine component increases	As above	As above
ICH NE to SW aspects (mesic to sub-mesic sites)	Periodicity: 150–200 years Size: 10–50 ha Character: finger-like burns to larger burns with few and small unburned refugia Veterans have small to no surrounding refugia patches	As above	As above	As above	As above
ICH (xeric to sub-xeric sites)	Periodicity: 150 years Size: 100–500 ha Character: similar to ESSF. Large intense fires. Single standing veterans with no refugia	As above	As above	As above	As above

Pests and pathogens

The impact of forest pests on the BFPA is thought to be relatively small, but still important. For example, endemic levels of root rot (*Armillaria ostoyae*) in the ESSF have resulted in frequent but widely dispersed small openings, while mountain pine beetle (*Dendroctonus ponderosae* Hopkins) outbreaks tend to produce fewer openings of more varying size, depending on the amount of lodgepole pine present in a stand. Both of these disturbance agents are important drivers of forest structural diversification at a landscape level.

Human disturbances

The current structure and function of the forests of the BFPA have been impacted by previous selection (high-grade) logging, which took place at least twice in the last 80 years. This removed the large veteran trees, which had previously escaped fire. The result is very few large veterans remaining, with these concentrated in small patches in wet, inaccessible areas. This has resulted in the loss of stand structural diversity from that which would be expected under natural conditions. Human-set fires in the early mining era and active fire suppression have imposed a new fire regime upon the landscape. The resulting differences in the frequency, size and intensity of the fires as compared to historical fire patterns have caused significant changes in the ecological structure and functioning of the forests.

Wind

Given the proximity of Bonanza Face to the broad valley and low pass linking the Arrow drainage and the Slocan River drainage, the area is likely to be subject to periodic high winds. The prevailing winds in the area are northwesterly along the axis of the Slocan Valley. Generally, the higher-elevation forests, particularly those located on or near the exposed ridge crests, are subject to higher-velocity and more sustained winds.

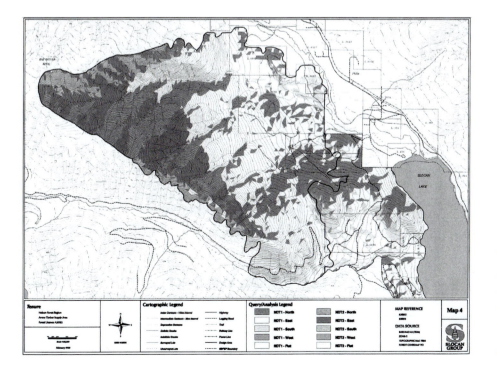

Figure 9.14
The map of disturbance types related to the landscape at Bonanza Creek. Source: Slocan Forest Products.

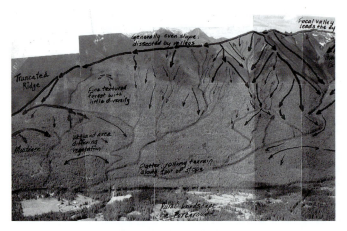

*Figure 9.15
Landscape
character analysis
on a sample
view at Bonanza
Creek. Source:
Slocan Forest
Products.*

Recovery and succession

The successional pathways for each BEC zone could be constructed from the various documents provided as part of the *Forest Practices Code Guidebooks* and the information accompanying the BEC handbooks. This showed that the forest is mainly currently in the mid-successional stages and that 20-year periods of intervention between adjacent harvest units would ensure that successional stage separation would be maintained.

Landscape character

With the ecological analysis complete and with a good understanding of the current landscape condition and ecological dynamics, the next step is to look at the character of the landscape.

Landform analysis

While the scale of Bonanza Face and Big Sister Mountain seems dominant in the immediate vicinity, it is typical of the scale of the surrounding area. The landforms of the Bonanza Face are the product of the geological and geomorphological history of the Ruby Range. The steep upper slopes, dissected by many avalanche slopes, gullies and small creek channels, gradually give way to more gentle undulating terrain on the lower slopes. Bonanza Face exhibits a number of distinctive glacial features including the *roches moutonées* and drumlinoid hills and a truncated spur on the toe of the ridge overlooking Shannon Creek. These landforms give the lower slopes a rather hummocky appearance that contrasts with the generally steep upper slopes and rugged ridgeline. This structure is reflected in the landform analysis.

*Figure 9.16
Landform
analysis on a
contour map at
Bonanza Creek.
Source: Slocan
Forest Products.*

Land features

The mature homogeneous forest matrix that covers much of the terrain dominates the landscape. Aside from a few sub-alpine meadows, avalanche slopes and rocky outcrops, the area is carpeted in the rich green texture of mature conifer forest. The subtly engraved, rounded creek drainages on the southern slopes of Bonanza Ridge gradually become more pronounced and rugged to the north, culminating in the steep angular valley emanating from the southern cirque of Big Sister Mountain. The agricultural lands, wetlands and deciduous forests to the east provide some contrast to the extensive even-textured coverage of the mature forest.

Bonanza Face also has a strong sense of *genius loci*. The combination of its location at the head of the scenic Slocan Lake, its proximity to locally significant recreational and aesthetic features and the long and varied nature of human history in the area combine to give Bonanza Face a very special character.

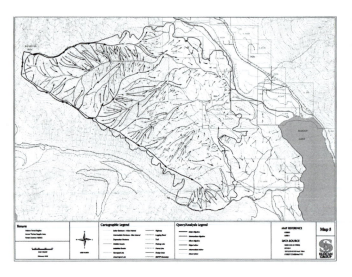

Constraints and opportunities analysis

This analysis was subdivided into two sections: higher-level and operational opportunities and constraints.

Higher-level opportunities and constraints

Higher-level (or top-down) issues include those which result from higher-order plans or objectives, government legislation and regulations, management guidebooks and policies or land and resource tenure factors. These are often based on very general ecological, social and economic assumptions and objectives. The following were analysed:

Visual quality and aesthetics
The Ministry of Forests has established Visual Sensitivity Ratings and recommended Visual Quality Objectives (VQOs) for the BFPA. The upper slopes are rated as high visual sensitivity and are to be managed to the partial retention standard while the lower slopes are rated as moderate and are to be managed to the modification standard.

Recreation
Although there were no formal surveys of actual or potential recreational use of the BFPA, the team did make use of considerable anecdotal evidence, observations, field encounters and consultation. The Recreation Opportunity Spectrum (ROS) ratings for the area highlight general activities which are known to occur or possibly occur in the areas delineated.

Archaeology
According to the Archaeological Overview Assessment there are two areas that have high archaeological site potential in or adjacent to the BFPA.

1   *Seral stage distribution* The *Forestry Practices Code Biodiversity Guidebook* recommends seral (successional) stage distributions for each Natural Disturbance Type (NDT) present in a given area. A comparison of these values for a larger landscape unit with the seral stage distribution of the BFPA provides an initial benchmark to help determine what rate of conversion could be ecologically sustained in the managed forest. These targets represent both a constraint on the timing and extent of forest harvesting and an opportunity to undertake forest harvesting to restore appropriate seral stage distributions in the BFPA
2   *Wildlife/biodiversity* Ungulate winter range areas have been identified in the southwestern corner of the design area on the southern slopes overlooking Shannon Creek. These areas may be subject to special management restrictions and objectives where the primary concerns are conservation of appropriate habitat conditions and maintenance of landscape connectivity corridors for the ungulate species.

Operational opportunities and constraints

Operational (or "ground-up") opportunities and constraints are those identified as important considerations through the inventory and assessment process. As a result of the inventories and analyses undertaken, there were a number of specific issues that were identified, each of which could be mapped:

Hydrology
Streams and wetlands and the adjacent terrestrial areas (collectively called hydroriparian ecosystems) provide unique habitat qualities and are generally the most biologically diverse areas in a landscape. In these areas large fires are rare and small stand-opening disturbances are more often the primary agents of change in the forest structure and composition.

*Figure 9.17
The map of
opportunities
and constraints
at Bonanza
Creek. Source:
Slocan Forest
Products.*

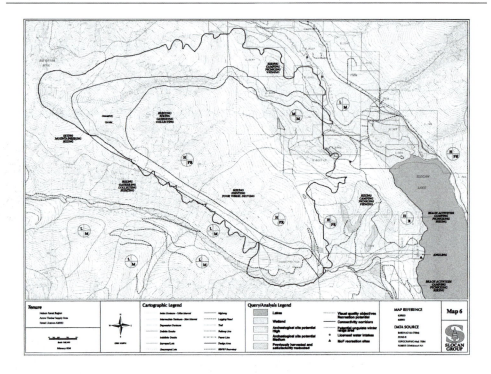

Equivalent Cut Area, or ECA, is a measure of the total area within a drainage that is non-forested through any type of timber harvest at any given time. These areas tend to behave quite differently from forested areas in terms of hydrological dynamics. It is therefore important to monitor and limit the ECA to appropriate levels within the various catchments of the BFPA.

Operability
Excessively steep, wet or rugged terrain often makes forestry operations impractical, ecologically questionable or very costly – as in the case of helicopter yarding. In the BFPA there are a number of areas that are considered inoperable including;

- all areas above 1,800 metres as this is the limit for reliable regeneration;
- the steep, exposed slopes above the Shannon Creek Forest Service Road at the southern end of Bonanza Ridge (6–8 km);
- the steep slopes of the *roches moutonnées* overlooking Slocan Lake; and
- the steeply incised gullies, avalanche tracks and riparian areas.

Soil resources
Localised steep and unstable (Terrain Instability Class IV or V) slopes may not be harvestable or may require special harvest systems. Future block- and road-specific assessments will be needed to identify operability and required measures. Gentle, well-drained lower slopes provide an opportunity to use conventional equipment with little adverse impact on soils.

Timber resources
The economic realities of forestry in British Columbia mean that higher operational and regulatory costs combined with lower returns, based on reduced harvest volumes and decreasing wood quality,

limit profit margins. The cost of harvesting timber from a site must be lower than the revenues generated. In the case of Bonanza Face, the operational costs are reasonable while the social and planning costs are high and the timber resource values are moderate.

Engineering systems

The combination of road construction and harvest systems is a critical element of practical planning when timber harvest is being undertaken.

Based on available geotechnical, hydrological and geomorphological assessments, a road system was planned for the BFPA to minimise the amount of road required while allowing access to timber resources.

There are a number of harvesting systems to choose from, including:

- ground-based yarding such as skidders, forwarders and crawler tractors
- cable yarding systems including high lead and skyline; and
- helicopter yarding.

In the case of the BFPA, all of these systems are potentially applicable but the final choice of system will be based upon a number of site-specific factors including:

- topography (slope steepness and variability)
- soil (composition, structure and sensitivity to disturbance)
- silvicultural system (level of retention, number of harvest entries)
- timber characteristics (log size and volume per hectare)
- potential road access and roading constraints; and
- yarding characteristics (distance, direction, slope and protection of other resource values).

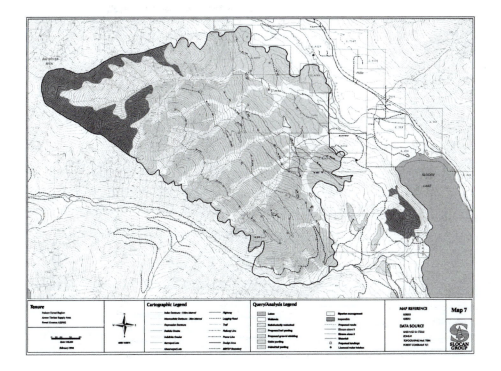

*Figure 9.18
The map of engineering and harvesting issues at Bonanza Creek. Source: Slocan Forest Products.*

Design strategy and concept

Although, in the processes described in Chapters 3 and 6, the desired future condition (DFC) is an end result of the ecological analysis, in this project it was deferred until the concept design stage where it was developed in combination with the outputs of the landscape character analysis and the constraints and opportunities. The forest ecosystem design for the BFPA allows planners to manage the working forest in a manner that will ensure the entire forest ecosystem maintains its health and functionality. In order to achieve the team's vision of creating a managed forest embodying a full range of ecosystem values in the spirit of sustainable forestry, a clear set of recommendations was needed.

Based on the results of the landscape ecological analysis and the landscape character analysis, the BFPA was subdivided into discreet DFC units, a kind of super-coupe or management unit at a scale above the normal coupe or cutblock, where distinctive natural disturbance regimes could be expected to occur. Each of these unit types has been assigned one of four DFC targets based on the expected disturbance regimes.

As a balancing act between human uses and a healthy and functioning forest ecosystem, the desired future condition will be achieved by:

1   Developing and maintaining an appropriate diversity of landscape structures, including matrices, patches, corridors and edges.
2   Ensuring a distribution of forest cover seral (successional) stages consistent with the recommendations of the *Forest Practices Code Biodiversity Guidebook*.
3   Limiting forest fragmentation and maintaining connectivity between old seral stage units, riparian areas and other important structures.
4   Managing for the conservation of biological diversity.
5   Protecting stream, wetland and riparian zone integrity.
6   Accepting the occurrence of endemic levels of forest pests and pathogens.

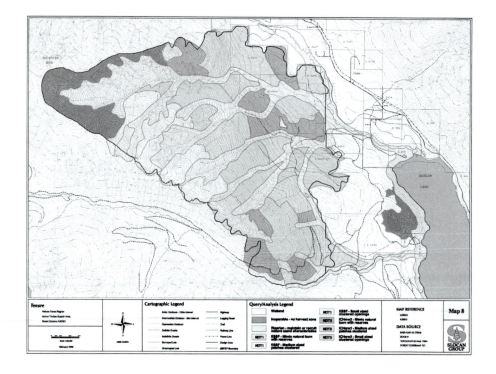

Figure 9.19 The desired future condition at Bonanza Creek. Source: Slocan Forest Products.

7   Modelling harvest patterns and systems on natural disturbance regimes.
8   Ensuring management activities do not degrade the aesthetic character of the area.
9   Maintaining existing human use patterns.

Sketch design and recommendations

Through the forest ecosystem design process a relatively detailed picture of the Bonanza Face landscape, including what it was like historically, what it is like currently and what we want it to be like (desired future condition) was developed. At the sketch design stage a landscape pattern was generated which combines the spatial organisation of the forest with the processes at work in it.

Based on the results of the ecological analysis and the landscape character assessment, the BFPA was divided into an intricate hydroriparian network and 60 individual design units or potential harvest cutblocks/silvicultural management units, varying in size from 47.6 ha to 6.9 ha. All of these areas were assigned a recommended management regime with the overall goal of achieving the DFC in terms of stand characteristics, forest structures and ecological functioning and their proportional representation within the landscape (Table 9.3).

The hydroriparian network was identified separately from the other design polygons and is to be developed under a special low-impact management regime. These areas comprise a critical component of the long-term landscape connectivity network, are recruitment zones for late seral stage forest and are important for protecting water and biodiversity values. Recommended management in operable areas would be single tree and small patch removal. With the exception of unavoidable crossings, no road development should occur, and to prevent soil degradation and sedimentation problems only minimal soil disturbance is recommended.

The remainder of the planning area is available for harvesting over a 300-year period. Forest harvesting is to be managed with the objectives of mimicking natural disturbance regimes, achieving seral stage distribution targets and conforming to equivalent cut area targets.

Design units
Each of the 60 discrete design units was assigned one of three management prescription frameworks with the intended goal of achieving an appropriate desired future condition for each individual unit and for the BFPA as a whole. The three management or treatment categories are intended to model forest harvesting activities on small, minor and large "catastrophic" natural disturbances.

The following are summaries of the silvicultural treatments designed for each of the disturbance types:

Only relatively *small openings* of less than 1 hectare will be prescribed to mimic single tree, small group or patch disturbances. These may be clustered or dispersed, depending upon the ecological nature of the block and the harvesting system involved. In more sensitive riparian areas

Table 9.3 Recommended disturbance treatments: proportions and total area

BEC	Disturbance scale	Proportion of design area (%)	Total area (ha)
(NDT 1)			
ESSF	Small (<1 ha)	40	178
	Minor (>1 ha <20 ha)	40	178
	Large (>20 ha)	20	89
(NDT 2)			
ICH	Small (<1 ha)	40	439
	Minor (>1 ha <20 ha)	30	329
	Large (>20 ha <40 ha)	30	329

*Figure 9.20
The final
cutblock design
at Bonanza
Creek. Source:
Slocan Forest
Products.*

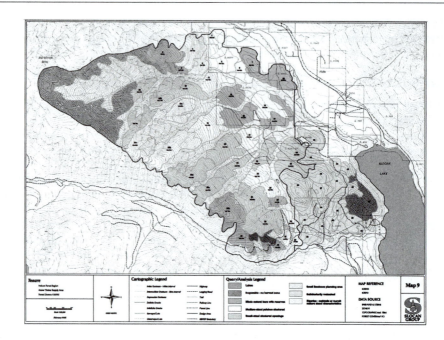

and designated corridor areas harvesting is to be very low impact. Management considerations for small disturbance openings include:

- use small openings to allow for the emulation of species succession
- vary opening sizes by removing individual trees or small groups
- locate these openings on wetter, cooler sites
- locate these events in younger stands
- complete reliance on these harvest methods will lead to excessive fragmentation.

Minor openings will be between 1 ha and 20 ha in size to mimic low-intensity, short-lived fires or other disease/insect outbreaks. These may be grouped or clustered as they often occur in this pattern in natural circumstances. The use of variable retention systems may be appropriate to leave veterans and reserves within some of the larger openings. Management considerations for minor disturbance openings include:

- use irregular opening boundaries
- vary opening sizes and cluster where possible
- locate these events where moisture deficits are not normally anticipated
- use variable retention to maintain single trees, snags and groups where possible
- complete reliance on these harvest methods will lead to excessive fragmentation.

Larger "burn mimic" openings or larger "catastrophic" openings of up to 40 hectares can be created using variable retention with reserves to mimic more extensive natural burns. Although large catastrophic openings of up 300 hectares could be expected under natural conditions, harvested openings larger than 40 hectares would be socially and ecologically unacceptable to the local community. Therefore, the creation of catastrophic-scale openings greater than 40 hectares in size will be accomplished through aggregation of blocks over time. Management considerations for larger "catastrophic" disturbance openings include:

- mimic mainly fire occurrence (does not include pest or pathogen disturbances)
- design openings with irregular boundaries following the local terrain
- leave retention patches, single trees and snags scattered throughout the openings
- site retention patches on wet areas or depressions that a fire might exclude
- locate these events on drier aspects where a moisture deficit might be expected
- create only a few larger openings at any given time
- possible to aggregate these openings into larger "catastrophic" disturbances once green-up is achieved.

Implementation and monitoring

Once the final design has been approved and accepted by the client, the implementation is to be carried out in phases. The Bonanza Face Forest Ecosystem Design is intended to serve as a long-term planning framework, subject to adaptive management revisions, to guide operations in the area for up to 300 years.

In order to make sensible decisions based on relevant circumstances and adaptive management feedback, tactical planning windows of 20 years and operational planning windows of five years are anticipated. For example, during the assembly of a five-year Forest Development Plan (FDP), planners and foresters will use the design as a planning framework to arrange the spatial distribution of operations and to schedule operations with consideration of new social, economic or ecological constraints, opportunities and understanding.

Furthermore, as part of the cutting permit application process, the design will help guide decisions relating to road systems, silvicultural systems and harvesting systems. In doing so the various management goals can be addressed while working towards the desired future condition for the area. Harvest blocks must be spatially and temporally distributed to ensure that the ecological and aesthetic integrity of the area is maintained.

During field layout and the development of silvicultural prescriptions the field forester will analyse the particular site characteristics through on-site investigation. This investigation will lead to the decision on which disturbance process is best suited to the site. The design units are distinguished by the likely natural disturbance regime anticipated within each stand. However, this does not preclude the introduction of a disturbance type other than the one defined in the design.

Adaptive management

Adaptive management (see Chapters 5 and 7) provides a framework for managers to learn from their actions, improve management and accommodate change. Implicit in this is the acknowledgement

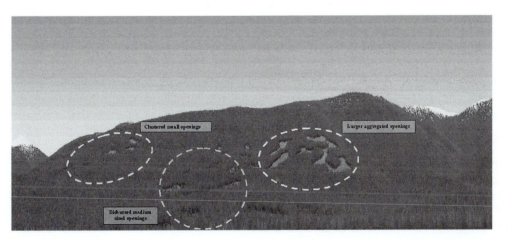

Figure 9.21 The silvicultural models developed for Bonanza Creek. Source: Slocan Forest Products.

of the limitations in understanding ecosystem response to management activities and the need to reduce these uncertainties through strategies involving operational experiments, monitoring and feedback. The framework of adaptive management involves five stages, of which the first three are fully developed as part of the forest design process. The five steps are:

1   Define management objectives and system parameters.
2   Develop a management plan.
3   Implement the management plan.
4   Monitor the implementation and outcomes.
5   Evaluate results, integrate feedback and adjust plans.

An adaptive management plan was developed for the BFPA and started immediately the forest design was approved.

Implementation

At the time of writing, quite a lot of the first five-year implementation of the design had taken place. The access road was constructed to the necessary high standard and in ways that were practical, cost effective and had low impact on the ecosystem or landscape. A number of harvest units have been completed using the silvicultural models described above. The units were laid out with little trouble, using some of the techniques described in Chapter 8.

Case study: Sutherlands Brook, Nova Scotia, Canada

Introduction

This forest design plan covered a relatively modest area of 5,100 ha of Crown land managed under licence by the forest products company StoraEnso. It is located in central Nova Scotia in Canada, in the area of the St Mary's River. The area comprises part of the drainage catchment of this river, mainly the tributary of Sutherlands Brook and its connecting streams. The landscape is gently sloping, flat or undulating, but nowhere of strong landform. The hydrological system forms a complex pattern of many streams (brooks), lakes, ponds and wetlands lying in hollows among the undulating topography, some of these forming the boundary to the area, and all generally flowing north to south. The St Mary's River is one of the most valuable salmon and trout fishing rivers in Nova Scotia and the project area is important for a wide range of recreation, including hunting and snowmobiling. The area is sparsely populated and there are no residents within the project area.

The forest is part of the so-called Acadian forest, a transitional type between the eastern hardwoods and the boreal conifer forest zones. It is mixed, with patches of hardwoods, such as maple and yellow birch, balsam fir, white pine and mixed stands set within a matrix of black spruce. The forest was by no means in a pristine condition, there having been several logging operations recently completed. These were not "standard" clearcuts but contained within them patches of retained trees and strips along riparian zones and were of an irregular shape. In fact, the coupe-level planning and design was good but there was no landscape-level plan within which to locate them.

Figure 9.22 Location map of the Sutherlands Brook project area in Nova Scotia.

The project was a pilot study for the company and was set up as a partnership between the company, based at Port Hawkesbury, the Nova Scotia Department of Natural Resources, the Canadian Forest Service and other forest users. It was also carried out with community participation (see Chapter 4).

Objectives

The following table sets out the objectives:

Inventory

Scoping produced much inventory information while the local knowledge filled in many gaps. The inventory materials were fairly basic maps,

Figure 9.23 An oblique aerial view of the Sutherlands Brook forest ecosystem design area.

aerial photos and reports. This is typical of the amount of information available for areas that have not seen much activity and therefore detailed survey work. It was also the case that there were significant gaps. The lack of detail was more than made up for by the extent of local knowledge available as a result of the participants in the design workshop. Professional expertise, including ecologists and recreation specialists, was available, which added to the quality of the work.

Analysis

The analysis process followed the fully developed process – landscape character, landscape ecology and constraints and opportunities – presented using the approaches described in Chapter 6. Each analysis produced some detailed outputs.

Landscape character analysis
The landscape of Sutherlands Brook is not highly visible and mainly experienced from internal viewpoints, for example from roads. As logging proceeds and the landscape becomes more open, then views are going to increase. However, the landscape character analysis offered more than that of the basic visual analysis for landscape design purposes, for example the landform analysis formed the basis for the concept options developed later on.

The landscape character analysis revealed a complex area despite relatively simple, subtle topography. The landform divided the area into three sections: the northwest and southeast areas were flattest, with fewest features, while the central section comprised more undulating topography with many well-defined knolls. This was to have a strong influence on the design. The land feature analysis was heavily influenced by some existing cutblocks as well as by the numerous small water features that characterise the area.

Landscape ecological analysis
The process followed was close to that described in Chapter 3. The landscape structures were defined as follows:

Matrix
There is one single matrix type identified for the area. It comprises between 30% and 100% black spruce together with varying proportions of balsam fir, larch, red and white pine and some intolerant hardwood (poplar, aspen and white birch). The ground layer comprises moss and ericaceous species such as *Kalmia* spp. There is no shrub layer, so it has a simple structure. It is even-aged, some 80–90 years old, single storeyed and is of fire origin. It has potential to continue as this forest type for some time unless disturbed, gradually becoming more open. It is vulnerable to few insect pests.

Table 9.4 Objectives at Sutherlands Brook

Resources	Objectives
Timber	To maintain a sustainable supply of timber while managing on an ecosystem and landscape level: • Balance the age class of the existing forest • Harvest maximum wood volume within the framework of project objectives • Actively manage productive areas (by promoting and encouraging natural regeneration) • Enhance productivity and quality of selected tree species by the use of silviculture • Maintain special species such as white pine or hardwood content
Visual landscape	To retain a high standard of visual integrity across the forest over the normal rotation: • Protect aesthetic aspects of forests, especially horizons, riparian zones and significant viewpoints • Locate roads to enhance views of landscape
Fish and wildlife	To protect and enhance fish and wildlife resources as specified by the province, provincial guidelines and company agreements (as a minimum standard): • Maintain wildlife habitat as outlined in the *Nova Scotia Wildlife Habitat Guidelines* • Minimise access to selected wildlife habitats to lessen disturbance to wildlife species • Protect critical habitat for wildlife (includes plant or animal) species, especially rare species
Old growth	To maintain a proportion of old growth forest over time and landscape: • Maintain 5% old growth forest
Recreation	To protect the existing recreational potential of the area and provide more recreational uses if appropriate for the ecosystem: • Utilise selected areas to enhance ecotourism opportunities • Enhance recreational opportunities e.g. hiking trails, snowmobile trails • Provide for segmented recreational opportunities to reduce conflicts between users • Establish designated parking for safety and convenience • Maintain recreational access for all-terrain vehicles (ATVs) and snowmobiles (retain established trails and old roads during harvesting)
Water	To maintain water quality: • Ensure the water supply and quality are protected for the St Mary's River system by maintaining feeder stream integrity • Maintain water quality in lakes
Forest health	To maintain overall health of the forest to ensure ecosystem function: • Design stands with respect to fire protection • Maintain a well-developed system of corridors and riparian zones to connect habitat types • Maintain or enhance biodiversity with a diverse network of cuts, stands and forest types throughout the area • Create a variety of cut sizes, shapes and tree species • Investigate disturbance history of the area and attempt to re-establish natural succession of ecosystems

(a) (b)

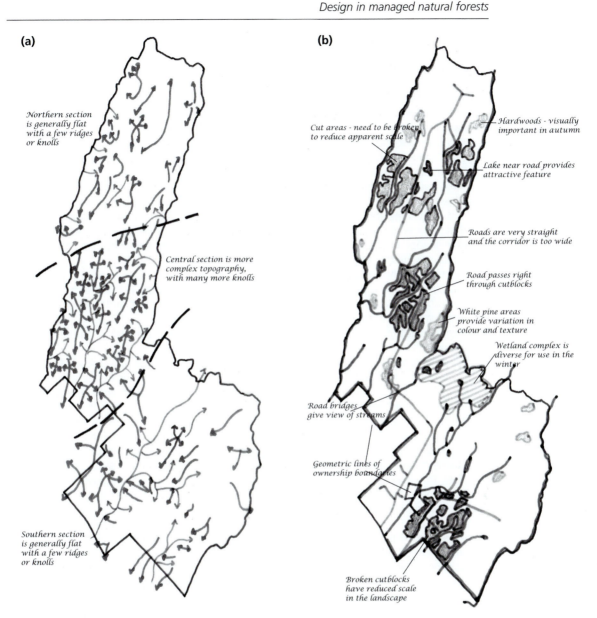

Northern section
is generally flat
with a few ridges
or knolls

Central section is more
complex topography,
with many more knolls

Southern section
is generally flat
with a few ridges
or knolls

Cut areas - need to be broken
to reduce apparent scale

Hardwoods - visually
important in autumn

Lake near road provides
attractive feature

Roads are very straight
and the corridor is too wide

Road passes right
through cutblocks

White pine areas
provide variation in
colour and texture

Wetland complex is
diverse for use in the
winter

Road bridges
give view of streams

Geometric lines of
ownership boundaries

Broken cutblocks
have reduced scale
in the landscape

Figure 9.24
Landscape analysis at Sutherlands Brook: a) landform analysis and b) land feature analysis. Source: Canadian Forest Service.

Patches

Most of the patch types are forest patches, some man-made, the rest natural. There is a wide variety of patch types showing that the area is very diverse in character.

- *White pine* stands range from areas of almost pure pine to patches with as low as 30% pine. They have varied ages, some as young as 30 years and some as old as 300 years. In the denser stands there is no understorey or ground layer, merely a carpet of needles, while in the more mixed stands a multilayered structure has developed with ericaceous shrubs.

They have arisen following fires but are stable and long-lived. They tend to be found on drier hummocks and vary in size from 1 or 2 ha to 20–30 ha.

- *Balsam fir* stands are associated with intolerant hardwoods and spruce. The patches are even-aged and may be over-mature and beginning to break up, with dead and rotten trees. Some may be of fire origin, other areas having colonised following some harvesting that took place in the 1930s. They have irregular shapes and vary in size from 5–10 ha in the south to 70–100 ha in the north.
- *Red pine* is found mixed with small percentages of larch and white pine. It was planted 25 years ago following a fire. It is infected with the pest *Sirrococus* and may have to be cleared. These stands have a high contrast with their surroundings as a result of being planted, though they have an irregular shape.
- *Mixed wood* stands are mainly intolerant hardwoods (white birch and red maple) with spruce and larch. They are full of herbs, ericaceous shrubs and lichens. They originated after fires that occurred 80–90 years ago and form a transitional stage, eventually succeeding to white pine or another conifer type. They also have an irregular shape and fairly high contrast, with a relatively narrow edge transition zone.
- *Cutovers* are harvested areas with early successional vegetation of bracken, grasses and *Kalmia*. They are unstable, being in the first successional stage, perhaps lasting for 5–10 years. They are more regular in shape than the natural forest patches, and tend to be grouped and larger than the average size for the natural patch types.
- *Planted cutovers* have been planted with black spruce but are otherwise the same as the cutovers.
- *Hardwoods on ridges* comprise mainly intolerant hardwoods but also some tolerant hardwoods (sugar maple and oak). There is more of a shrub layer, with fewer ericaceous species, and is one of the most floristically diverse patch types. It is also of fire origin and unstable because of the intolerant hardwood content. However, these stands are likely to succeed to tolerant hardwood and become very stable. They have a high contrast and strong edge, are often linear in shape reflecting the ridges where they are found, some 7–15 ha in area.
- *Hardwoods along streams* mainly comprise intolerant hardwood, especially red maple. Stands can be wetter or drier. The wetter variants have cinnamon fern as part of the ground layer. They may have arisen from fires or be fire refugia and are stable because of the edaphic conditions. They have a high contrast, a strong edge and narrow linear shapes, being quite large, some 1–20 ha in area.
- *Borrow pit* This is an area where road-making materials were extracted. It has pioneer vegetation of lichens and mosses. The shape is regular, the edge hard and the vegetation succession is likely to be slow. It is 2 ha in area.
- *Barrens* are areas where forest succession has failed to occur after fire. They have become dominated by ericaceous vegetation, mainly *Kalmia*. They are very stable, since succession is extremely slow. They have a high contrast with their surroundings and a hard edge. They are between 2 and 10 ha in size.
- *Lakes* provide open water and are meso-oligotrophic in character. They are of glacial origin, mainly kettle holes. They are very stable, quite regular and simple in shape (round or oval) and can be linked to other wet features. The average size is 4 ha.
- *Open bogs* comprise sedges and mosses and contain waterlogged peat and, if they have areas of open water, leatherleaf. They originate from shallow ponds that have gradually filled up with mosses and peat. They are now very stable and present a high contrast with their surroundings. Edges can be abrupt or more gradual as trees spread into the moss. They have a regular circular or oval shape.
- *Treed bogs* are waterlogged areas of sedge and peat but being slightly drier have become colonised by black spruce, larch and even fir and red maple on the drier hummocks. They represent a successional stage beyond the open bogs. They have a lower contrast and graded edges.

- *Fens* are less acidic and more fertile in character and also contain sedges and peat but are colonised by red maple and alder. They tend to be part of the riparian system and are flooded, receive sediment and are associated with the edges of water features. They are linear in shape and of high contrast.
- *Ponds linked to streams* are meso-eutrophic in character because they receive silt and nutrients from the stream system. They originate from small kettle holes and are round or oval in shape. They are less stable than lakes because they are smaller and shallower and often have a lot of weed and emergent vegetation.
- *Isolated ponds* are also of kettle hole origin but, being isolated tend to be eutrophic as there is no flushing of nutrients. They have similar plants to the other ponds but are less stable and more likely eventually to fill up and turn into bogs.
- *Beaver pond* This is an isolated example on a stream, dammed by beavers. It is extremely unstable and could burst or be washed away any time. It is shallow and 1 ha in size.
- *Graveyard* This is a small area used for the burial of local people. It is mown grass and stable while maintained as it is.

Corridors
- *Roads* These are strips cleared out of the forest with a gravelled centre section and mown edges kept at an early seral stage of growth. All are kept in use and maintained, so they are stable. They have a high degree of contrast with the edges and neighbouring stand types. There are approximately 35 km of roads in the area.
- *Streams* These include perennial and seasonal watercourses. They have a low to moderate gradient and flow at moderate rates. They are mainly floored with pebbles, gravel or some silt. They are stable and offer a high contrast with their surroundings. Some channels were straightened in the past to make log driving easier. They tend to flow from the north to the south – there is a distinct pattern to them. They range in width from 5–7 m to 0.5–2 m.
- *Riparian zones* These can be composed of any forest type depending on where the streams flow, and form a transitional zone from the wetter edge of the streams and ponds to the upper edge of the shallow valleys through which the streams flow. When left after logging they present a hard outer edge and strong contrast, but when part of a larger patch there is very little contrast, if any. The width of the corridor depends on the width of the stream valley.
- *Snowmobile trail* This is a narrow opening created in the forest. It has no constructed surface and creates barely any contrast with the surrounding forest.
- *Road right of way* This is a temporary corridor created to make a road but not yet surfaced. It is like a road but lacks the gravel surface.
- *Boundary strip to reserve* This is a strip cut through the forest to mark the boundary of a nature reserve. It is narrow and presents almost no contrast with the surrounding forest.

These were mapped and immediately a pattern of the occurrence of many patches began to emerge.

Landscape flows
Landscape flows were chosen as recommended in the description of the landscape ecological analysis process and included the typical categories of biotic and abiotic flows.

- *Water* is an important flow in this landscape, related to many structures and sensitive to forestry activities.
- *Deer and moose*, taken together, are large mammals using much of the forest at different successional stages throughout the year.
- *Song birds* as a group comprise many species, some with different habitat requirements and also varying seasonal use patterns.
- *Black bear* is an important high-food-chain species, which is also sensitive to human activity.

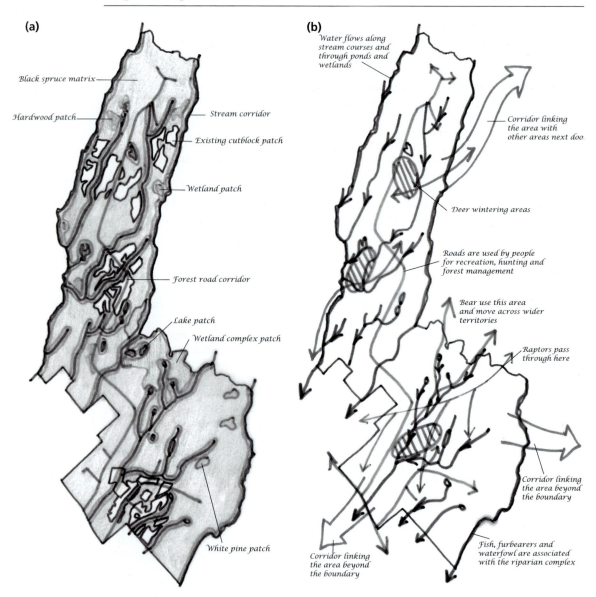

(a)

Black spruce matrix

Hardwood patch

Stream corridor

Existing cutblock patch

Wetland patch

Forest road corridor

Lake patch

Wetland complex patch

White pine patch

(b)

Water flows along stream courses and through ponds and wetlands

Corridor linking the area with other areas next door

Deer wintering areas

Roads are used by people for recreation, hunting and forest management

Bear use this area and move across wider territories

Raptors pass through here

Corridor linking the area beyond the boundary

Corridor linking the area beyond the boundary

Fish, furbearers and waterfowl are associated with the riparian complex

Figure 9.25
Ecological analysis at Sutherlands Brook: a) ecological structure and b) landscape. Source: Canadian Forest Service.

- *People use* the landscape for timber production but also make a lot of use of it for different types of recreation at different seasons.
- *Fish* can be found in most of the streams, especially the perennial ones, and other water features and are sensitive to water quality and to levels of shading by riparian forest.
- *Raptors* are also a group of high-food-chain species with seasonal habitat requirements.
- *Furbearers* are a group of animals of some economic value and tend to favour forests with a high matrix component of mature forest. They also use large-scale territories.

- *Water fowl* are users of all the water and wetland patches and corridors.
- *Wood turtles* are an especially important species with specific habitat requirements, mainly in the riparian zones and wetlands.

In this case study there is not room to present the spreadsheet of interactions between structures and flows. What emerged from this analysis was that, as expected, the streams, ponds, lakes, wetlands and riparian forest areas, as a network, were particularly important for many functions and flows. Areas of old forest, especially mixed species types, tolerant hardwoods and wet forest areas, were also important. Maintaining areas of mature conifer to act as winter range for deer also proved to be a significant requirement. Early successional forest is important because of the browse it provides for ungulates and as habitat for rodents, which are prey for raptors and some furbearers.

Linkages between the study area and the surrounding landscape were also considered. The study area is not large and forms part of a wider range for large mammals such as moose and bear. It only forms part of a larger hydrological system. Maintenance of corridors and late successional habitat connectivity with adjacent areas needs to be considered. Forest management practices in these areas may cause problems for the ecological functioning of the study area but this is outside the control of StoraEnso.

Disturbance and succession

The interactions between disturbance and succession in this forest type are rather interesting. Depending on the effects of disturbance the main forest type, black spruce, may develop different successional pathways leading to very different replacement stands.

Disturbance

The analysis of natural disturbance factors for Sutherlands Brook proved interesting. The disturbance agents considered to be significant were fire, insects and fungi, wind, sleet, snow and ice, fauna, such as porcupine and deer, beaver, moose and larch sawfly (as a specific insect pest). Human disturbance, such as clearcutting and road construction, was also included so that a comparison between these and natural agents could be made.

Succession

Several successional routes were evaluated for this project. Since the main forest type is black spruce, succession pathways were developed for it depending on different factors, such as scale of disturbance and available seed sources. A succession pathway from clearcutting was also included for comparison purposes.

There are distinct contrasts between, for example, the post-fire black spruce, which proceeds through five successional stages to become old growth at age 250 years, and managed black spruce after clearcutting, which only proceeds to around 80 years before being ready for harvest. This highlights the differences between the natural and the managed stands, although it is doubtful that much of the post-fire black spruce grows beyond 100 years due to the likelihood of repeated fire. Old growth conditions tend to occur in the places least likely to burn such as riparian zones, which will be retained in any case. Thus, the apparent discrepancies between the two forest types may not be as great as they first seem.

The conclusions about the role of fire and the chances to create mixed stands as well as the potential similarity between naturally disturbed and managed spruce suggest that moving away from some kind of modified clearcutting system may not be entirely appropriate from an ecological point of view, as long as the features left in stands after fires are also retained after harvest. This provides some interesting possibilities for developing the desired future condition (see design concept stage below).

Table 9.5 Analysis of disturbances at Sutherlands Brook

Type	Scale	Duration	Frequency	Structure
Natural				
Fire	Large fires: 1,000s ha. Small fires: 1–100 ha	Months. Days–weeks	100 years	Scatter of large old pine, wet and low-lying areas survive and some tolerant hardwoods left unburnt, creating a broken and patchy area
Insects and fungi	Minor – individual trees or small groups affected throughout the forest	Continuous effects	Continuous	Small holes, especially in mature conifers, leading to a multi-storeyed forest canopy
Wind	Patches of 1–5 ha may be blown in mature conifer stands, especially where holes have already been created by fungi or insects. These may keep enlarging over time	Hours	Every few years, but no known pattern	Patches which gradually grow and allow regeneration to develop. Pines left standing
Sleet, snow and ice	Individual trees, especially hardwoods, in the north of the area	Hours	Every winter	Individual trees break or fall over under the weight of snow and ice, opening small gaps and allowing shade-tolerant regeneration to develop
Fauna, porcupine and deer	Fraying and browsing on tree bark kills individual trees or small patches, especially fir and hemlock	Over a winter in range areas	Every winter	Patches of dead and dying standing trees
Beaver	Creates ponds of up to 5 ha in area, felling trees in the vicinity of such areas	Years	Continuous where they occur	Flooded areas, standing dead trees in wet areas, suckering of aspen after beavers have felled them
Moose	Browse trees in wintering areas, 1–2 ha in extent	Over winter	Annually	Affect forest structure but no mortality
Larch sawfly	All larch trees	Several seasons	50–60-year cycles	Groups and patches of standing dead trees
Human agents				
Clearcutting	From 30–50 ha (upper legal limit)	Months	Around 90–100-year rotations for spruce	Open areas with some snags left, also patches of trees on wet areas. Lots of slash and the removal of much of the biomass
Road construction	Variable	Permanent openings	Once there, are permanent	Hard edges, linear openings, maintain early seral stage vegetation

Table 9.6 Analysis of succession pathways at Sutherlands Brook

Stage	Successional stage	Age	Structure and composition
	Post-fire black spruce: succession back to spruce		
1	Stand initiation	1–20	Ericaceous species, bracken and herbaceous species colonise and dominate. Birch and spruce seed blown in and become dominant by the end of the period
2	Early stem exclusion	20–30	All vegetation becomes shaded out
3	Late stem exclusion	30–80	Vegetation continues to be shaded out, natural thinning takes place. Mosses eventually form a ground layer. Duff builds up. Stands become increasingly at risk from fire
4	Mature, stand reinitiation	80–150	Canopy starts to open and some insect, fungi and wind damage allows some regeneration to begin. Gradual development towards old growth condition. Fire risk continues to increase
5	Old growth	150–250	Old growth. A more open canopy develops and advance regeneration leads to a multi-storeyed forest. Indicator species of lichen and fungi may also appear. Dead wood builds up, providing more fuel for fires
	Post-fire black spruce: succession to intolerant hardwood or mixed wood		
1	Stand initiation	1–10	Invasion of the site by intolerant hardwood species – seed blown in from neighbouring stands. Fire was so intense that most of the spruce seed was destroyed. Grey and white birch, pin cherry and maple are the main species
2	Stem exclusion	10–20/50	Pin cherry and grey birch are shaded out, the white birch is thinned out and maple persists. The rate of progress depends on the fertility of the site
3	Regeneration initiation	50 onwards	There are two possibilities: 1 On the wetter and less fertile sites spruce and pine eventually invade and gradually grow through the canopy, resulting in a mixed wood with maple as the main hardwood. This forest type may be prone to fire but the chance of large-scale fire is lower because of the hardwood element 2 On well-drained and fertile sites the intolerant hardwood persists and develops. Black cherry may invade. Some softwood species may invade gradually over time. This forest type is less prone to fire
	Clearcutting of black spruce with natural and artificial regeneration		
1	Stand initiation	1–10/15	Either natural regeneration of black spruce or, if this does not occur quickly, artificial regeneration by planting black spruce or Norway spruce. Crown closure of black spruce by year 15 and of Norway spruce by year 10
2	Stem exclusion	10–45 or 15–80	Understorey and ground vegetation is either shaded out or artificially removed. Self-thinning of natural regeneration of black spruce. Planted stands may be thinned once trees reach merchantable proportions
3	Maturity	45+ or 80+	Norway spruce is harvested sometime after it reaches 45 years, depending on growth rate. Black spruce is harvested after 80 years but before it reaches 100 years
	Small-scale distribution of black spruce within the fire cycle		
1	Stand initiation	After 80 years	Small-scale disturbances by small fires, wind, pests, etc. create small openings, up to 5 ha in size, allowing regeneration to commence. This may be of shade-tolerant or intolerant species, both hardwood and softwood
2	Stem exclusion	After 20 years	The regeneration reaches canopy closure and shades out the ground vegetation. Intolerant hardwoods begin to be shaded out
3	Understorey reinitiation	After 80 years	The regenerated areas start to open up and allow some understorey reinitiation. Leads to multilayered stands and, in the absence of large-scale disturbance, could create a very different forest

Opportunities and constraints analysis

The opportunities and constraints analysis, being done in a participatory way, yielded a large amount of information that was both recorded on a map and presented in a table. As well as analysis in relation to the resource values identified in the objectives, other factors were selected, such as ownership, logging techniques, land-use history and the location of the area. What became clear was that there were many opportunities presented by the easy terrain and the nature of the forest to protect features of value and to incorporate complex harvesting methods to emulate the effect of natural disturbance, which would be of great benefit to wildlife.

Design concept

The design concept stage is based on the identification of all the critical, irreplaceable habitats and a network linking them all together. This incorporated the riparian zones, corridors, deer wintering areas, old growth and all the water features. Also, the tolerant hardwood stands were added as a second element. This resulted in a complex pattern that served to subdivide the area into many large and small patches of operable forest. The cutblock design reflects the variability in the landscape character, by including some larger-scale blocks to the northeast and southwest, where topography is flatter and the landscape holds few barriers to the spread of a fire, while in the more diverse central portion, some smaller blocks would naturally fit.

One difference of the Sutherlands Brook plan compared with the other case study areas is that it only covers a period of ten years, with part of it already cut. This is unusual, but reflects the forest management system and the inflexibility of some aspects of harvest planning and wood supply in that part of Canada. While this results in significant disturbance over the next ten years, after that the forest will be undisturbed for a century, allowing large-scale habitat development, recreation and water quality protection. Such is the value of the forest ecosystems network provided by the critical habitats that this was also considered to prevent the scale of harvest being considered too much of a problem. The plan area is relatively small for the scale of ecosystem processes. Fires could naturally burn the entire area, so that the rate of harvest can be argued to be within the natural range of variability. However, there continues to be the problem of lack of integration with neighbouring land and the effect of large-scale cutting all over the landscape.

The company presented the plan to the Provincial Integrated Resource Management Committee for approval, which was obtained without any problems.

Implementation

The plan was completed and approved in 1995 and implementation has proceeded since, reaching completion recently. This involved setting out some complicated shapes and also carrying out some cutblock level of design to incorporate smaller-scale features of stream zones and wildlife tree patches which were not considered at the scale of the whole plan. This had the effect of reducing significantly the visibility of the larger blocks from any given viewpoint, and provided a good deal of retained habitat within the blocks. In fact, so much material is left behind that they present what might be regarded as a scruffy appearance, untidy and badly finished. In fact, the scruffier they are, the more "natural" they are, so public perception of tidiness needs to be adjusted to accommodate the needs of emulating natural disturbance.

In 1999, a visit was made to Sutherlands Brook by Simon Bell and Tom Murray (of the Canadian Forest Service) to inspect progress on the ground and to obtain feedback from the managers and stakeholders. According to the planning forester Derek Geldart, the company was then in the fifth year of implementation, i.e. halfway through, and the integrity of the plan remained intact, although a right of way for a gas pipeline had been put through part of the area. Between 1996 and 1998, 400 ha had been harvested yielding 64,000 m^3 of softwood and 1,000 m^3 of hardwood. 150 hectares were scheduled to be cut in 1999 to yield 20,000 m^3. Natural regeneration on most stands has turned out not to be adequate and so in-fill planting is taking place.

Table 9.7 Opportunities and constraints at Sutherlands Brook

Factor	Opportunity	Constraint
Roads	Already built and allow operations to be moved around freely Provide access for recreation Provide access for hunting There is a chance to review road plans and revise the routes for new ones to fit the plan	The alignments cannot be changed The roads are straight and not optimal for wildlife, recreation or amenity Roads give too much easy access, which could lead to over-hunting or over-fishing Roads split the area and may present barriers to wildlife
Logging techniques (use of forwarders)	To reduce the road amounts needed by increasing forwarding distances To create more variable shapes to cutblocks Less ground disturbance than from skidders Combination of roads and forwarders enables fast responses to blowdown or fire salvage Few limits on size of cutblocks – small blocks are economical There is lots of wood, so cutting can take place anywhere and be spread around the landscape	Soft ground is inaccessible to equipment, leading to unlogged areas and loss of potential timber
Maturity of existing forest	More flexibility of products possible Big potential for deer winter range To create early seral stage habitat and set off succession to widen age classes/successional stages	The old forest is at greater risk of catastrophic disturbance The large proportion of old forest limits habitat values over the area The old forest as winter range limits some harvest opportunity
Balsam fir	To develop recreational fishing	This is old and starting to collapse, so needs to be cut soon to avoid wasting timber
Hydrological system	To retain high visual quality and diversity Potential for greater biodiversity The widespread distribution of water in the area gives opportunities for habitat development To develop and connect riparian corridors To improve fish habitat, such as structure in the streambeds To break up the likely spread of fire	Causes operational access limitations because riparian areas are no-go areas for machines Some wetter sites restrict tree growth and timber productivity Water quality and fish spawning areas downstream need to be protected if the fish are to continue to reach the project area
Crown ownership	No limits on access by the public, therefore potential for recreation throughout the area and all year The province can exert influence for the land to be managed for a wider set of objectives	All-year-round open access can lead to disturbance of wildlife at breeding seasons, etc. More limitations placed on forest management activities

Table 9.7 (Continued)

Factor	Opportunity	Constraint
Private land	May be a contrast in management with Crown land The opportunity to incorporate private land into the plan depends on the attitude of the owner	Management of private land may cause problems if habitat connectivity is broken The shape of the private land boundary is very geometric
Protected area	To use it as a benchmark for monitoring purposes To become an old growth component To link with other ecological corridors To represent a reasonable size of undisturbed habitat To prevent disturbance by people	It has been removed from the lease and therefore is unavailable for cutting It is an island so needs to be linked to other areas Its shape is somewhat artificial It has restricted public access – no trails
Topography	To develop a diversity of structure related to topographical variations Very suitable for winter operations Most of the area has a southerly aspect, so good for deer wintering	Not many views are available, so the landscape can be monotonous
Geology	To develop mineral exploitation Sandstones give moderately fertile and free-draining soils Fast and cheap road building No need to import non-local road materials	The simple geology means fairly uniform soils and sites, so diversity potential is limited Permanent environmental damage could be incurred by mining
Soils	Firm, stony soils present good running for machinery and low risk of sediment yield Stable soils offer chance for off-road vehicle use with low risk of damage	Fairly limited species choice due to soil type Slow growth rates affect timber productivity, silvicultural flexibility and rotation length Thick duff layer affects regeneration Compaction risk in some areas
Vegetation	Fire may be a management tool to maintain natural character To use some introduced species such as Norway spruce to increase timber yield To maintain present levels of diversity	The dominance of black spruce limits diversity in products and habitat Difficult to achieve natural regeneration Less desirable to remove the vegetation diversity, i.e. non-black spruce species

	The high proportion of white pine gives diversity Some unusual tree combinations can be protected and enhanced by management The natural large scale of the vegetation pattern presents opportunities to develop large patch sizes May be a prime area for new discoveries of fungi, mosses and lichens (poorly surveyed to date).	Fire risk Low percentage of natural regeneration over the area *Kalmia* grows well and is not conducive to spruce – pine is better
Fauna	To improve habitats by creating a range of sizes and shapes of cutblocks The undisturbed nature of the area could benefit larger mammals compared with adjacent land areas	Logging needs to be modified in deer wintering and riparian areas
Surrounding landscapes and land-use history	To introduce characteristics of similar areas to east and west into the plan consideration and to respond to the contrast with north and south landscapes, that is the grain of the wider landscape Therefore there is no need to provide everything everywhere	Cutover areas nearby put pressure on mature areas within the plan area Red pine of poor performance on the barrens may affect preferences for the use of species and affect the landscape at the bigger scale
Location of area	There is low pressure on the area, so it could remain remote and wild If it is of low ecological interest in Nova Scotia terms, more intensive management could relieve pressure elsewhere To link with the trans-Canada trail passing to the north of the area	It is a long way from settlement and difficult to get there by non-motorised means Pressure from motorised recreation access Monitoring of recreational use difficult because of location
Existing logging	Already diversifying age structure. To change species on areas logged but not yet replanted To learn from what has been achieved so far in terms of regeneration, damage, stand treatments etc.	Logged areas cannot be used for timber production for a rotation Visual problems of shape and scale Shape and scale may not be best suited to biodiversity Adjacency rules constrain where next logging can take place
Timber	To supply wood to local mills	Already committed to produce certain timber volumes in the short term

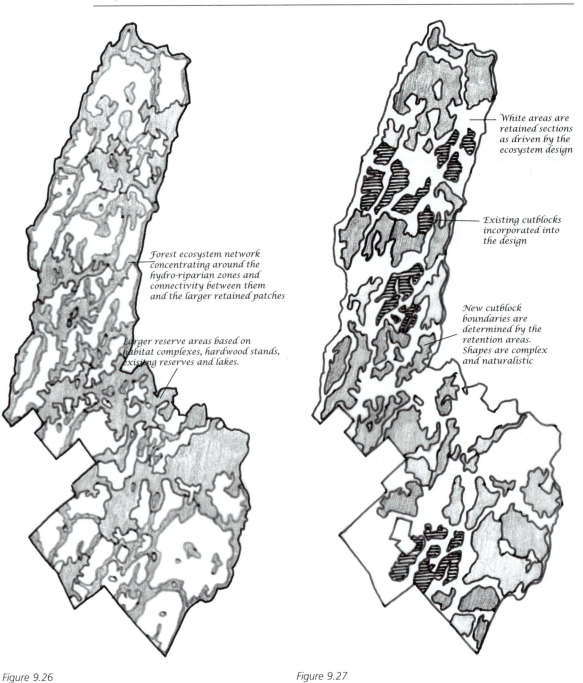

Forest ecosystem network concentrating around the hydro-riparian zones and connectivity between them and the larger retained patches

Larger reserve areas based on habitat complexes, hardwood stands, existing reserves and lakes.

White areas are retained sections as driven by the ecosystem design

Existing cutblocks incorporated into the design

New cutblock boundaries are determined by the retention areas. Shapes are complex and naturalistic

Figure 9.26
The map of the ecological structure concept at Sutherlands Brook. Source: Canadian Forest Service.

Figure 9.27
The map of the final cutblock design at Sutherlands Brook. Source: Canadian Forest Service.

Figure 9.28
A view of one of the
implemented cutblocks.
Source: Canadian Forest Service.

Case study: Vuokatti, Finland

Introduction

This case study took place in a forested area in north-central Finland not far from the town of Kajaani, where the landscape is a prominent part of the scene but where there are many owners, each of whom owns a relatively small part of the area. Thus, because no single owner has control over the way the forest as a whole is developed, it was important to try to develop a strategic plan for the whole area within which individual owners could carry out their own management activities. The project offered an ideal opportunity to pilot the approach on a different landscape type with different ownership patterns. Stakeholders included Sotkamo municipality, local land-owners, Forestry Centre Tapio (the private forest owners association), Vuokatiopisto Forestry College and Metsahallitus (Forest and Park Service). The project was also carried out with a public participatory project at the same time.

Landscape-level management objectives

The first task was to set management objectives. This followed the approach described in Chapter 5, identifying resource values and then describing a number of objectives for each in terms of criteria or indicators of success. These objectives were as in Table 9.8.

Inventory

The next stage was to review the information available. This was organised as site factors, management factors and cultural and aesthetic factors. The information included:

- 1:10 000 maps of the area with contours and land features
- Aerial photos showing all the vegetation, settlements and other features in false colour at a scale close to 1:10 000
- Reports on the geology and geomorphology of the area
- Maps of the region, showing recreation areas and routes
- Other reports on rare plants and cultural aspects.

Figure 9.29
Location map of
the Vuokatti
project area in
Finland.

Table 9.8 Objectives at Vuokatti

Objective	Criteria and indicator of success
Wood production	To gain income from the forest by harvesting timber while safeguarding special features in the forest
	To manage to sustainable forestry principles and to balance the age structure across the landscape
	To develop the timber production capacity of the lower slopes so as to reduce the demands on the upper, less productive and sensitive areas
	To manage the upper slope areas for multi-age structure in order to maintain their values for forest vitality and landscape
	To use native species in regeneration which are best suited to the sites
Recreation	To develop recreational uses within the carrying capacity of the landscape and avoid site degradation
	To manage and develop the use of the long-distance hiking route
	To maintain the natural qualities of the environment to provide the best recreational experience. This includes both close views from paths and more distant scenes
	To maintain the forest structure to keep berry and mushroom picking available
	To manage the landscape to keep game species in sufficient numbers to be able to hunt
Tourism	To develop tourism based on nature to continue all year round, especially during the summer
	To develop recreational opportunities on the lakes for summer use
Agriculture	To save the farms which are active at present and ensure they continue into the future for their value as parts of the cultural heritage, including farm buildings
	To maintain fields in an open condition by agricultural operations of a conventional form and by alternative means and grants
	To add multiple means of livelihood on farms, including different kinds of services
Landscape	To maintain the different characteristics of the area that distinguish it from neighbouring environments
	To maintain or to open important views out to the landscape
	To maintain the balance of open land to forest in the characteristic pattern of its distribution across the landscape
	To manage the scenery and views along public roads
	To manage the landscape of small-scale pathside views and to ensure it is kept clean and tidy
	To improve past activities which have spoilt parts of the landscape, such as clearcut areas, and to protect landscape quality when planning future harvests
Biodiversity	To maintain and protect valuable habitats of human origin such as old pastures
	To maintain and protect valuable natural habitat types
	To promote the wider use of broadleaved tree species as part of the forest habitat
	To develop the landscape structure to reflect the natural range of plant communities and successional stages
Water	To maintain the quality of water where it is used for human consumption by safeguarding the stream and water flow structure and character
	To protect the character of lakes and other water bodies used for recreation
	To protect the stream pattern and structures in the landscape, especially during road construction and timber harvest

However, much valuable information is not available in written or map form, but is held in the heads of the participants, as local knowledge or expertise in various subject areas. The stakeholder group was very well endowed in this respect and it made the project much more realistic.

Analysis

The next stage was to embark on a series of analyses.

Landscape character analysis

This was carried out in two parts, analysis of landform and of land features, as explained in Chapters 3 and 5.

Landform analysis

The landform analysis picked out the grain, texture, scale and shape of the landform and proved to be important for relating land use and vegetation patterns to landform and for designing patches of forest to blend into the landscape. A definite landform character emerged of the main ridge area being more broken and irregular, flanked by lower and simpler areas of less complex structure. This tied in very closely with the geology of the area and the glaciated landforms. There was also a hierarchy of the landforms, the main convex forms being subdivided into a series of smaller convexities, so increasing the complexity of the whole pattern.

Land feature analysis

The next stage consisted of describing and accounting for the patterns of vegetation and land use in the landscape, which here was closely related to the landform and geomorphology.

The study area was divided into three character areas on the basis of landform and land use.

1 *The central ridge system* consists of the main bedrock exposure of ancient quartzite rocks, which were resistant to erosion by the ice. The landform consists of a complex structure of rounded, smooth, convex forms now covered in dense forest in which spruce is the main species. Seen from viewpoints either within the area, such as at the summit of the main hill, Vuokattivaara, or from external views looking back to it, such as from the lakes or a hill

Figure 9.30

Views of the Vuokatti project area: a) from the neighbouring village, b) from a lake and c) from a farm in the area.

some distance away, called Naapurinvaara, it presents a dominant dark-coloured, medium-textured landscape. On the northern flanks the ski slopes and some existing clearcuts of geometric shape spoil the natural, wild character.

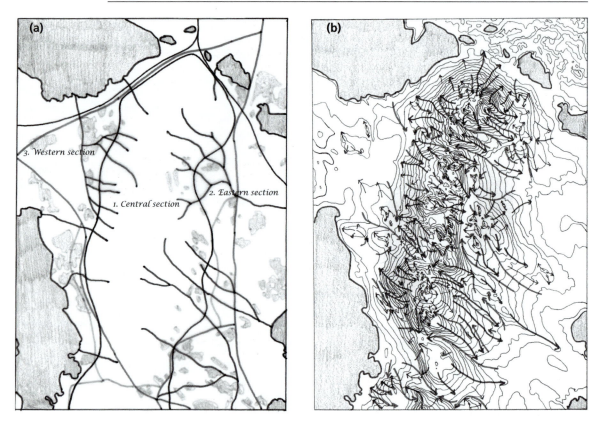

Figure 9.31
Maps of landscape character at Vuokatti: a) the wider landscape and its setting and b) the detailed landform analysis.

There are small, enclosed areas within the zone seen from roads passing through it or from footpaths. Some of these exhibit a strong sense of *genius loci* as a result of the combination of *roches moutonées* and the stunted pines growing among them. The sites of ancient Lapp (Sami) sacrifices add a further aspect to this quality. The overriding character is of a wild, natural area where economic activity has not been particularly important in recent years. The dominant presence of spruce is a result of the heavy snow conditions during the winter which are not favoured by pine at higher elevations.

2 *The eastern section*. This area is undulating and the landform is of glacial depositional origin. It contains some linear moraines, old beaches and terraces from a former ice-dammed lake. The area contains a few farms of small size set among a heavily managed forest. The farms tend to be on the better-drained soils and their open land gives opportunities for views and a sense of variety in the landscape. The forest has been managed so that it contains a wide range of ages and species, especially a high proportion of broadleaved trees. The edges of the fields are somewhat irregular in shape and have a varied structure. The farm landscape is also varied by patches of trees remaining in the steeper gullies. The farm buildings and houses are mainly traditional in style, which gives a strong sense of unity and contrast with some nearby areas.

3 *The western section*. This part of the landscape also lies on lower ground but lacks the depositional landforms. There are fewer variations in landform so that the main sites of better,

freer-drained soils are under agriculture of a larger size where several farms are joined to give much larger-scale, open landscapes than the eastern section. The forest is also managed and altered. Part of the area is flat and boggy. The forest contains fewer broadleaved trees than the eastern part. Some of the farms now have more modern houses, which has changed the character from wholly traditional to more modern and prosperous. The larger scale of the open landscape affords more views up to the wooded ridge or down to the lake.

Constraints and opportunities analysis

This analysis followed the recommended approach and looked at all the practical issues, many of which relate to areas or features of the landscape and either restrict the freedom of action in pursuit of the objectives or provide greater freedom of action: they are constraints or opportunities. The analysis was presented in both plan for those which can be spatially related (not illustrated), and as in Table 9.9.

Landscape ecological analysis

As explained in Chapter 2, the first step of the analysis is to define, map and describe the existing landscape structure. This was divided as usual into the matrix, patches and corridors/pathways.

Structure

Matrix
The matrix was divided into two types. The first was defined as the extensive areas of mature spruce forest, containing trees of 140 or more years old, called by the group the "upper spruce forest zone". These trees are large, the forest is very dense and it has a clean forest floor because of the heavy shade cast by the dense spruce. This forest lies on the upper slopes where the spruce dominates because it is better adapted to the heavy winter snow at the higher elevations.

The second matrix type was defined as the "natural mixed forest zone", found at lower areas, comprising spruce, birch and pine of less than 100 years old. It has a very dense stand structure but, because of the lighter canopies of pine and birch, there is more vegetation on the forest floor, including herbaceous plants.

Patches
Relatively few patch types were identified, showing that the landscape is simple in some ways, but more complex in others, since there were large numbers of many of these patch types, such as the regenerating forest or the clearcut areas.

- *Rocky forests*: These are areas of rocks with spruce trees growing among them, including some large trees; there are lichens on the rocks. These patches are found right on the top of the ridge, in the highest and most exposed areas.
- *Regenerating forest*: Up to 20 years old, consisting of young trees, herbs and grasses. The age reflects the fact that the forest cutting has been more intensive over the last 20 years. There are many examples of these patches.
- *Clearcut areas*: Newly felled areas, with slash and debris lying on them and some standing trees retained on the most recent examples, in compliance with recently introduced regulations.
- *Fields*: Open areas, both cultivated and under grass, with some clumps of trees dotted around them. The field size varies according to the landscape character. On the eastern side of the hill the fields tend to be 2–3 ha in size as islands in the forest, while on the western side they are amalgamated together to form extensive areas of 60–70 ha. This reflects the better soils, flatter terrain and lower elevation.
- *Lakes*: There are several large lakes surrounding the area, forming extensive patches of open water, surrounded by forest. The water tends to be brown in colour because of the peat areas also found nearby, and of low nutrient status.

Table 9.9 Constraints and opportunities at Vuokatti

Factor	Constraint	Opportunity
High rocky areas	Low timber production Erosion risk from too much access in some places Skyline views need to be protected	To allow good hiking To protect biodiversity To permit views out over the landscape To show the history of glaciation
Mature spruce forest on ridge	Visually prominent Steep slopes are difficult for access, difficult to reforest and present a risk of erosion Difficult for public to gain access in some places and at some times because of reforestation	Lots of wood To give logging and silvicultural work To maintain mature forest communities To demonstrate mature forest To pick berries and mushrooms To see snow-covered spruce trees
Agricultural areas	Tourism can disrupt farm work Traditional farming is difficult to keep going economically Tourism can put extra loading on the farm environment and resources The farming population is aging	To develop tourism on farms To sell farm products to tourists To get more grants because of the location of Vuokatti To repair and reuse old farm buildings for tourism To give jobs to local and young people To sell the advantages of a clean, unspoilt environment
Skiing	Capacity and scale of downhill ski area is limited Visual appearance of the facilities and slopes Rocky ridge is difficult to cross-country ski over	To maintain and develop good cross-country skiing in the forest areas To promote the skiing centre as a top-class place for international competitions To build a novice skiers route
Land ownership and local inhabitants	Worries about change and new things Ownership boundaries may conflict with landforms	To use and build on local knowledge and interest in the development of the area To promote management co-operation among landowners
Summer tourism		Wide range of possibilities offered by terrain To develop berry and mushroom picking, summer camps and routes for all-terrain cycles
Managed lower-elevation forests	Forests are already very diverse and of mixed ages and species Difficult for public to gain access in some places and at some times because of forest operations Geometrical edges of felled areas spoil the landscape	Lots of berries to pick To hunt, especially birds To use a range of silvicultural techniques To improve timber production for the future by silviculture To improve the design of future harvest areas

- *Ski area*: This is a single area comprising several ski slopes cut out of the forest and maintained as grass, with ski tows, buildings and a ramp.
- *Abandoned fields*: These are areas with rough vegetation, grasses and herbs and may start to revert to forest after a short time.

Corridors
The corridors were typical of those found in many projects.

- *Streams* are found, but these are very small and are possibly not perennial. They tend to lie within dense forest areas.
- *Major roads* run through the area. These are paved public roads running in corridors cut through the forest, with mown verges.
- *Minor roads* lead to houses and also serve forest management purposes. They can be either paved or gravel, are narrow and less well maintained than the major roads.
- *Power lines* also run through the forest, along straight corridors of cleared vegetation.
- *Recreational routes* can be found as 8–10 m wide path corridors with narrow gravel surfaces or as narrower routes with no surfaces, and they may be waymarked.
- *A railway line* runs through part of the area. It forms an open linear space with rails and ballast and is separated from its surroundings. There is some vegetation maintenance along its edges.

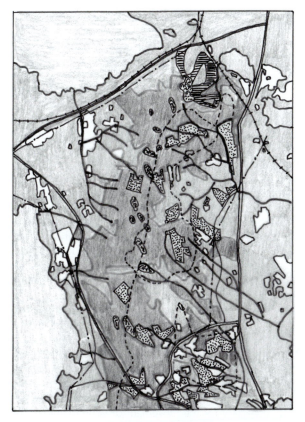

KEY

Mature spruce matrix

Mixed forest matrix

Fields

Felled and regenerating forest

Rocky forest

Ski area

Lakes

Streams

Major roads

Minor roads

Recreational routes

Railway lines

Electricity lines

Figure 9.32
The map of landscape structure at Vuokatti.

Flows

The next task was to choose the flows to be representative of the ecosystem. As described earlier, ones that are helpful in determining whether an ecosystem is reasonably healthy are those that are indicative of a function. The examples used are typical of those found in other projects. The flows chosen were:

- *Water* enters as rain or snow and leaves the area, so being captured, stored and released. Because of the landform there is not much water capture or storage at higher levels, accounting for the minor character of the streams, although a considerable amount of snowpack can accumulate during the winter. Most water is found in the lake systems at lower elevations.
- *Elk* are large mammals that roam widely through the landscape and require a number of plant communities for their survival.
- *Flying squirrels* are a special animal of the Finnish forests that have exacting habitat requirements and so need to be managed in specific ways.
- *Raptors* such as hawks and owls are at the top of the food chain and indicate the health of the ecosystem.
- *Hares* are less common than formerly and are important prey species for a number of carnivores. They live in the farmland areas.
- *Black grouse* are typical forest birds with special habitat needs, especially for mating. If the forest has not got the correct structure they will leave it.
- *Lynx* are carnivores at the top of the food chain and are also susceptible to disturbance by humans.
- *People* work and make a living in what is largely a cultural landscape. They also come to visit to partake in recreation and to enjoy the scenic qualities that are sensitive to careless development or management. Tourists represent an economic resource to local people.

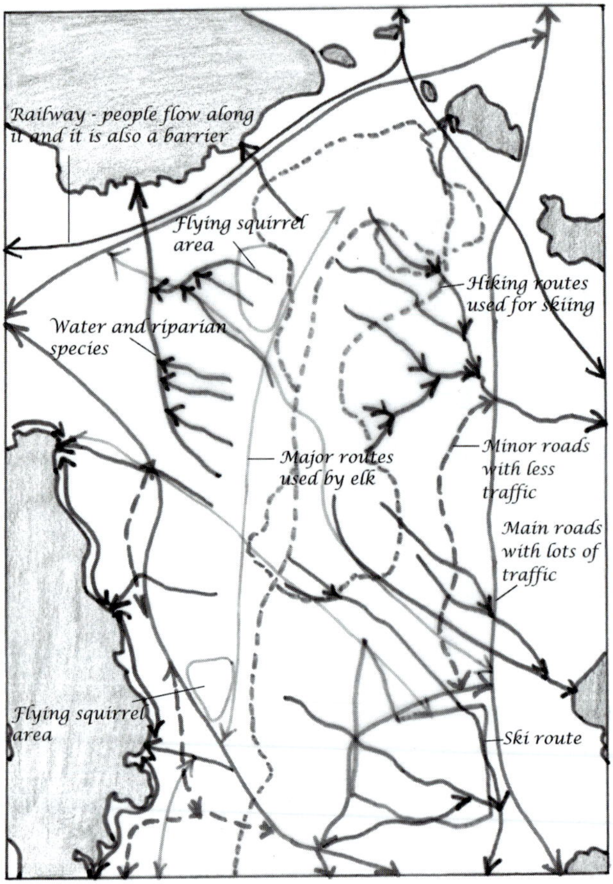

The structures and flows were mapped to explore the interacting patterns. An analysis in some detail was also carried out to explore these interactions more closely. This analysis, using the tabular spreadsheet format as described in Chapter 2, helps to show the relative importance of various categories of landscape element in their contribution to the functioning of a healthy ecosystem. Owing to the relatively limited number of structures and flows, this spreadsheet was relatively simple compared with other projects.

From the analysis it can be seen that the mixed forest matrix is one of the most important elements of the landscape, yet this is becoming more fragmented as the timber production levels increase. The regenerating forest is important too, reflecting the need for different successional stages to be present. Corridors such as streams are important too, especially for the movement of large mammals. As well as the forest, the old abandoned fields are important habitat. These are acting as old flower-rich

*Figure 9.33
The map of
landscape flows
at Vuokatti.*

Railway - people flow along it and it is also a barrier

Flying squirrel area

Water and riparian species

Hiking routes used for skiing

Major routes used by elk

Minor roads with less traffic

Main roads with lots of traffic

Flying squirrel area

Ski route

Table 9.10 Interaction of structures and flows at Vuokatti

	Water	Elk	Flying squirrel	Raptors	Hare	Black grouse	Lynx	People at work	People at play
Matrix									
Mature spruce	Intercepts rain and snow, controls water release and flow into the landscape	Thermal cover against wind and rain	Uses the edge	Hawks nesting	No use	No use	Major habitat	Timber values, maintenance of trails	Hiking, skiing, mushroom collecting, scenic viewing, hunting
Mixed forest	As above but more evaporation. Bogs used to hold water but have now been drained	Forage, cover, mainly summer use. Breeding and calf rearing	Some use for breeding Edges preferred	Nesting in older areas if the trees are big enough	Some use	Quite a lot of use. Feeding on birches	Hunting of hare	Important timber-producing area	As above, but fewer views out, mainly internal experience
Patches									
Rocky forest	Water runs off, earlier snow melt, drier in summer	No use	No use	Nesting in big pine trees and in rocks (eagle owl).	No use	Feeding on bilberries	Some use, denning	Making trails	Hiking, views, very important for *genius loci*
Regenerating forest	Similar to matrix but less important	Primary food area. Some cover in older sections, resting	No use	Hunting, especially in younger stands and more open areas	Breeding and feeding	Some feeding and short-term residence	Hunting hare and small mammals	Future wood supply. Management and maintenance work	Creates views from roads. Elk hunting, some skiing, landscape diversity and autumn colours

Table 9.10 (Continued)

	Water	Elk	Flying squirrel	Raptors	Hare	Black grouse	Lynx	People at work	People at play
Patches									
Clear-cut areas	No interception, faster run-off, risk of soil erosion, snow melts early	Avoid	No use, disturbance during forest operations	Hunting, perching on trees left on site	Use if some cover and food (felled aspen) left on site	Nesting in slash	Avoid	Wood production	Seen as unattractive, opens up views of landscape
Fields	Pollution risk, erosion risk to exposed soil Snow melts early	Forage	No use	Hunting, some nesting or perching by hawks	Food	Forage in autumn after harvest	Avoid	Food production, tourism on farms	Attractive cultural landscape Staying in cottages on farms
Lakes	Store water, collect silt, control flows into rivers. Some evaporation losses Hold ice in winter	Drinking, forage on water edge plants, cooling and escaping from insects in summer	No use	No use	No use	Frozen lakes used as leks in spring	Avoid	None	Attractive feature, used for swimming, boating, fishing
Ski area	Managed snow storage. Erosion risk in summer, pollution risk Artificial snow used in autumn	Avoid	Disturbance	Some hunting	Some food in summer	No use	Avoid	Skiing as commercial venture and jobs in winter and summer	Skiing, views, visually intrusive to some people

Abandoned fields	As for fields but lower risk of pollution or erosion	Forage, less disturbance	No use except where there are edges with old broadleaves	Hunting	Main habitat	No use	Hunting hare	Not used	Look abandoned, grow wild flowers, keep views open
Corridors									
Streams	Variable flows, heavier in rain storms and during ice and snow melt	Main movement routes, foraging, drinking	Stream valley sides used for forage	No use	Some use of bank-side vegetation	Food for young birds	No use	No use	Too small to be significant
Major roads	Run-off and pollution risk, some sediment risk	Danger to cars and drivers from accidents	Barrier and danger	No use	Danger	Dangerous to young birds	Danger, barrier to movement	Travel and transport	Travel, views of landscape by tourists
Minor roads	As above, but less so except in spring when erosion risk	Travel, some accident risk	As above, less so	No use	As above, less so	As above, less so. Stones for bird's crops, sand baths	Low risk	As above	As above
Power lines	No effect	Some travel	No use	Some hunting, perching on poles	Food	Lines can be dangerous	Some hunting	Power	Visually intrusive
Recreation routes	Erosion and silt risk	Some use for travel	No significance	No use	No use	No use	Avoid – disturbance	Building and maintaining trails	Major use for hiking and skiing, and using ski-mobiles
Railway line	Erosion, some pollution risk	Dangerous	No interaction	Some hunting	Forage	No use	No interaction	Travel and transport	Travel, views of the landscape, noise

meadows and need to be maintained as such, and not allowed to regenerate to forest. Old forest with large trees is also an important component for raptors and other species and needs to be retained in the landscape.

Recently harvested areas, arable fields and the ski area are the least valuable ecological components and may be contributing problems, for example affecting water quality and disrupting the movement of animals. Other barriers are the major roads and railways, although the area is not heavily dissected by them.

Now that the understanding of the current landscape as an ecological system is known, the next step is to look at the dynamics of the landscape in terms of disturbance and succession.

The following disturbance types were identified:

Upper forest spruce zone

Natural disturbance
- *Fire* The average size of a fire is around 100 ha, with a range of 10–500 ha. Fires arise from lightning striking trees on prominent ridges or hilltops and burning down ridges. The size of burnt areas is limited by topography, open and wet areas. Storms able to generate enough lightning to result in serious fires tend to move into the area from a southwesterly direction, on a roughly 200-year cycle. The structure resulting from a fire includes an irregular shape, with patches of forest and snags left in rocky or wet places, also possibly any broadleaves that were present on the site.
- *Wind* This tends to produce smaller patches of disturbance, from below 1 ha to 3 ha in size. Usually wind disturbance affects the exposed upper parts of southwesterly facing slopes, because of the prevailing wind direction. These can be considered to range from minor events, which occur almost annually, to major events every 10–15 years. Trees are mostly uprooted, leading to soil disturbance and upturned root plates, allowing some regeneration to start in the denser stands. Wind snap is less common.
- *Insects* These tend to be more common in the oldest spruce and their presence, for example bark beetles, often follows storm events, resulting in the death of some living trees to produce dead wood, both standing and downed. These insects are endemic to the forest.
- *Snow* Every winter, some individual trees snap under the weight of snow. This creates snags and dead wood, especially on hilltops around 28 m above sea level, where the snow is heaviest. Rot can start from the tops of broken trees.

Human disturbance
- *Timber cutting* Older harvest practices tended to be clearcutting on a large scale (over 10 ha), which produced regular shapes, or followed stand boundaries, leaving occasional patches standing. These often became larger as the wind eroded the edges.

Newer harvest practices result in smaller areas – 3 ha on average – and leave a percentage of trees scattered about to retain a light canopy. Salvage of dead/dying trees removes dead wood from the forest.

Natural mixed forest zone

Natural disturbance
- *Fire* Fires tend to be smaller than in the upper forest spruce zone. They produce areas very irregular in shape, containing many more unburnt areas, such as swamps and wet areas. Fires tend only to burn small patches, a few hectares in area, because of the broken terrain. In this zone, at lower elevation and less at risk from lightning strikes, fires tend to be less frequent, perhaps at 300–400-year intervals.

- *Snow* Individual trees, such as pine, can be snapped by heavy snow. Younger trees and smaller plants can be damaged in spring by icy snow, over extensive areas. Trees can be broken, branches damaged or young trees flattened.
- *Wind* Small patches, around 0.5–5 ha, are damaged by wind on cycles of around ten years, affecting some 2–5% of vulnerable stands.
- *Insects* These tend to follow damage caused by other agents, for example after storms (bark beetles). They are always present, waiting to attack weak trees in older areas.

Human disturbance
- *Timber cutting* There was more cutting in the past, leading to a fragmented forest. Cutblocks range in size from 2–10 ha. There are also more thinnings carried out, which has an effect on forest structure, composition and succession processes. Shapes of cutblocks tend to be regular, following ownership and stand boundaries.
- *Clearance for agriculture* In the past, clearances were made for agriculture following the practice of swidden. Such areas were smaller (a few hectares) in the east, larger (10s of hectares) in the west, constantly kept at an early succession by farming. Swidden is no longer practised but the forest structure reflects something of the pattern. It is believed that the species composition included more birch when swidden was practised.

The successional pathways were also worked out for the two zones:

Upper forest zone
- *Primary succession post fire* Birch seed blows in, as do those of grasses, while the seeds of herbs and berry plants are introduced by animals and birds. A birch forest then develops. After 20–30 years spruce re-enters the canopy. Pure spruce stands will have developed after around 100 years, having reached an early mature/late pole stage. The mature phase of stand development occurs by 160–200 years, at which time the canopy starts to break up through small-scale disturbance, unless a fire has affected it in the meantime.
- *Artificial regeneration post cutting* Spruce is planted and established by year five; natural birch regeneration within the planted stands is allowed, thinned out if need be (if it becomes too dense). By 30–40 years the stand will have become, through thinning and management, pure spruce. Thinning takes place at 40, 60 and 80 years with the final cutting taking place around 100 years.
- *Secondary succession spruce* Spruce seeds into holes created in the canopy by wind or insect damage; there is too much shade to allow birch to regenerate. Mixed-age stands of spruce result. It takes around 150 years to replace mature stands completely by such small-scale regeneration.

Natural mixed forest zone
- *Primary succession* After a fire or after fields have been abandoned, the first phases of regeneration are similar to the upper forest zone but with more herbaceous plants, grasses, shrubs and with faster growth of birch, which tends to colonise everywhere. Spruce regenerates in wetter areas while pine prefers sandy, drier sites and rocky places.
- *Artificial regeneration* The process is much the same as for the upper spruce. There is a need for more weeding of the young trees as they become established, cleaning to remove unwanted species such as too much birch, and more thinning at shorter intervals of both spruce and pine during a 100-year rotation because of the richer sites.
- *Secondary succession after patch disturbance* Pine tends to regenerate mainly into sites suitable for it, such as the sandier soils, and spruce on the wetter sites, when seeding into patches created by either natural or human disturbance.

After completing the description of the agents of disturbance and succession a picture of the landscape as an ecological system has been assembled. It became clear that the spruce ridge was more intact and continuous than might be expected in nature, due to fire suppression, while the lower forest was more fragmented, because of human management and cutting for timber as well as clearance for agriculture. It was also possible to see that the move away from larger clearcuts is mainly beneficial but should not be universal, as larger openings, perhaps modelled on natural disturbance patterns, have definite benefits for many of the wildlife species examined earlier.

A useful way to sift through the amount of ecological information is to consider what elements need to remain in the landscape in more or less fixed locations. These represent critical natural/ cultural capital, in that they may be irreplaceable elements. They provide a good starting point for developing a network of habitats as part of the design concept. There are also landscape elements that should always be present at any given time but, since their location is not so critical, can move from place to place over time. These represent constant natural/cultural assets. Successional stages of the forest matrix are examples of such features.

Finally there are some landscape elements that are liabilities ecologically, visually or economically and need to be removed, repaired or replaced.

These three categories of elements were identified by the participants and are listed as follows. They include features identified during the landscape character analysis well as the landscape ecological analysis:

1 **Critical natural/cultural capital**
   - Lakes and the edges of big lakes
   - Streams
   - Rock ridges
   - Old buildings of historical value
   - Wetland areas
   - Sites of threatened (immobile) species
   - Sacrificial sites (of Lapp/Sami origin)
   - Archaeological sites
   - Sites with strong *genius loci*, e.g. key viewpoints and views
   - Agricultural areas in the eastern portion
   - The larger village of Vaarankyla
   - The habitat reserve of old spruce (115 ha owned and managed by the state and municipality)
   - Old spruce in wetlands (fire refugias) in valleys, on hills and next to hiking routes.

2 **Constant natural/cultural assets**
   - Mature spruce on caps of ridges: whole landform units over 200 m in elevation
   - Mixed forest: large connected areas with patchy openings
   - Larger openings to give larger areas (in the natural range of successional stages) for habitat purposes
   - Agricultural fields (arable)
   - Meadows (much more important than arable fields)
   - Forest grazing areas on eastern side (bounded by stone walls).

3 **Liabilities**
   - Ski area and ski jump tower: visually unsightly
   - TV tower: detracts from the natural character, but little can be done about it
   - Blocked views – to lakes in Vaarankyla and other places, which need to be restored
   - Derelict, ugly, shabby buildings, which need to be renovated or demolished
   - Unused fields: to be brought back into agriculture or turned into forest
   - Ugly clearcut areas, which need to be reforested and if possible redesigned in shape.

Concept/vision: "desired future condition"

Now it is time to consider how all the information, the analyses and the objectives come together to lead to a vision of how the landscape could be in the future. This was started by considering the zones of landscape/ecological character and then using the list of critical natural/cultural capital as a place to start developing a long-term or permanent framework into the landscape. After that it was a relatively easy step to divide the landscape up and develop design criteria for each component of constant natural assets within each zone. Finally the liabilities were added, as areas for short-term improvement. The landscape character analysis is used to ensure that main boundaries between sections are designed to fit and to specify the kind of design to be adopted at the next level of single ownership plan.

The concept vision of the desired future condition for Vuokatti comprises the following categories:

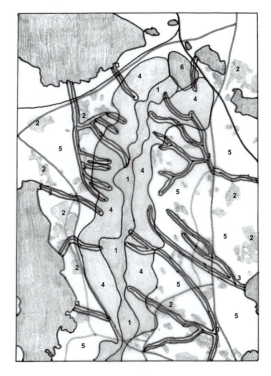

1 Ridgetop forest

Rocky forest and mature spruce: this is visually important, especially on the skyline; it contains important areas of strong *genius loci*, viewpoints, ancient forest, hiking trails, it performs the function of water interception, provides cross-country skiing opportunities, lynx habitat and moose cover. It is of low timber value.

An area encompassing this zone should be designed to relate to landform and connect with other areas of long-term features in the landscape.

2 Settled landscape

Farmland, village: these are the traditional places where people live and contain the inheritance of the rural culture. The places fit into the landscape and people generally keep the landscape in good condition. Open views and the visual contrast with the forest should be maintained. Forest edge habitats and wildflower meadows are important elements. There is some tourism use. Grants to repair buildings, walls and other cultural assets may be available. Abandoned fields should be re-used. Old village roads should be maintained for tourism use.

3 Stream corridors

These include small lakes, lakeshores, wet areas, some rocky areas and some very old spruce patches that would have escaped burning. They provide a series of NW-SE connections across the landscape, connecting the upper and lower parts of the landscape. They are also important for protecting natural capital. They should be maintained in a largely natural condition to provide wildlife habitat, especially for the flying squirrel, water protection and elk routes.

4 Spruce on lower ridge slopes

This is the area between the ridge cap and the lower mixed forest. It is important for timber production. Cross-country skiing, elk habitat and hunting by raptors are all functions of this type. It is also visually sensitive so timber-felling areas should be designed to blend into landform (shape, size, edge, texture). Silvicultural choices include creating larger-sized cuts to mimic fire patterns and smaller patches that reflect the size and shape of wind-blown patches, in order to reflect the natural character. There may also be the potential to make corrections to poorly designed existing cuts. There is also the need to protect small key biotopes during detailed harvest planning.

The matrix character should be maintained over time, aiming to minimise fragmentation and maximise connectivity with the ridge cap and stream corridors.

Figure 9.34 The design concept at Vuokatti: zone 1 is the ridgetop forest, zone 2 is the settled landscape, zone 3 is the stream corridors, zone 4 is the spruce on the lower ridge slopes, zone 5 is the lower mixed forest and zone 6 is the ski slope.

5 Lower mixed forest

Mature, cut and regenerating spruce, pine and birch. This is the most productive timber area. Moose habitat and raptor nesting and hunting are important functions, while the forest also provides visual diversity and autumn colours. Berry and mushroom collection also takes place here.

There is a need to reduce fragmentation and to increase the degree of connectivity and improve its matrix character. Some areas can be maintained as "porous matrix" by carrying out small group or patch fellings while a range of cut sizes can be allowed elsewhere. Cutblocks should be designed to be more natural in shape with edges to be varied in structure.

6 The ski area

There needs to be visual improvement of ski slopes and equipment. It is close to settlement but cut into the forest. The edges of cut ski slopes could be made more varied, vegetation on slopes made more natural, equipment and buildings improved (graffiti removed) and wear and tear repaired. Natural materials should be used around the site.

Consultation

At this stage a meeting was held to present the ideas to the local people. It was held one evening in a local school and was well attended. The concept plans were presented and questions invited, leading to a lively discussion. The meeting was very positive and the plan ideas quite well received. This was partly because three local landowners had participated in the workshop discussions and knew how the ideas were developing as well as acting as the representatives of the community on the design team.

A number of useful comments were received regarding the need for co-operation between owners when new roads and trails are planned, the need to repair buildings, concerns about numbers, siting and design of new houses expected in the area over the next few years and expectations of more demand for skiing.

At this point the design plan was felt to be complete, as it was never intended that the detailed design of a coupe pattern was going to be feasible due to the large numbers of landowners. The mode of implementation was anticipated to be through each landowner agreeing to follow the principles laid out in the plan and, importantly, to protect any of the critical capital elements that happened to lie on their land. For example, if an important stream corridor was identified, it was necessary for each landowner along the corridor route to protect their section so as to maintain the connectivity, migration routes, riparian forest character, etc.

Conclusions

Managed natural forests probably account for the majority of the temperate forests where wood production is a major objective of management. In some parts of the world, no harvest has yet taken place but owing to demand pressures and the need for economic development these remoter places are unlikely to escape. Given the fact that in many countries the better-quality land on easy terrain has been taken for agriculture, leaving the more mountainous areas under forest cover, the visual impact of logging in these landscapes can be severe unless carefully planned, designed and implemented. The ecological impact can also be very great and if not properly understood at the time of planning the damage, in those areas still undisturbed by management or logging, still has the potential to be severe.

Ownership, climatic conditions, legal frameworks and market factors vary from country to country but even so it is possible to apply a well-developed approach to the design of forests to ensure that ecosystem integrity, visual quality, sustainable timber production and many other benefits can be secured. In this chapter the general process presented in Chapter 6 has been explained in more detail in terms of its application to design in managed natural forests. The three case studies, each of which represented fully developed model applications as well as real, practical plans, ably demonstrate that the process is flexible enough to be applied with subtle variations to reflect local conditions. The depth of the presentations of these case studies should enable readers to gain some confidence in the application of the process for themselves. The case studies were completed as initial plans some time ago, when the development phase of forest ecosystem design was at its height. They are now in the implementation phase, which of course lasts for several decades and the results are only now beginning to show themselves, such is the nature of forestry and the time-scales over which it is carried out.

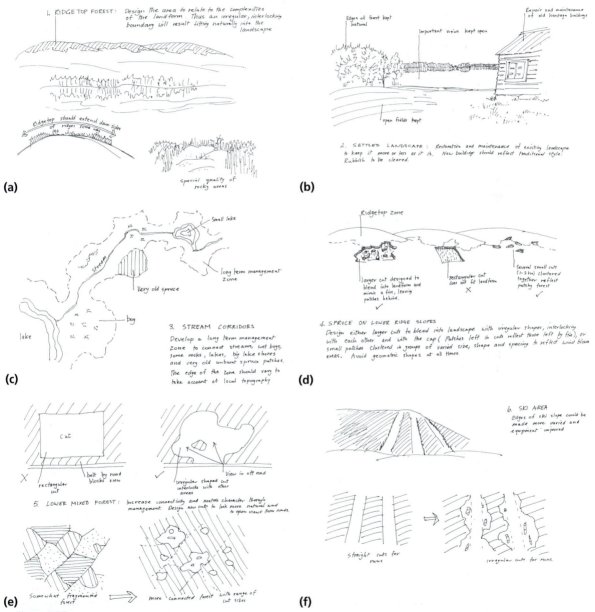

Figure 9.35

Sketches showing how the design concept should be implemented in the different landscape/forest zones at Vuokatti: a) zone 1 is the ridgetop forest, b) zone 2 is the settled landscape, c) zone 3 is the stream corridors, d) zone 4 is the spruce on the lower ridge slopes, e) zone 5 is the lower mixed forest and f) zone 6 is the ski slope.

Chapter 10
The design of plantation forests

Introduction

The previous two chapters have covered aspects of managing forests within the context of natural ecosystems. While a large proportion of wood production will continue to be provided by such forests, in many countries, such as Chile and New Zealand, it is plantations that are expected increasingly to supply significant volumes. These plantations may be of native species, possibly of superior genetic origin, planted as monocultures in order to boost production, or they may be of tree species not native to the country but suited to the climate and growing conditions. Intensive silvicultural practices to hasten regrowth after logging and to boost productivity may use plantation techniques dealt with in the previous chapter. In this chapter, the larger-scale and generally non-native varieties of plantation will be considered.

Tree plantations can be found all over the world, from temperate to tropical zones. This book covers mainly temperate or Mediterranean conditions, but some of the issues and design principles should be applicable to tropical situations. Major examples of plantation forestry include the extensive use of conifers from the Pacific Northwest of America and Canada and in Britain and Ireland, using species such as Sitka spruce (*Picea sitchensis*), Douglas fir (*Pseudotsuga menziesii*) or lodgepole pine (*Pinus contorta*) planted on former rough grazing or moorland. In Australia, New Zealand, South Africa, parts of South America such as Chile and Argentina as well as Spain and Portugal, large expanses of Monterey pine (*Pinus radiata*) have been used, native to a narrow coastal belt of California. Eucalyptus species, native to Australia, are used in southern Africa, South America, Portugal and Chile. Even within Australia, eucalypts native to one area are planted in other parts of the country where they are not native. This extension of the use of native species off site is another variant of plantation silviculture. It has been a feature of the use of Norway spruce (*Picea abies*) in Scandinavia and parts of the former Soviet Union, black spruce (*Picea nigra*) and jack pine (*Pinus banksiana*) in eastern or central USA and Canada or Douglas fir in the Pacific Northwest of the USA. Also in the Pacific Northwest, hybrid cottonwood plantations (a mix of western and eastern cottonwood) have become widespread along floodplains, displacing low-value pastureland and hayfields.

Plantation forests are characterised by species that grow fast and produce higher levels of timber volume per hectare/acre using simplified stand structures. These plantations are typically of one species, although mixtures sometimes feature and are frequently planted all at more or less the same time (or within a few years) so that they form even-aged stands. They are usually managed on short rotations, calculated either on the basis of the age of maximum mean annual increment (MaxMAI), the time at which volume production peaks before falling off as the stand matures, or the age at which the most economic return is obtained, that of maximum net present value (MaxNPV) according to the forestry economic model using the technique of discounting. Frequently,

Figure 10.1
Some examples of plantation forests from around the world.
Most are based on non-native species:
a) Sitka spruce in Glenummera, County Mayo, Ireland
b) Monterey pine in South Island, New Zealand
c) Monterey pine near Valdivia in Chile
d) Monterey pine in the Basque country of Spain
e) Japanese cedar between Kofu and Mt Fuji, Japan.

the intensive silvicultural management of plantations seeks to exclude non-productive species, and employs pesticides, fertilisers, artificial drainage, pruning and other activities to ensure the best growth rates and productivity, as long as it can be proved that there is an economic benefit to carrying out these investments.

Historically, plantations have often been laid out in geometric patterns to simplify management. These patterns comprise a series of rectangular compartments, possibly laid out on a strict grid, separated by open strips and fire breaks, the trees being planted in straight rows to facilitate machine operations such as ploughing, draining, weeding or thinning. Records of growth rates, silvicultural activities and harvest products are easy to keep when the plantation is laid out in this way, so such forests can be likened to wood factories geared to the single objective of the production of a standard, predictable product of known performance. Harvesting usually also follows the same grid pattern, whole compartments being felled at a time, possibly following a time sequence so that in the ultimate plantation forest, every year the same area is felled producing the same amount of timber. This is replanted and eventually comes to be felled again. The total number of compartments equals the rotation lengths (10, 20, 30, 40, 50 years, etc.) so that a sustained yield of wood can be guaranteed, assuming there are no catastrophic problems such as fire, disease, windthrow, etc. This concept is known as the "normal forest" but it is rarely, if ever, realised because there is not the degree of predictability in a forest plantation that there is in a pig farm or a steel factory. However, the notion as an ideal has been highly influential and has had a major impact on forestry practice over the past several decades.

From the description above, it is easy to see that a plantation forest differs markedly from a natural one in a number of attributes. These differences are often cited as predominantly negative aspects of plantations, in that they are rightly seen as ecologically deficient in comparison with natural forests. A key challenge is to understand how much a plantation should be modified towards the conditions of a natural forest in order to satisfy the requirements of sustainability. Table 10.1 below summarises these differences.

Sustainability issues in plantation forests

There are several issues that have to be addressed if plantation forests are to achieve long-term sustainability. These are landscape context, site productivity, disease risk, habitat, water quality and aesthetics.

Landscape context

Clearly, plantation forests, particularly those with non-native trees, do not offer the same suite of ecological services as do native and natural forests. If the native forests of a country or region are completely or substantially replaced by plantation forests, then some important ecological services are likely to be lost. However, plantation forests can be part of a national or regional strategy for sustainability if they are planned as part of a larger landscape network. Conservation biologists such as Reed Noss and Larry Harris have promoted regional-scale forest networks with

Table 10.1 Comparison of natural and plantation forests

Natural forest	Plantation forest
Species composition	Mixed stands of site-native trees, bushes etc.
	Single species stands, undergrowth mainly absent
Layout	Irregular stands in size and shape, blurred edges, patterns related to soil, aspect, microclimate
	Regular, geometric stands with sharply defined edges, site amelioration may be used to even out the influence of site variation
Rotation length	Forest passes through all successional stages, affected by natural disturbance such as fire, insects, wind
	Forest felled at age probably equivalent to late stem exclusion, active measures used to prevent fire, disease, wind etc.
Stand structure	Complex stand structure may form one to many layers; dead trees, dead wood etc form the oldest trees in the stand
	Simple stand structures, one canopy layer, dead trees may be removed, little or no dead wood, dead trees only due to self-thinning and are not old
Visual impact	Natural forests are part of the landscape, they express its character. Negative impacts occur at regeneration harvest
	Plantations stand out from the landscape and contradict its character during all stages of growth and harvest
Ecological impact	Natural forests provide complex habitat with multiple niches
	Plantations provide very limited habitat and few niches

three types of forest: natural forests as core reserves, lightly managed forests as buffers and intensively managed forests to provide the necessary materials for human culture. This system is not unlike the medieval European forest system described in Chapter 1. There are various conceptual arrangements for these forest types.

Clearly, each country or region has to debate and decide for itself how much of each forest type to plan for, and how these are arranged will depend greatly on past land-use legacies. For example, core reserves will almost always be on public lands (wilderness, national parks and roadless areas). Intensively managed lands may be in either private or public ownership.

This model works well in the North American West, where there is ample public ownership and substantial amounts of wilderness, but works less well in Europe and eastern North America. New Zealand has in effect adopted this model by clearly distinguishing plantation forests from natural ones, and prohibiting any further conversion of natural forests to plantations. The assumption is that sustainability is achieved by allowing very intensive production in suitable zones in order to take the economic pressure off remnant natural forests.

Site productivity

As is the case with intensive agriculture, intensive forestry tends to be demanding on soil fertility. Single-species plantations, particularly conifers, can reduce soil fertility and increase acidity over time. Many forest ecologists, for example Chris Maser, have argued that this practice will not be sustainable for more than a few rotations. Conifer needles either break down quickly, adding little in the way of nutrients to the soil, or do not break down at all and produce acidic conditions in the upper layers of the soil. Lack of broadleaved trees, particularly nitrogen fixers such as alder, means that some key nutrients may become exhausted on certain soil types. Even hybrid cottonwood plantations in the Pacific Northwest quickly use up soil nitrogen. Lack of large wood residues left behind on the forest floor results in a gradual loss of wood fibre, a critical soil component. The forest ecologist David Perry considers that "the importance of soil to forest productivity cannot be overstated." Not only is soil critical for the present growth of trees, but also is the legacy from which future trees will grow. In other words, soil outlives the present rotation, and must be considered over the very long term for sustainable plantations to be possible.

In addition to the potential for plantation forests to exhaust soil fertility, site preparation, planting and harvesting can make further impacts on the topsoil. Most plantations are on marginal farmland, or low-elevation forest land, which are characterised by rolling to steep topography and thin topsoil. Cultivation may thus expose the soil to erosive forces. Mechanical harvesting results in compaction if done while the soil is wet. In hybrid cottonwood plantations of the Pacific Northwest, harvest is done throughout the year (including the wet season) in order to have a sustained flow of fibre to local paper mills. Decaying logs, a common product of natural forests, are usually absent or quite scarce in plantations. These logs have been found to be critical centres of biological activity. They reduce erosion, store water during droughts and provide cover for small mammals that spread mycorrhizal spores. According to David Perry, in natural Douglas fir forests in the Cascade Mountains, as much as 30% of the upper soil layers are composed of decaying logs.

Thus management and replenishment of the soil resource is crucial if plantations are to be sustainable. In New Zealand, plantation managers have begun to cultivate nitrogen-fixing groundcovers under *Pinus radiata* plantations as one strategy to retain fertility. They have also adopted best management practices to protect soil during site preparation and harvesting. Nevertheless, it is uncertain how many generations of pine can be grown on a given site before significant soil depletion occurs. Again, according to Perry, plants and animals of the forest are as critical to the health of the soil as the soil is to the plants. In simplifying a forest system to where only one or a few trees are grown on continuous rotations, much of the cycle that links the plants to the soil is lost.

Disease risk

It is generally assumed that plantations run a higher risk of being attacked by disease or infestation, as is the case with any monocultural crop. New Zealand and Great Britain have very stringent plant health quarantine regulations to protect non-native plantations from being exposed to diseases. Attention to genetic variability is another hedge against disease. When a nation becomes reliant on one or a few species of tree, these are intensively studied, and prevention and treatment methods become tested and distributed fairly quickly. Nevertheless, the high productivity of plantations is associated with an increased risk.

Habitat

Plantation forests provide some habitat, but in nearly all cases must be considered inferior to native forests in a number of ways. Firstly, by definition, plantations do not provide complex structure, thus the available ecological niches are fewer. Secondly, intensive management means frequent habitat disturbance. Thirdly, short rotations generally mean that the forest is in its least biologically diverse stage for most of its existence. Early and late successional stages typically have far more species than are found in a closed canopy stage forest. Lack of snags means lack of tree cavities for nesting birds. Lack of species diversity means fewer food sources, and so forth. In addition, older forest stages have become increasingly rare worldwide as commercial forestry has expanded. As a consequence, even if forest plantations provided adequate habitat for the large number of species that rely on early successional structure, those that rely on older forest are less well provided for.

Wildlife habitat in plantations can be improved substantially by conserving small special habitats within them. For example, wetlands, rock outcrops and remnant groves of older native forest can be left as undisturbed islands. This is a common practice in the UK at present.

Water quality

Since plantations are intensively managed, they can have a greater impact on water quality by generating sediment and chemical pollution. Aquatic buffers, careful road construction and best management practices (BMPs) are the most common ways to mitigate water quality impacts from plantation forestry.

Aesthetics

Plantation forests clearly have strong impacts on the aesthetics of landscapes, both at the creation stage (afforestation) and the harvest stage, since clearfelling is the most common technique. In fact, it can be argued that the whole field of aesthetic forestry essentially developed in response to the challenge of plantation forestry in Britain and the USA (see Chapter 3). We usually do not think of aesthetics and sustainability in the same context. Sustainability is normally addressed in economic and ecological arenas. But forest aesthetics contribute or detract from local and regional quality of life, and can have profound effects on local economies. For example, the tourism economy of British Columbia is just as land-based as is forestry. It is a rapidly growing segment of the overall economy of the province, and may eclipse forestry in total economic value within a few years. The retirement economy is also highly dependent on landscape aesthetics. The Ponderosa pine forests of central Oregon and some parts of the Oregon coast have attracted a rapidly growing number of wealthy retirees in large part due to the aesthetic quality of nearby national forests. So in order to contribute to regional "sustainability", aesthetics of plantation forests must be successfully addressed.

Chapter 9 demonstrated how to manage design in natural forests so that their essential qualities are maintained while at the same time producing timber and other products. Achieving the same in plantation forests is a more complex task. Firstly, these forests represent a high level of capital investment. In most cases, someone has bought the land and invested money in planting it, and expects to get a return from it. Secondly, the plantation may be part of a landholding dedicated to supply a mill, so that any reduction in volume or change in quality may be difficult to justify. Thirdly, non-native trees cannot re-create a natural ecosystem, but conversion to native species may not be an option due to the timber productivity that might be lost.

There are ways that plantations can be manipulated so as to address the issue of sustainable forest management. These relate to plantation layout, stand composition and structure, and rotation length. Experience has shown that adopting principles to design forests to fit into the landscape better from an aesthetic standpoint can also improve ecological functioning and water protection.

The key challenge is to be able to do all of this while minimising the loss of timber production, or not increasing the operational cost to the point where the capital investment is no longer viable. Support for modifying plantations by means of grants, subsidies or tax incentives can be used to help landowners improve their practices, as can forest certification, which may improve net value and/or access to markets.

The rest of this chapter will demonstrate tried and tested principles of plantation design, illustrated with case studies. The first section will discuss afforestation design, and the second section will examine ways of modifying existing plantations so as to achieve better sustainable forest management.

Figure 10.2
An example of a very open, deforested landscape in Scotland where some isolated plantations of conifers appear to float randomly in the scene. View from Ben Kilbreck, Sutherland.

Afforestation design

Afforestation is the act of planting a stand of trees on land not formerly forest. In some cases, an area may have been forest once upon a time, but for a long period has been in some other land use. Most often this is agriculture or grazing, in a somewhat degraded state, perhaps with the soil fertility exhausted. Afforestation can be a means of restoring the ecosystem (see Chapter 8), but in this case we are primarily concerned with plantations established where wood production is the main objective, although in one of the case studies this is not a significant factor.

The first issue to address is that of land acquisition and the overall scale of the forest in the landscape. As noted earlier, intensively managed plantation forests should, where possible, be developed as part of a larger regional scheme that includes lightly managed native forests and unmanaged wild ones. Before deciding on a location for a plantation, a large-scale ecological analysis should be done to identify suitable potential locations. This analysis should consider whether a particular area might better be afforested with restoration objectives at the forefront.

In many cases in Britain, Ireland or New Zealand, the landscape into which a plantation forest is to be established is open and almost treeless due to previous extensive clearance for agriculture. A plantation inserted into this type of landscape will be a large patch floating and isolated from other areas of forest. This sets up strong ecological and aesthetic contrasts. For example, birds that nest in large expanses of open land may be predated by raptors which find convenient perches in the new plantation. A treasured visual prospect from a local village or hiking area may be negatively affected. In other settings, such as Tasmania or some parts of New Zealand, a new plantation fills open patches within a matrix of existing forest, possibly of native origin. Design is much more challenging in the first case because there is nothing to link the plantation to, either visually or ecologically.

In Chapter 5 an overall forest landscape design process was outlined. It emphasised the need for survey or inventory and analysis prior to concept generation. In plantation afforestation design the following issues need to be addressed as part of that process:

- *Objectives*: while timber production is likely to be the primary objective, secondary ones may include improving the ecology of the area by diversifying habitat, protecting water quality in local streams or lakes through good soil management and ensuring that the forest fits into the landscape aesthetically in order to support local tourism.
- *Survey/inventory* of the area should at a minimum include: soils, topography, microclimate, hydrology, cultural, archaeological or historical features, visibility in the landscape, wildlife habitat, access, the overall landscape character and recreational use.
- *The analysis* should look for opportunities to incorporate elements that help satisfy the secondary objectives while aiming to reduce as far as possible the constraints to achieving

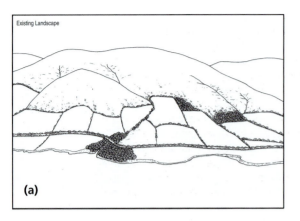

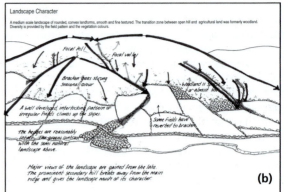

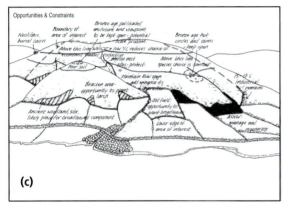

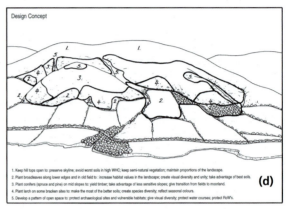

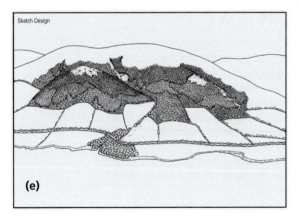

Figure 10.3
The sequence of design for an afforestation project, based on British experience: a) the original landscape before afforestation; b) landscape character analysis; c) opportunities and constraints analysis; d) design concept; e) sketch design. Source: Forestry Commission.

the primary objective. Landscape character and landscape ecological analyses must examine the potential effects of the significant changes to the landscape that will happen as a result of plantation afforestation.

- *The concept design stage* is extremely important because, not only can we expect a drastic change to the landscape that is likely to be of long duration, but the investment into the forest may be large and is expected to be ongoing over many generations. It is vital to ensure that all key aspects are fully incorporated so as to avoid costly adjustments later on.

Certain economic and practical aspects of plantations have tended to become embedded in forest managers' psyches to a degree that standardised methods of doing things may have to be challenged. Two items in particular are accepting the need to give up areas of perfectly plantable and productive land in deference to other objectives, and laying out compartments in non-geometric patterns in order to mitigate aesthetic impacts. These measures are likely to result in lower total production, increased operational cost or both. However, it is important for forest managers to recognise that the old ways of plantation management are no longer acceptable under sustainable forest management objectives or certification systems, and that if properly assessed at the concept stage the costs of these can be minimised and practical problems largely overcome.

Large-scale landscape change is nearly always a major issue with the public, and past experience with some very unattractive plantations may raise suspicions and lower confidence in the desirability of any new proposals. This is to some extent the opposite side of the same issue raised when considering logging in natural forests in Chapter 9. A key design challenge is the completely different appearance that will result from a plantation forest when compared with existing open ground, whether farmland, moorland, grassy steppes or fells. In extremely open landscapes there is nothing to tie the new forest edge into, such as clumps of trees, hedgerows or field patterns. In these circumstances the landform and, possibly, semi-natural vegetation patterns can be used to guide design. Thus, a full landscape character analysis is vital, with a thorough examination of the landform (see Chapters 4 and 6). If there is a stronger pattern in the landscape provided by fields, hedges, trees, woods or other features, this should be used to help achieve visual unity.

One of the problems often encountered with new plantations is that of ownership boundaries. These may often be set as straight lines running at right-angles to the contours, so that if the whole plot of land were to be planted the outline would probably be highly unsympathetic to the landscape by presenting a hard, unnatural edge. Solving this may require greater or lesser amounts of land left unplanted to allow for a more sympathetic shape to be designed. The external shape of the planted area is one of the keys to successful visual design, and devoting attention to solving this problem is very worthwhile.

As in all aspects of designing shapes to fit into different landscapes, the principle of fitting to landform applies. However, there are different segments of the external boundary or margin that require different treatments. These are upper margins, skylines, side margins and lower margins.

Upper margins

Upper margins are where the planted area terminates part way up a hill or mountainside. Economic parameters may suggest that a certain elevation marks the cut-off point for viable planting due to the drop-off in growth rates with elevation and poorer soils. This contour, if planted, would result in a severe and unnatural horizontal line. However, detailed examination of the landscape will usually show that better conditions persist in the gullies where shelter, soil and moisture provide more favourable conditions than on ridges that are more exposed, drier and have thinner soils. Therefore, a more subtle approach is to reflect the topography by raising the planting up into the hollows and dropping it on ridges to different degrees depending on the relative strengths of the feature. This produces a much more natural-appearing shape that is easier to unify while having little effect, if any, on the economics.

Developing an upper margin in this fashion emulates the natural pattern found at tree lines in moister areas. However, this planted upper margin usually lies at a much lower elevation than would a natural tree line. Once the overall shape has been designed it is often possible to emulate a natural tree line further by developing a density gradient where the solid canopy becomes broken into patches, small clumps and then individual scrubby trees, or even an artificial krummholz (by planting naturally shorter species). It is important for this zone to be sufficiently deep, since too narrow a zone will fail to provide the ecotone valuable for ecological purposes and may result in an unfortunate-looking thin strip along the contour left at later harvesting.

As well as the shape of the upper margin, the scale must be well designed. A narrow sliver left unplanted looks out of proportion. The discussion of scale and proportion in Chapter 3 provides

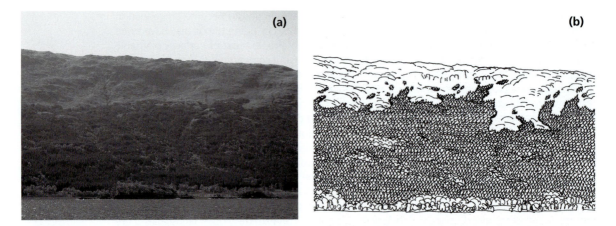

(a)   **(b)**

*Figure 10.4
a) an example
of a poorly
designed upper
margin to a
plantation in the
Great Glen in
the Highlands of
Scotland, which
merely follows
the contour line;
b) shows how it
could have been
designed to
reflect the
character of the
landform.
Source: Forestry
Commission.*

guidance on this, for example the rule of thirds is particularly useful. To achieve a good scale may mean again losing some plantable ground, but this should be of marginal value in any case at upper margins. An alternative, if the land is available, is to plant over the top and completely clothe the skyline.

The character of the upper margin should reflect different qualities in the landscape. Smoother, rounder landforms should be reflected in more flowing shapes while more jagged, rocky topography can be used to develop a much more fragmented pattern. The opportunities presented at the upper elevation transition may be ecologically valuable so that as well as developing a naturalistic edge to satisfy the landscape character, it may be possible to create a valuable new habitat with little impact on wood production. Frequently the species of tree or shrub changes at the natural tree line, perhaps into a different conifer, such as sub-alpine fir (*Abies lasiocarpa*) in British Columbia, mountain pine (*Pinus mugo*) in Central Europe, and dwarf birch and willow in Scandinavia. Thus, there are opportunities to introduce native trees and shrubs into this zone, perhaps increasing its depth, to reflect the natural character even though the adjacent plantation trees are non-native.

Side margins

The problem of ownership boundaries running straight up a hill has already been mentioned. Side margins are not usually found in natural forests so there are few precedents to emulate. Instead, the topography and vegetation patterns have to be used with the aim of trying to create a generally designed line so as to avoid any sense of unnatural geometry inadvertently developing. Since the lower-elevation areas are likely to be fairly productive, loss of ground here is more expensive than at higher elevations. Where the topography is varied and steep, perhaps with rocky ridges on knolls, it is much easier to run the margin around them than if the slope is smooth and featureless. If the plantation has to be fenced to keep grazing and browsing animals out, any unplanted land inside the fence will tend to develop a different colour and texture from that outside, which may serve to draw attention to the straight line. Thus, the design of the fence line may also need attention in order to avoid this.

Where there are other features, such as clumps of native trees, hedges or patches of woodland near the plantation boundary, these can be visually and ecologically linked into the plantation, either by planting or by encouraging natural regeneration. These drafts of woodland and trees should use the topography wherever possible and reflect the pattern of the existing features to impart a sense of unity and continuity to the landscape.

One of the places requiring careful design is the junction of the upper and side margin. Any sense of a right-angled corner at this point immediately produces a geometric effect. Short sections of straight line are acceptable as long as the overall shape is that of a generally diagonal, curving line.

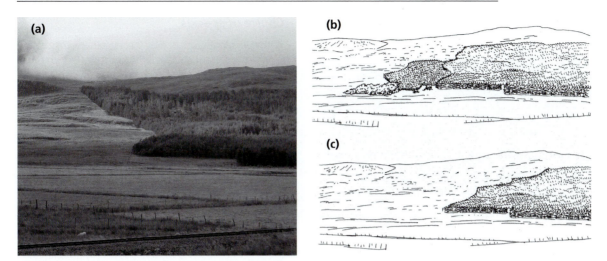

(a)

(b)

(c)

Figure 10.5
a) an example of
a poorly
designed side
margin, where
the forest stops
at an ownership
boundary;
b) presents an
alternative for
extending part of
the forest; this
solution would
require the
cooperation of
the neighbouring
landowner;
c) resolves the
problem by
retreating the
forest in places,
within the
ownership
boundary.

Lower margins

The lower margin of a new plantation is frequently the easiest to cope with, because at lower elevations the plantation often tends to meet agricultural land, which often has a strong pattern of fields, hedges or small woodland elements. If this pattern has integrity there may be no need to alter the plantation shape. Instead, the existing network of hedges and wooded elements can be pulled up into the lower portion of the plantation to increase interlock and establish a firm sense of unity. This often also has habitat benefits by linking the plantation forest to existing woods, thus diversifying and enlarging available habitat and increasing connectivity.

If the plantation lies on a slope and extends down to a valley floor with a river, it is usually a good idea, and may be a regulatory requirement, to develop a transitional riparian area using native species to buffer the stream from the plantation. This pattern should then be connected to any smaller streams or swales flowing through the plantation, which themselves may need special treatment (see below).

In completely open landscapes, without any vegetation structure of trees or woodland, it is likely that landform will be the most dominant visual influence. In these cases, the lower margin should follow or reflect the landform. Again, it may be advantageous to plant native trees for riparian value if there are streams within these gullies.

Along the edges of the lower margin it is often possible to create more ecological and visual diversity by planting clumps and patches of native trees. Where space allows this can be developed into a broader gradation resembling a natural ecotone, similar to the upper margin. If there is a perimeter fence included around the plantation, it may need to be installed as a series of straight stretches for cost and practicality. The edge can be developed within this fence line, which may benefit from the lack of grazing pressure, thus allowing for a richer ecotone that is of greater value for wildlife.

If the plantation is on flatter terrain the external margin shape is not as aesthetically important unless the forest is seen from elevated viewpoints in the surrounding landscape. In this case, it is the edge structure that has most impact. Such a plantation may be in an open landscape or one with a strong pattern of hedges and hedgerow trees. The edge zone may abut intensive agricultural land, so that the ecotone can be an element in increasing or maintaining local biodiversity. The spatial structure across the edge zone should be matched by a corresponding vertical structure. A layered effect of parallel bands of shrubs, small trees and larger trees should be avoided in favour of irregularly shaped and spaced clumps of varying size, thus resembling in appearance and structure a naturally developing colonisation zone. This zone can then be maintained together with patches of retained mature forest while the plantation behind is harvested.

*Figure 10.6
A good example
of a lower
margin to a
forest, where the
plated area
interlocks with
the field pattern
and broadleaved
trees extend up
into the conifers.
Ireland.*

Choice of species

Within a plantation it may be necessary to plant a range of different species in order to reflect changes in soil, moisture, elevation or exposure. In other cases, there may be no diversity of species whatsoever. Species variation can be valuable for promoting both visual diversity and biodiversity, but if not done sensitively it can also create problems.

In British plantations on mountainsides, soil variations can be identified by noting vegetation changes such as heather (*Calluna* spp.) in podzolic conditions, grasses (*Molinia* spp.) on gleyed soils or bracken (*Pteridium aquilinum*) on richer upland brown earth soils. It was an early tradition among foresters to match these with suitable species such as pine (*Pinus* spp.) on the heather which can cope with podzolic soils, spruce (*Picea* spp.) on the gleyed soils under the grass and larch (*Larix* spp.) on the brown earth soils beneath the bracken. If the natural site boundaries were followed, an intimate and diverse pattern was created that to some extent reflected the pattern of the surrounding landscape (especially the larch, which is deciduous and becomes golden yellow or orange in the autumn as does the dying bracken). With more intensive silviculture (ploughing, drainage, fertilising) and due to the poorer prices received for larch timber, this practice went out of fashion. However, recently the development of the Ecosystem Site Classification (ESC), based on the British Columbian biogeoclimatic ecosystem classification system, has resurrected the principle of the approach so that species variation, if introduced, can be better related to productivity (see Chapter 8).

The ESC uses a climatic zone based on factors including oceanicity, elevation, rainfall, exposure and warmth, within all of which soil type and site variables of soil moisture and soil nutrients interact to produce a grid. A map of sites related to this grid, which itself is linked to species suitability, provides a good starting point. Research has also shown that many of the sites are related to topography (such as the podzolic sites lying on knolls and the surface water gleys in the hollows).

Using this or similar methods appropriate to other countries, a pattern of species can be developed that provides the basis for greater unity and diversity of the forest landscape and to some extent can increase potential biodiversity. The design of these species patterns should reflect not only the soil pattern but also the topography (which should go together) and may need to be simplified for ease of planting and management. To be avoided are simple geometric layouts following grid-like compartment boundaries, even when they loosely reflect changes in site conditions.

To mitigate aesthetic impacts the proportions of different species should reflect asymmetric balance according to the rule of thirds. For example, the layout could appear in the landscape as two-thirds evergreen conifers, one-third other species of which two-thirds could be larch, one-third broadleaves and so on.

*Figure 10.7

An example of planting different species to reflect soil types. In this case larch was planted on bracken (brown earth soils), spruce on grass (gleyed soils) and pine on heather (podzolic soils). Achray Forest, Scotland.*

Species can be planted singly within compartments or as mixtures. Single species are simpler to manage but may appear too highly contrasting. Mixtures, as long as they are intimate or of varying diversity, work well but stripes, bands or chequerboard effects can be highly intrusive in visually significant areas because of the geometry they introduce. Junctions between species can be blended to reduce any hard or abrupt boundaries, but only when the correct shape has been designed, since this is no substitute for poor layout.

As well as introducing a variety in the productive species in the plantation (all of which may be non-native), it is important to consider including at least a proportion of native species. If this proportion is small, it should be concentrated lower down the slope of a plantation on a hillside due to the scale variation in the landscape (see Chapter 3). It is particularly important to link edges with native woodland components outside the plantation. Attention should also be paid to stream sides or around water bodies where water quality and fish habitat protection may require a riparian zone of native woodland. These zones should not be merely the minimal protection zones required as a basic measure by various jurisdictions, but should vary in width to reflect the topographic variation along the watercourse. Generally, the idea is to build onto or link with existing natural remnants in order to augment their aesthetic and habitat value.

In single-species plantations such as Monterey pine in New Zealand or Tasmania, one of the major ways of uniting the plantation into the surrounding landscape and introducing diversity is by incorporating native forest or bush, especially along watercourses. Other sites that may be favoured for such treatment include rocky knolls or cliffs, wetter areas unsuited to pine, and skyline ridges, as long as the scale can be respected. If the plantation lies within a matrix of native forest, the introduced native woodland elements can provide valuable linkages helping the habitat requirements and movement of a number of animals and birds that might otherwise avoid the plantation itself.

Open space

Open space habitat is another component that may be needed within plantations. If not adequately provided through the opening of rotational felling (which may or may not resemble natural disturbance patterns – see below), it may need to be included within the initial layout. Even when temporary open space habitat is adequately provided by felling, there is a period of a full rotation before the first felling in a plantation takes place. This means that some interim provision may be needed. Open areas can fulfil a number of purposes: the protection of features such as archaeological sites, retention of non-woodland habitats, places to control deer, recreational use areas (picnicking, camping) or as viewpoints. Open space in British forests is particularly important due to the value of many non-woodland habitats as compared with that of dense stands of exotic conifers, particularly until the plantation matures and is more diversified in structure. These open spaces can take the form of isolated areas, but it may often be better to link them together in order to provide greater connectivity for their colonisation or use by wildlife. A network of open areas can be developed that becomes a long-term or even permanent structure within the forest. Linear open spaces may also be used to divide the plantation into compartments for management purposes, although geometric shapes should be avoided in favour of organic ones following landform as described in Chapter 3. These linear spaces can provide useful, wind-firm boundaries to demarcate possible future felling coupes.

Figure 10.8 Open space within a plantation forest: this shows the types of spaces needed under British conditions when using non-native conifers.

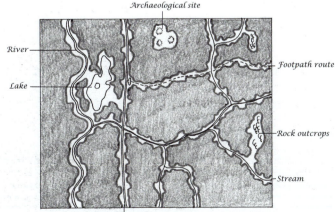

All of the open spaces should be designed into the fabric of the plantation, varying their shape and scale, so as to maintain as natural an effect as possible. Patches of native woodland can also be incorporated to provide further diversity, while the edge structure of all open spaces can be developed as described above. This will help them attain greater habitat value and blend them better into the landscape.

In forests where the main plantation species choice is limited to one or two species, the components of remnant or introduced native forest and open space become even more important for providing continuity of habitat and connectivity across and through the landscape. It may be necessary to devote a considerable proportion of the area, up to 20%, in order to satisfy the non-timber requirements in differing proportions of open space and native forest, depending on the circumstances. In conventional plantation layouts the linear spaces used as compartment boundaries, and sometimes as informal access routes, can represent as much as 12% or more of the land area. Using open space in a more multi-functional way as described here need only add a small extra amount to that percentage. There are also likely to be open patches left where sites cannot be planted due to streams, open water, wetlands, rock outcrops, poor soils and other physical obstacles. These should be incorporated and their value towards the forest ecosystem maximised, since such places usually have habitat and landscape value and may also be useful for recreational use.

Integrating plantations into native forest areas

In many countries, areas of native forest have been converted into plantations capable of yielding a much greater wood supply. These areas need not be wholly detrimental to biodiversity or landscape values as long as they are well integrated into their setting. In other cases, abandoned farmland within a forested landscape may be re-afforested using plantation techniques.

The shape of a plantation created from or within an existing forest should be designed according to the principles described earlier in this chapter, following the dictates of site and landform where appropriate. There is likely to be much less of a contrast between the plantation and the surrounding matrix forest compared with the examples set in open landscapes described earlier, although textural differences between conifers and broadleaves (e.g. Monterey pine and eucalyptus) can be striking. In order to achieve a better design where the shape of the proposed plantation is unsatisfactory, perhaps due to it having been a relatively geometric field, it may be necessary to cut into the surrounding matrix forest, plant part of the area with native trees or both.

The fact that the plantation lies immediately adjacent to native forest means that wildlife will have an easier time colonising the planted area. The early years of a plantation, until the canopy closes (a shorter time in intensely managed forests than in most natural ones), provide valuable habitat. Colonisation of the area by other forest plants is also likely to be quicker as those sites which have been forest in the past will have a seed bank of forest plants while those on former farmland will tend to have arable weed species of lower ecological value.

An internal component of mature forest should also be included, especially along any watercourses flowing through the area. Other habitats such as relict patches of mature forest on sites unsuitable to intensive plantation silviculture, wetlands, water holes, rock outcrops and cultural or historical relics, should be protected and incorporated into the layout as described above.

Figure 10.9 In Tasmania, Australia, plantations of non-native species tend to be located in a matrix of native forest and retained elements of the former vegetation can more easily be incorporated into the layout, for example along riparian areas.

Managing plantation forests

Once the plantation has been established a number of stand-tending operations usually take place, such as weeding, brushing/cleaning (the removal of woody species that interfere with the planted trees) and eventually spacing, pre-commercial thinning and commercial thinning. Some of these activities can be used to promote or maintain biodiversity and landscape values.

Once the planted trees have reached thicket stage they close canopy, cut out the light and prevent much undergrowth from surviving (not much different from the stem exclusion stage of a natural forest succession). Up to this point plants that interfere with the achievement of a full stocking of productive trees are removed by cutting or herbicide. However, there is scope, at least around the edges and close to elements of native forest and open space, to permit a small proportion of non-productive trees and shrubs to remain in the canopy and, hopefully, to persist if they can compete for light. It is also possible, if the planted stand is spaced to reduce the number of stems per hectare/acre, to deliberately reduce the density further in order to help to naturalise the edges. This reduces the abrupt junction between planted and native forest/open space and introduces another ecotone into the forest structure.

The thinning stage (whether commercial or pre-commercial) provides another opportunity to incorporate natural structures or elements into the plantation. At this stage a proportion of the stand is removed in order to reduce the total number of trees and to concentrate wood production on the remaining trees. Normally this is carried out to maintain a complete canopy so that no potential timber productivity is lost. However, there are a number of ways thinning can be carried out. If the trees were originally planted in straight lines it is possible to remove complete rows (such as every third one), some rows with selection from amongst the trees between the rows, or completely selective thinning. Row thinning is a purely mechanical process and is easy to carry out, requiring little supervision or skill. The results emphasise the geometry of straight lines and do not allow for the retention of all the trees of best form in the stand. The removal of some rows (sometimes two together) permits easier access by machines, the operators of which can then select the trees to be removed from the adjacent stand. The result is a less formal arrangement of trees that become more varied in spacing as the number of times the stand is thinned increases. Completely selective thinning requires the greatest skill and supervision but produces both the optimal quality in the stand and the most naturalistic effect. Gradually the canopy becomes somewhat more open and it is possible for ground vegetation and, eventually, some woody understorey to develop. Towards the later growth stages of the plantation, well-thinned stands can develop a valuable structure, but they tend to lack the larger-sized dead and decaying wood that would be found by this stage in a natural stand.

Plantation forests mature for harvest at an age when they either reach maximum mean annual increment (MaxMAI), that is the point when the rate of growth averaged across the life of the stand starts to level off, or at the point when they reach maximum net present value (MaxNPV), where the difference between total discounted costs and discounted revenue, at a particular test discount rate, is at a maximum. The later calculation is the one used in most true plantation forestry. Age at final harvest may vary from

Figure 10.10 This plantation of black spruce in New Brunswick in Canada has been selectively thinned. The results start to develop a less formal, more natural structure and appearance. J.D. Irving Ltd land.

20–30 years in eucalyptus, 30 years in southern pine in the USA and Monterey pine in New Zealand, 40–45 years for Sitka spruce in Britain and Douglas fir in the Pacific Northwest, or 70–80 years for Norway spruce or pine in Sweden. These ages are clearly not the natural old age of the stands which, in a natural forest, could reach up to several hundreds of years. In fact, a plantation is harvested at or before the late stem exclusion stage of most natural equivalents of their type and species.

It has long been normal practice for plantations to be harvested by clearfelling for two reasons. Firstly, clearfelling is generally inexpensive from a layout and operational standpoint. Secondly, it facilitates cost-effective site preparation and replanting, which allows the productive cycle to be repeated. If the plantation forest was planted more or less all at once it is likely to mature over a short period of time and thus to be most economically harvested quite quickly. This, if followed to strict economic principles, results in the rapid clearance of the forest and its complete reversion to the establishment phase.

Under the requirements of sustainable forest management, such as expressed in the UK Woodland Assurance Scheme (UKWAS), SFI, FSC and other certification protocols (see Introduction), such rapid clearance is not permitted. Instead, patches of the forest plantation are removed at intervals, with the intention of restructuring the forest from being largely even-aged towards a more multi-aged structure. Normally it is desirable that at least two metres of growth, preferably rather more, on each replanted harvest area is achieved before any adjacent patches are removed, thus creating a pattern of different ages across the forest landscape. It is also desirable to retain certain stands on a longer-term basis in order to maintain some mature forest in the landscape, and to allow a small proportion to develop the stand characteristics of old forest, including standing dead trees, large-sized dead wood and a multilayered structure.

The planting and layout of this pattern of different harvest patches needs a good deal of care. If the forest was originally planted in a grid of compartments, such a geometric pattern, while achieving a new structure, would hardly fit into the landscape aesthetically. Thus, felling design in plantation forests is a crucial stage and must be done well. Opportunities exist not only to diversify the age structure but to eliminate visual and ecological problems associated with the original layout, such as geometric external or internal shapes, lack of open space or native forest elements, damage to watercourses, valuable habitats or archaeological sites, and also to improve the quality of timber grown in the future through replacement of poorly performing species or varieties. Since the new structure developed by patch clearfelling will be difficult to alter once it is set in place, it is important to ensure that it fits into the overall landscape and ecological setting while at the same time is practical and keeps any extra costs of implementation to a minimum.

The process of designing the felling of plantations to achieve multiple design objectives is probably most developed in Britain, so the approach to be described here will be illustrated with British examples. A British case study later in this chapter will also show the main elements of the complete process. The steps to be followed are those already presented in Chapter 5, with an emphasis on the redesign of plantations that possess poor layouts.

Figure 10.11 A view of Ennerdale Forest in Cumbria, England, which has been in the process of redesign for around 27 years. The phased coupes and the various areas of replanting can be seen.

As is the case with afforestation, the design of a cutting pattern should begin with a thorough inventory of the land and its aesthetic character. In older plantations, there could be a number of elements that need to be incorporated into the plan. In hilly or mountainous landscapes it is normally the landform that will be used to drive the design of the felling coupes. The logical starting point is an analysis of landform using the principle of lines of force described in Chapter 3, and presented in other chapters. The lines of force principle is particularly important in plantation forests because there are often no other strong existing patterns available to borrow from. A complete pattern of coupes can be designed that follows topography, using the lines of force as an underlying template. This is achieved by, firstly, locating coupes in hollows or on convexities, generally occupying the main extent of such features. If the forest covers the skyline, coupe boundaries should only cross it at saddle points to prevent exposed and intrusive visual edges. Coupe boundaries should reflect the character of the landform, whether rounded and flowing or more broken and angular, for example. Some key design principles are as follows:

- Coupe edges should rise in hollows and descend on convexities.
- Coupes should be larger higher up hillsides and smaller lower down, to reflect the scale and proportion of the landform.
- Avoid straight sections of boundary, particularly horizontal or vertical in direction.
- Coupe shapes should not be symmetrical, nor should coupes of the same size and shape appear in the same design.
- Interlocking shapes should achieve unity.
- Areas to be retained should be designed into the plan, generally located in sheltered places in fertile soil on lower slopes.

Once the initial pattern has been designed around the landform it is necessary to check that each coupe can be harvested using the equipment expected to be used, from the road system and landings in place or to be built. Adjustments should be made, if necessary, to ensure that no parts of a coupe are inaccessible and that excessive costs are not going to be caused by the design.

The next step is to consider how to phase the timing of felling for each coupe. It is usually desirable to aim for as long an interval as possible between the felling of adjacent coupes so that the maximum height and age differences can be achieved. This time interval depends greatly on the expected growth rates of the replanted stand, the key requirement being a minimum of two-metres height growth.

There are frequently also economic considerations to be taken into account. In order to spread the felling over time in an even-aged plantation, it is necessary to fell some coupes earlier and some later than the economic or growth optimum. This normally reduces revenues (discounted over time). Although some cost is inevitable, it is a good idea to try to keep it to a minimum by not spreading the harvest out over too great a time period. Generally, it is more expensive to advance felling than to delay it, and also more costly for the more productive species or stands. However, it costs a negligible amount to delay or advance felling for up to five years from the optimum, so that in areas where good growth of replanted stands is achievable, it is possible to develop at least three age classes for little cost.

The general scope can be as follows: fell around 70% of the forest at the optimum age or five years either side, delay up to 20% for another five years and advance the rest by a further five years. In the case of slower-growing plantations these periods may have to be stretched to seven or even ten years apart, but delaying or advancing such stands is not as costly as for faster-growing species.

Figure 10.12 An example of a well-designed coupe reflecting the landform in its shape and also the scale of the landscape. Sutherland, Scotland.

The proposed felling sequence can be generated with the help of a GIS by overlaying the coupe design on the forest inventory. Some decision rules, such as the need to achieve two-metre height growth on adjacent coupes and the economic constraints, can then be applied to establish the optimal phasing.

Assuming that the coupe pattern is well designed and will lead to a new structure to the forest that will last for a long time, it is sensible to use this as the basis for the replanting design. There is a major opportunity at replanting to improve the layout of the forest for visual, biodiversity, water, recreation and heritage reasons. Where there is a lack of open space or native woodland, or where sites have been planted or awkward visual effects have been created, then once the existing stands have been removed these can be reconstructed. The following aspects should be considered; all follow the design principles for new planting:

- The external margins of the plantation forest where it meets open, unplanted areas or native forest. Areas can be left unplanted in order to correct geometric shapes and allowed or helped to revert to natural vegetation.
- Streams can be opened up and native woodland elements introduced, and connections made to features outside the forest.
- Internal open spaces for wildlife habitat, archaeological site protection, recreation or hydrological protection reasons.
- Tree species can be changed by planting the whole or part of a coupe with an alternative to vary colour, texture, habitat value, to rebuild depleted soil or to respond to disease issues (such as root or butt rot).

So far the discussion has concentrated on clearfelling and replanting, with small areas retained to grow and develop more ecological maturity. There is also the possibility of managing some of the coupes by non-clearfelling systems, sometimes known as continuous cover because they are not characterised by large cleared expanses. There are a number of systems, traditionally described as seed tree, shelterwood or selection systems, that can be applied (see also Chapter 9). Some of these are more suited to certain species than others, depending on the tolerance of the regenerating seedlings to shade. Generally, these silvicultural systems aim to protect the sheltered microclimate provided by the tree canopy and to prevent changes to soil moisture (water tables often rise following clearfelling), to suppress weed growth and to allow stands to regenerate naturally from seedlings produced by the mature trees. Each silvicultural system needs considerable skill to manage and requires a good deal of continuity and stability of management for the stand to develop. There is no space here to discuss the detailed aspects of each system, but the salient ones that are appropriate in this context will be briefly described.

Shelterwood systems involve the removal of a proportion of the stand to facilitate regeneration, often timed to coincide with a season of good seed production. This removal can be uniformly spaced over the stand (uniform shelterwood) or concentrated in small openings (group shelterwood). As the regeneration becomes established and grows, further canopy removal is undertaken in one or more places to allow more light into the stand. Finally, all the remaining overstorey trees are removed. The group system is implemented by the progressive enlargement of the original openings until they amalgamate into one stand. In leave shelterwoods, some portion of the overstorey is permanently left for habitat.

Selective systems operate on a finer scale. The group selection system resembles the group shelterwood except that instead of a more or less even-aged stand, the addition of new groups over time results in an uneven-aged stand. The single-tree selection system involves a mix of all ages of trees across the entire stand and is managed by controlling the numbers of trees in each size class according to the set of ideal proportions.

Once implemented, each system cannot be reverted to simpler forms of silviculture very easily, so it is essential that the coupes are well designed in the location, extent and shape, so as to fit into the wider pattern of the forest.

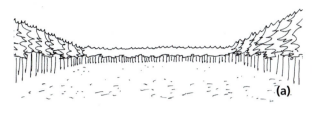

Figure 10.13
Design of felling coupes on flat land. a) shows how uninteresting a geometric shape can be, displaying no mystery and being able to be seen all at once. b) shows a coupe design with an irregular, organic shape, retained stands and a depth of composition which makes it much more interesting. Only part of the coupe is visible at one time, so reducing the apparent scale.

In flatter terrain where landforms are not so strong, coupes are usually seen from within, from viewpoints along roads or paths. This does not mean that geometry is not visible. In fact, the internal shape of a rectangle is very bland and uninteresting. Organic shapes, especially those that form a strong interlocking pattern, are preferred because they provide more visual stimulation and richer edge habitat niches. Highly curvilinear forms also invoke a sense of mystery because not all of the coupe can be seen at once. Retained clumps and other features provide visual diversity and habitat values as described for natural forests in Chapter 9. The same principles of phasing and replanting apply with particular importance given to roadside, pathside and other edges.

If there is some minor relief such as low glacial deposits, these can be used to emphasise the shape, by using small ridges as locations for retained stands that maintain or increase the degree of enclosure and control of spatial scale.

The case studies presented in this chapter focus on two different examples from the UK. The first is a new planting project using native broadleaves in Kent, England, that is similar to an ecological restoration in some respects but contains a strong design element. The second is the management including felling and replanting of a plantation conifer forest on a private estate in the Highlands of Scotland.

Case study: Victory Wood, Kent, England

Introduction

This project was a design for a new area of woodland to be planted on land at Lamberhurst Farm in Kent. The Woodland Trust, a charity dedicated to protection, creation and restoration of woodland, acquired the land in order to develop native woodland on it in commemoration of HMS *Victory* and the 200th anniversary of the Battle of Trafalgar which took place in 1805. This is one of a series of woods being planted across England, each named after a ship of the line present at the famous battle. The design was completed in late 2005 and the planting is underway at the time of writing.

The site before planting was some 140 ha of farmland but until the mid-twentieth century was woodland, part of the Blean complex, an extensive area of ancient woodland in the North Kent Weald, so that in many ways this was a landscape and ecological restoration project rather than an entirely new woodland, although the design reflected the need to meet certain objectives such as public access.

England

London

Victory Wood

Figure 10.14
Location map of
the Victory
Wood project.

Location, context and background

Lamberhurst Farm is located in northeast Kent, north of Canterbury and not far from Faversham or Whitstable, near the small villages of Dargate and Yorkletts. It occupies a ridge top, a north-facing slope and an area of flat, lower ground. The soils are clay and somewhat poorly drained, and the area has recently been used for arable production.

The location is part of a series of ridges which, together with the coastal plain, is known as the Blean. This has been identified in the Kent landscape assessment and its character described (see below). Historically, the farm, at least the slopes and ridge-top part, were wooded. Old Ordnance Survey maps from the 1870s and 1930s clearly show the area as wooded, forming part of the continuous matrix of woodland that characterises the Blean.

Landscape character

The countryside character programme assessment for Kent compiled by the Countryside Agency and English Nature identifies the Blean as part of the North Kent Plain landscape unit. Kent County Council has also compiled a more detailed landscape assessment for the area: The North East Kent Landscape Assessment. In this study the Blean is identified as a local character unit. The key aspects of the landscape noted in the assessment are:

- Densely wooded, rounded hilltops with sparse nucleic settlements and few roads within the woodland
- Flat coastal plain
- Haphazard seaside and leisure development
- Neglected pasture near the coast
- High proportion of unfarmed land.

The landscape character analysis includes some landscape management guidelines. The main aim of these is to restore it to reflect the character of the Blean Woods. These guidelines are:

Figure 10.15
A photo of the
Victory Wood
site. Source:
Woodland Trust.

- To restore native broadleaf woodland to link with Ellenden Woods (a nearby area of ancient woodland)
- To improve ecological and wildlife potential by restoring or creating suitable networks, linking with existing hedgerows and woodlands where appropriate
- To avoid inappropriate large-scale or obtrusive elements on the visually sensitive ridgeline.

Ecological character

The Blean has been identified in the Kent landscape assessment process as part of the most extensive area of nearly continuous woodland on London Clay in the southeast of England. The woodland area is mostly ancient and has traditionally been under coppice with standards management. The main species is sessile oak (*Quercus petraea*) with areas of beech (*Fagus sylvatica*), sweet chestnut (*Castanea sativa*) and hornbeam (*Carpinus betulus*). Areas of ash (*Fraxinus excelsior*) and hazel (*Corylus avellana*) and some pedunculate oak (*Quercus robur*), as standards, can also be found. The ecological value has been recognised by the designation of several Sites of Special Scientific Interest and National Nature Reserves with two areas, Blean Wood and Ellenden Wood, being considered as Special Areas for Conservation. These two woods lie either side of Lamberhurst Farm and would once have formed a continuous wooded matrix. These woods are important habitats for several Red Data Book species, mainly invertebrates, including the heath fritillary butterfly. Several populations of breeding birds such as nightingale, nightjar and golden oriole are also present. Restoration of native broadleaf woodland on Clay Hill is seen as a desirable option to reconnect the existing areas and to re-create a large single element of woodland instead of the two separated pieces of Blean and Ellenden Woods.

Objectives of the proposal

The objectives of the proposal to create woodland are as follows:

- To restore the woodland across an area that has been cleared relatively recently and to reconnect the woodlands in the wider landscape of the Blean Woods Complex
- To maintain and improve the biodiversity potential for the existing woodland and non-woodland habitats within the site
- To improve habitat values for a range of woodland and non-woodland plant and animal species
- To provide, over 50% of the site, a significant area of open-space scrubby grassland to be managed as a semi-natural habitat and grazed
- To allow the expansion of site-native broadleaved woodland through natural regeneration developing from the existing woodland and on land neighbouring it
- To provide for inclusive public access
- To improve and restore the landscape as seen from external viewpoints
- To create an attractive landscape as experienced from within the woodland
- To provide interpretation linking HMS *Victory* with the Trafalgar Woods Scheme to commemorate the Battle of Trafalgar and The Woodland Trust's Corporate Objectives.

Survey

As part of the planning and design process a number of surveys were undertaken:

Site survey

This is a basic survey of the features to be found on the site, soils and drainage, previous land use, power lines and pipelines, tracks and roads and any other elements. In summary, there are a number of features that affect the site. To the northern end are some power lines and also a proposed sewer route that might interfere with part of the development. The power lines run across the area most accessible for parking, etc. A gas pipe runs close to the southern boundary but happily is outside the ownership. Steep slopes occur on the site which may cause problems for access.

Landscape

The area was visited, viewpoints identified and a series of panoramic photographs taken from each. These were used for analysis of the landscape and to illustrate the design. In summary, the area is highly visible and panoramic views can be obtained from the hill tops. In the recent past most of the hilly section was wooded and in the landscape character assessment restoration of the woodland is a priority. The landscape is sensitive, being a prominent ridge visible from many

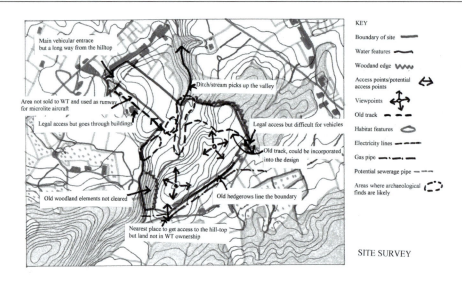

surrounding roads and settlements. There are also views to be obtained from within the site which need to be taken into account.

Ecology

The site has been assessed to see what species, especially invertebrates and birds, currently use the area. This information was used to help develop the appropriate habitats and to ensure that no valuable habitats or species will be damaged by the creation of the woodland. In summary there were some important features to be retained and protected, such as the ditch known as Hawkin's Dyke and its overgrown pond, the southern boundary hedge, Blean Wood, Triangle Wood, the northern meadow with its old veteran oak tree and the woodland edges.

Archaeology and historic landscape

The site was surveyed and extensive archival and other desk-based research was undertaken to chart the long history of land use and landscape in the area. From this it was clear that the landscape has been used by people since prehistoric times, with the woodland areas (cleared in the twentieth century and now to be restored) managed for coppice to provide fuel for local industries and the fields once divided by hedges as part of farmland also with a long history. While most features of the historic landscape had disappeared some remain and the recommendations of the survey, to restore the landscape as close to its historical pattern as possible, were taken into account in the design. This included restoring the woodland area to its former extent, reinstating old trackways, supplementing the remaining veteran tree with new oak pollards and replanting some of the field boundary hedges.

Soils

A soil survey was carried out to assess any problems for tree growth or any polluted areas. In summary, while the climate of this part of Kent is quite dry, the soils in this area, being clay, retain moisture and present few problems for tree growth. Some soil was removed from the area so that topsoil is lacking and there has been some dumping of material that presents some very localised phytotoxicity problems.

Analysis

The next step in the process was to analyse the information found in the survey. Two aspects have been analysed; the site itself, in terms of the interaction of the different factors found there, and the landscape as seen from external viewpoints. The first of these surveys presented the information as a set of opportunities and constraints, while the second presented an analysis of landscape character. The opportunities and constraints are presented in a table, while landscape character is presented in maps and perspectives.

Opportunities and constraints
Table 10.2 presents the opportunities and constraints affecting the design in terms of the effect on meeting each of the project objectives.

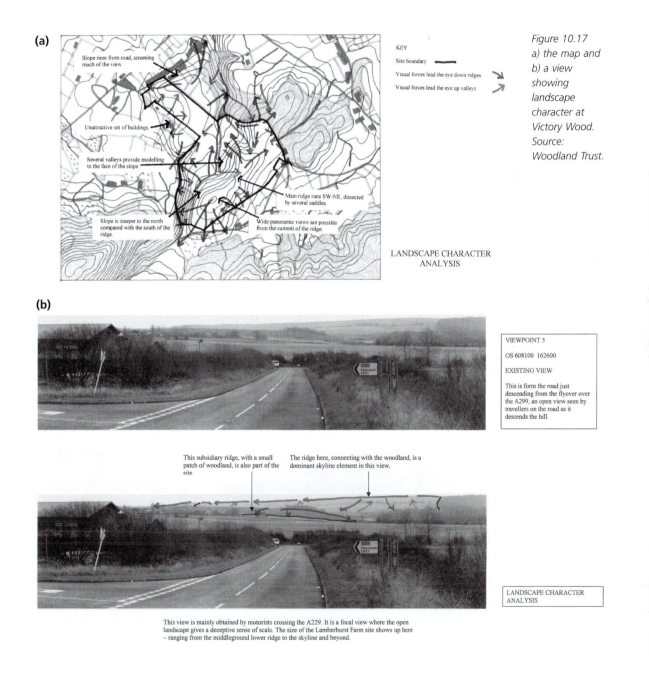

Figure 10.17 a) the map and b) a view showing landscape character at Victory Wood. Source: Woodland Trust.

Table 10.2 Constraints and opportunities at Victory Wood

Objective	Opportunity	Constraint
To restore the woodland across an area that has been cleared relatively recently and to reconnect the woodlands in the wider landscape of the Blean Woods complex	To plant across the cap of Clay Hill as recommended in the landscape assessment, more or less to the original extent To use the same species of trees as found in the local woodlands	The lane leading from Dargate now presents a physical obstacle to the connection between Blean Wood and Clay Hill Nutrients in the soil and the presence of ruderal species will affect the development of ground vegetation
To maintain and improve the biodiversity potential for the existing woodland and non-woodland habitats within the site	To manage through thinning and singling of coppice the 7.6 ha of existing woodland to create high forest To start building up deadwood habitat within the wooded areas and to conserve other deadwood features (trees, stumps) To provide a larger water feature centred around the Hawkins Hill Ditch by enlarging the current small boggy area at its eastern end on the site	Most of the existing woodland is on a slope making timber extraction more difficult The existing woodland is small in size compared to the whole site, thus the impact of this mature woodland habitat will be localised Hawkins Hill Ditch only flows after heavy rainfall, so no vast or even small permanent lakes could be created Woodland edges that receive light and have a valuable structure need to be protected
To improve habitat values for a range of woodland and non-woodland plant and animal species	To establish trees over at least 50% of the area because there is little conservation value in the arable crops To build a range of habitats into the design, especially on the lower areas and in association with the features that remain	
To provide, over around 50% of the site, a significant area of open-space scrubby grassland to be managed as a semi-natural habitat and grazed	To provide this marginal habitat of grassland mixed with areas of developing scrub woodland (covering up to 15–20% of this open-space area) bordering significant areas of mature woodland To provide a good habitat for birds and invertebrates and small mammals including bats	Grazing animals may (albeit in small numbers) be off-putting to the visiting public Potential usefulness of this habitat for ground-nesting birds might be jeopardised by dog-walkers with uncontrolled dogs Could look unsightly if left rough

Objective	Approach	Potential problem
To allow the expansion of site-native broadleaved woodland through natural regeneration developing from the existing woodland both on our land and neighbouring it	To develop a mix of areas including glades within woodland and open areas with patches of trees	Potential problem with noxious weeds
	To enable colonisation of site-native species to occur To reduce the area to be planted and therefore a perceived cost The edges of Victory Wood, alongside the existing wooded areas, to be more natural looking with widely spaced planted trees to be "filled in" by natural regeneration	Natural regeneration could be slow in coming or non-existent, so some planting may be required
To provide for inclusive public access	To provide an all-access path/route around the site	The terrain means that it may be difficult for people with mobility problems to get to the top of the ridge from the bottom
To improve and restore the landscape as seen from external viewpoints	To restore the skyline as a wooded feature in the landscape To reflect the strong topography in the layout of edges and woodland types	The edges of new woodland meeting the existing woodland may need to be kept apart for ecological reasons, which may appear strange
To create an attractive landscape as experienced from within the woodland	To maintain an open viewpoint at the summit To create pleasant glades along paths To design the edges for close viewing, with lots of diversity of species and structure To use strategically placed groups of trees to break up the unattractive farm buildings on the boundaries of the site	The tree growth on the summit may make keeping panoramic views open difficult

Source: Woodland Trust

Design concept

From the development of the objectives, surveys and analysis, it was possible to develop an overall concept for the creation of Victory Wood. The concept was divided into three main, interconnected elements:

1 Woodland and non-woodland habitat creation that fits into the landscape and historical character (discussed below)
2 Access around the site
3 Interpretation and commemoration.

Woodland and non-woodland habitat creation
The landscape and ecological character provided strong guidance for the overall pattern and distribution of different habitats. The historical land cover was clearly defined into the woodland and non-wooded fields. The wooded areas were mainly on the slopes. The current objectives are to create woodland on around 50% of the area and have a mix of other habitats on the remaining 50%. The original land cover was much more than 50% woodland, so a simple restoration was not envisaged. However, for visual reasons, in order to look balanced from key viewpoints and to restore the character, it was important for woodland elements to extend down as far as the original boundary, although the type of habitat to be adopted on the lower southwestern portion of the slope would not be continuous woodland but of a patchy character, interspersed with open areas. Equally, the lower fields, which historically were divided into smaller fields, would not be left as they are but some boundaries would be restored and they would be developed into a mosaic of grassland, scrub, wetland, small patches of woodland and individual trees. This was designed and laid out to allow for a degree of grazing as part of the management. The area was divided into a series of zones.

Zone 1 Woodland matrix: This was the main area for woodland creation, mainly through planting using transplants grown from locally collected seed but also including some direct seeding and natural regeneration, especially close to the existing woodland edges. The planting mixes were developed from the recommendations in the ecology report. They included mixes aimed at

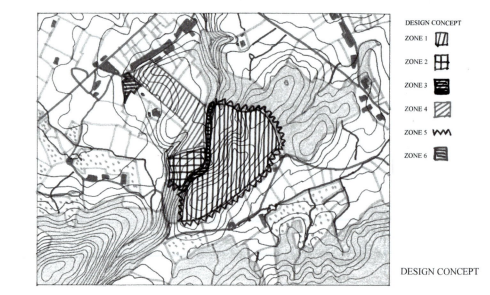

Figure 10.18 The design concept for Victory Wood. Source: Woodland Trust.

producing both coppice-with-standards and high forest. The largest remnant existing woodland patch was incorporated into the matrix and is to be managed to open the canopy and to convert the coppice stems into single trees. There were some open spaces left in the matrix, especially where small glade habitats were to be encouraged but also to provide recreational space, views and a setting for a sculpture.

This area contributes to meeting the objectives as follows:

- It will restore the woodland across the area that was cleared relatively recently and reconnect the woodlands in the wider landscape of the Blean Woods Complex.
- It will maintain and improve the biodiversity potential of the existing woodland and non-woodland habitats within the site.
- It will improve habitat values for a range of woodland and non-woodland plant and animal species.
- It will allow for the expansion of site-native broadleaved woodland through natural regeneration developing from the existing woodland and on land neighbouring it.
- It will improve and restore the landscape as seen from external viewpoints.
- It will create an attractive landscape as experienced from within the woodland.

Zone 2 Woodland/scrub/grassland mosaic: This area provided a mixed type of habitat comprising patches of woodland of various sizes interspersed with areas for scrub to develop and eventually become woodland over a long time period and grassland to be managed on a low-intensity basis. This would be concentrated in one area but there would be some other places where it would be developed. The wooded elements would be planted using transplants grown from locally collected seed, in mixes developed from the recommendations in the ecology report. This would be at varying density, including areas of scrub species at wide spacing which would remain ungrazed. The planted areas would be fenced to prevent grazing damage with some grazed areas running between.

This area contributes to meeting the objectives as follows:

- It will maintain and improve the biodiversity potential of the non-woodland habitats within the site.
- It will improve habitat values for a range of woodland and non-woodland plant and animal species.
- It will provide some of the open-space scrubby grassland to be managed as a semi-natural habitat and grazed.
- It will improve and restore the landscape as seen from external viewpoints.
- It will create an attractive landscape as experienced from within the woodland mosaic.

Zone 3 Riparian: This area is based along the stream and associated wet areas and former pond. It would be developed and managed as a partly open, partly scrubby woodland habitat. The appropriate species would be taken from the recommendations in the ecology report. Grazing animals will be excluded.

This area contributes to meeting the objectives as follows:

- It will maintain and improve the biodiversity potential of the non-woodland habitats within the site.
- It will improve habitat values for a range of woodland and non-woodland plant and animal species.
- It will create an attractive landscape as experienced from within the woodland mosaic.

Zone 4 Grassland: This area provides the semi-natural grassland habitat that is valuable for a range of species. It would be established using a mix of grasses and wild flowers appropriate to the site and grazed by cattle. These mixes would be developed from recommendations in the ecology report. Some areas that are not convenient for fencing or need to be accessible and open would be mown occasionally. Hedges with hedgerow trees would be planted and along the boundaries.

This area contributes to meeting the objectives as follows:

- It will maintain and improve the biodiversity potential for the non-woodland habitats within the site.
- It will improve habitat values for a range of non-woodland plant and animal species.
- It will provide a significant area of open-space scrubby grassland to be managed as a semi-natural habitat and grazed.
- It will improve and restore the landscape as seen from some external viewpoints.
- It will create an attractive landscape as experienced from within the woodland.

Zone 5 Woodland edge: This zone is a relatively small one, comprising the edges of the wooded area, especially against neighbouring land but also against the other existing woodlands. Here the composition and structure of the woodland edge would be established by planting appropriate tree and shrub species and by leaving open spaces along the fences to allow an ecotone to develop. The species mixes suitable for the edges would be developed from recommendations in the ecology report. There would be differences where the edges lie against fields and where they were near other woodland edges. The latter would be able to rely on natural regeneration and colonisation much more than the former.

This zone contributes to meeting the objectives as follows:

- It will be part of the restoration of the woodland across the area that has been cleared relatively recently and reconnect the woodlands in the wider landscape of the Blean Woods Complex.
- It will maintain and improve the biodiversity potential of the existing woodland and woodland edge habitats within the site.
- It will improve habitat values for a range of woodland edge plant and animal species.
- It will allow the expansion of site-native broadleaved woodland through natural regeneration developing from the existing woodland and on land neighbouring it.
- It will improve and restore the landscape as seen from external viewpoints, especially the close ones to the south.

Zone 6 Recreation and interpretation zone: This zone is the developed area to the north, where the main access, car park, specific elements to commemorate the *Victory* and associated interpretation will be developed and managed.

Design

The concept was developed into a comprehensive design showing the different areas of planting with trees and shrubs, the open spaces within the wooded areas, the fencing, the paths and other access routes and the interpretative elements. This was presented in plan and supported by a number of perspective views. The idea was to create a unified design where everything fitted together harmoniously.

A range of habitat types would be developed, all linked and arranged according to the site. The main woodland would include both high forest and coppice. Woodland edge zones would create a network of corridors throughout the site and allow for natural development of the edge ecotone. Some internal open spaces would let light in and create more diversity along the paths. The woodland/scrub/grassland mosaic would be established with a series of grazing exclosures until the growth has developed sufficiently to enable the fences to be removed and the area opened to low-intensity cattle grazing, after which time a complex structure will develop. Wetland would also be developed along one section of the stream and different types of grassland would develop through a combination of seeding and pasture management.

The design will protect the areas likely to contain archaeological remains as identified in the survey by keeping them open and under grass.

The visual appearance of the wood is demonstrated in the views. The result is one of restoration of the Blean landscape, with woodland once more clothing the skyline and the slope. From inside

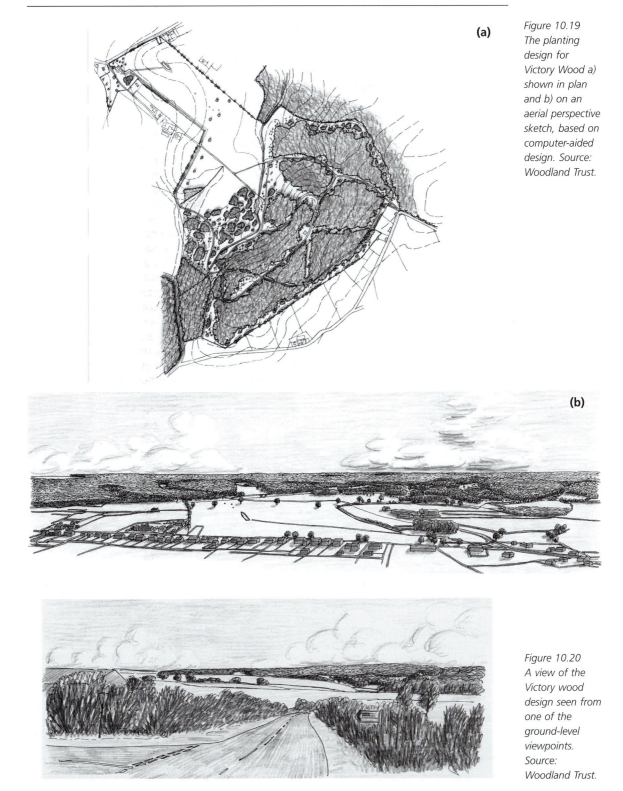

(a)

*Figure 10.19
The planting
design for
Victory Wood a)
shown in plan
and b) on an
aerial perspective
sketch, based on
computer-aided
design. Source:
Woodland Trust.*

(b)

*Figure 10.20
A view of the
Victory wood
design seen from
one of the
ground-level
viewpoints.
Source:
Woodland Trust.*

*Figure 10.21
Location map of
the Dochfour
project.*

the site the landscape will be diverse, including open and enclosed areas with different views ranging from the immediate landscape of the woodland to broader vistas out to the countryside beyond.

Case study: Dochfour, Inverness-shire, Scotland

Introduction

Dochfour Estate is a mixed estate of approximately 3,450 ha lying to the northwest of the Great Glen, above Loch Ness and the River Ness, 2 miles to the southwest of Inverness. It is privately owned and managed, making it different from most of the other case studies presented here. The need for public consultation remains important, but public participation in the plan was not undertaken as it is not yet part of the accepted way of managing private land in the UK. However, a scoping meeting with representatives of a number of agencies took place and a scoping report used to identify the various issues that would need to be taken into account. When the plan was prepared there was no automatic right of public access to the woods, although the legal position in Scotland was rather complex. Since then the Scottish Parliament has passed legislation granting the public right of access to forests, moorland and rough ground. This means that during the plan preparation recreational values only played a small role.

The estate is prominent in the landscape, especially those areas rising from the valley floor of the fertile flood plain of the River Ness onto the steep slopes and hilltops above. As well as forest there are farms, some moorland and a designed parkland landscape of considerable historic interest.

The forest occupies approximately 1,500 ha of the total land area of the estate, which is made up of approximately 50% conifers, planted mainly for commercial reasons, and 50% mixed woodlands managed for their amenity and conservation value. These include the extensive designed landscape and semi-natural woodland. The forest is bordered to the north by open heather hills, and other, mainly commercial plantations in a variety of ownerships, both public and private.

The forests vary in character from areas of semi-natural woodland, which have not been managed intensively, to commercial conifers of mainly introduced species, which were planted in the 1950s on agricultural land (mainly rough grazing) and on forest areas that had been felled for their timber during the Second World War. The commercial forest areas have been thinned where they are accessible, and contain stands which are approaching financial maturity.

In 2000, the owner wanted a long-term forest design plan and as a scheme of financial support for the preparation of such plans by private owners had just been established by the Forestry Commission, the plan was developed. Simon Bell worked on the design aspect along with another consultant, Steve Conolly, who manages the woodlands for the owner.

*Figure 10.22
One of many
views of the
landscape of
Dochfour Estate.
Source: Cawdor
Forestry Limited.*

Objectives of management

Objectives were set for a number of forest resource values. These arose from a combination of those of the estate owner, national forestry policy requirements and pertinent issues raised at the scoping meeting.

Table 10.3 Objectives at Dochfour

Resource	Objective	Indicator
Timber	To produce high-quality timber To maintain or enhance productivity	Maximise percentage of high-quality sawlogs Maintain or improve average yield class of commercial woodlands Utilise genetically improved planting stock where appropriate
Financial	To maximise sustainable long-term income	Felling takes place as close to maximum Net Present Value as possible Increasing positive net cash flow that can be demonstrated to be sustainable in the long term
Sporting	To maintain and enhance sporting values, without compromising the long-term environmental and silvicultural objectives	Maintain and expand the sporting enterprise Generation of income, whilst achieving successful tree regeneration
Landscape	To maintain the current level of landscape diversity To protect and manage the designed landscape	Landscape perspective views can be demonstrated to be compatible with good design and with the landscape The character and quality of the designed landscape continues
Biodiversity	To protect and enhance the nature conservation values, particularly of the native and policy woodlands	Evidence of natural regeneration of semi-natural woodland Species and age diversity are maintained or enhanced Rare wildlife populations are maintained
Deer control	To reduce the deer population to allow forest regeneration	Evidence of natural regeneration and restocking without resorting to fencing Reduced deer damage to standing timber
Recreation	To provide specific managed public access	Implementation of specific schemes such as the Great Glen Way

Source: Cawdor Forestry Limited

Survey

Information for the development of the plan was gathered from a number of sources. A GIS database was already in existence, so that a lot of the analysis was able to be carried out using it.

Geology
The underlying solid geology of the area varies from Old Red Sandstones, conglomerates and Moine-series schists and gneisses. Only part of the lower valley floor is covered with a depth of morainic material and this is mainly occupied by the designed landscape.

Soils and site types
Soils vary from fertile brown earths on the lower areas through surface water gleys on the mid-slopes through gleys to podsols and deeper peat on the more exposed, upper areas.

The sites most suitable for growing commercial forest are generally at the lower elevations, less than 280 m, and fairly sheltered.

Windthrow hazard

The low elevation and freely draining soils result in the windthrow hazard class being between 2 and 4 for most areas. This was calculated using DAMS, however individual coupes were modelled using Forest*GALES* software (see Chapter 5). Most stands have been thinned and there has been no damage to the edges of an existing coupe, suggesting that the assessment of wind damage risk is correct.

Forest stands

Good-quality forest stand information exists, which was produced using air photo interpretation and ground checking in 1999. This information comprises species, age (from planting year), yield class and management history.

Archaeology

A desk survey of archaeological sites was carried out by a commercial archaeology firm, which supplemented information obtained from the Highland Council Sites and Monuments Record.

Nature conservation

Site designation information (Sites of Special Scientific Interest, Ancient Semi-Natural Woodland, Ancient Woodland Sites) was obtained digitally from Scottish Natural Heritage and incorporated into the GIS.

Landscape

A full landscape survey of viewpoints was carried out and an extensive suite of photographs taken.

Analysis

The main types of analysis carried out were those of landscape character and constraints and opportunities. It was not possible to carry out a full landscape ecological analysis as the means of this had not been developed for plantation forests.

Landscape character analysis

Dochfour Estate lies mainly on the western slopes of the northernmost section of the Great Glen. From south to north the topography changes quite markedly, producing a transition from a steeply sloping, rugged landscape wild in character to a series of rounded hills of lower elevation and less broken landform. The forest changes too, from a naturally fragmented pattern interspersed with scree and rocky outcrops rising to bare summits, to completely clothed hills of a smooth, almost continuous texture. The forest on the rising slopes is intimately connected to an extensive designed landscape occupying the main valley floor to the north and west of Loch Ness and the River Ness. For planning purposes the landscape has been divided into seven zones. This zonation is used for landscape character analysis and the design strategy (see strategy map below for the spatial extent of the zones).

Lochend
This zone lies on the steepest, most rugged portion of the landscape, bounded to the north by a deep ravine and to the south including land outside the estate. Loch Ness lies immediately below these slopes. Views from the A82, the main road passing along the Great Glen, are limited but the zone is highly visible from Loch Ness, the Caledonian Canal, from roads on the eastern shore and from numerous roads south of Inverness, particularly the two military roads. The landform is steep with rock outcrops and gullies dominating the upper slopes and scree and other deposits below. Heather, bare rock and scattered birch form the semi-natural component of land cover.

The forest can be divided into two halves:

1   The southern portion is highly fragmented, irregular in shape, broken by scree and rock. It comprises mainly birch on the lower slopes and scattered Scots pine above. This is quite

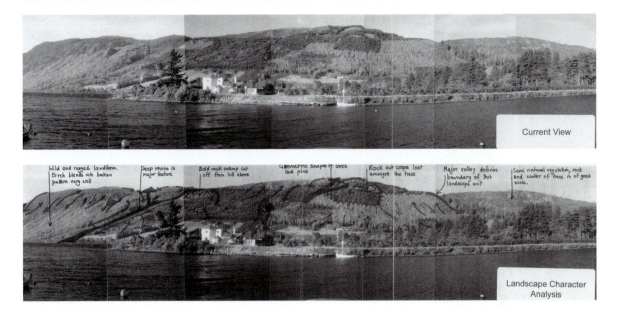

Current View

Landscape Character Analysis

attractive in the landscape, besides being uneconomic and physically difficult to harvest (see Constraints and opportunities analysis below) so is ideally left to its own devices, perhaps with some management to the lower, more easily accessible areas and removal of lodgepole pine where feasible.

2   The northern portion lies above some fields. It comprises larch below and lodgepole pine above. The pine lies somewhat awkwardly on the landform and the species boundary with the larch is geometric.

Figure 10.23 A second view and a perspective showing landscape character at Dochfour. Source. Cawdor Forestry Limited.

Dochfour Hill/Blackfold

This zone wraps around a major ridge from the north of Loch Ness up into a deep valley. The lower-elevation slopes include some steep sections and very complex landform. This area is less visible but still prominent in some wider views and from selected viewpoints along the A82. The forest on the lower slopes of Dochfour Hill is extremely diverse, with a range of species, open areas, lower-density patches and a naturalistic grading of forest to mountain. The area blends into the woods of the designed landscape and forms a backdrop to Dochfour House. The higher-elevation forest up in the valley (Blackfold) is mainly lodgepole pine on wetter, colder soils.

Craig Leach

This is the most dominant zone from many views both distant and from the A82. It is, however, the simplest in landform and forest cover. It comprises a simple ridge of relatively smooth slopes, steep lower down, rising to a flatter top. This zone forms the backdrop to the majority of the designed landscape and there is scope to retain some of the lower area of forest as a transition between forest and park. This zone is entirely afforested, predominately with even-aged commercial conifers. An existing felling coupe is a significant feature that is not sympathetic in either shape or scale, and could be reshaped to provide landscape improvements. Other geometric species shapes show up strongly during autumn, winter and spring.

Dunain Hill

This zone lies on a very prominent, steep-sided, rounded hill with rocky outcrops to the east, which is a landmark for miles around. It is flanked by two lines of electricity pylons, has a radio transmitter on its summit, both of which detract from its appearance. The forest is predominantly of even-aged mixed conifers, which due to a variety of species, site types and areas of rocky terrain is perhaps somewhat too diverse in its pattern.

The designed landscape

This zone is the largest and lies all along the valley floor, and is prominent mainly in close-up views from the A82. The woodlands comprise a number of avenues and tree clumps set in a series of open spaces, carefully composing views and vistas.

Many of the woods occupy knolls that both remove difficult terrain from agriculture and increase the sense of enclosure of the landscape. A range of exotic trees provides diversity of colour, texture and height throughout the landscape. Maintenance of the landscape structure is a priority here, mainly achievable through forms of continuous-cover silviculture.

Lentran Hill

This zone comprises a gently rolling summit area, part of the ridge which divides the Great Glen from the Beauly Firth. It is only visible as a distant though prominent part of one of the views, from the Black Isle. The forest comprises even-aged conifers, although inaccessible due to the currently poor road infrastructure, to the north, and open native Scots pine, which is an important black grouse habitat, to the south. The area is surrounded mainly by commercial forestry in other ownerships.

Constraints and opportunities analysis

This analysis explores the range of physical, practical, economic, social and cultural issues that affect the area and what can be achieved there. Table 10.4 sets out these issues and describes the constraints and opportunities in relation to each.

Design strategy

Following the analysis, the next stage is to develop the design strategy leading to the sketch design. The overall design strategy has been developed from a combination of the information gathered for the plan and the overall objectives and analyses. This provides long-term goals for the plan and enables each successive phase of activity to be planned in detail thereby contributing to the overall success in achieving those goals.

The design strategy has been developed in relation to each of the landscape zones, which have distinct attributes and characteristics that need to be developed in specific ways. For each of the zones the key features are followed by the strategy and then a reference back to the objectives.

Lochend

This area is highly visible, scenically dramatic, steep and difficult to work, with limited access and low production potential apart from the lower slopes above the fields.

The design strategy is to:

- Remove the uneconomic lodgepole pine and to replace it with a native woodland which would blend into the commercial forest areas below. This would be composed of native Scots pine, birch and open ground
- Fell the larch and spruce area over a series of phases and restock, mainly with larch. Some small areas will be retained, and the overall broadleaved component will be increased.

This will:

- Increase biodiversity values by expanding native habitat
- Improve the landscape by improving the upper margin

Table 10.4 Opportunities and constraints at Dochfour

Factor	Constraint	Opportunity
Ownership boundaries	Eastern boundary conflicts with landform As coupes are felled the boundary may become more prominent	To reduce the impact of the boundary To co-ordinate plans with neighbouring owners to improve boundaries
Powerline wayleave	Severely conflicts with landform	To reduce the impact of the wayleave at replanting by use of open ground and species variation
Telecom mast	Prominent	None
Age of existing stands	Many of the crops will become mature in a short time period Large areas will reach maturity within 20 years of each other This will necessitate some areas being felled before or after the optimum rotation	To commence restructuring to spread felling ages as far as possible, resulting in the second rotation having a wider spread of ages
Timber market	Limit areas which can be harvested economically – specifically removal of poor-quality lodgepole pine stands on inaccessible sites	To have flexible felling permission in place, allowing harvesting of inaccessible areas to take advantage of market fluctuations
High deer population	Limits regeneration Neighbouring forest provides significant cover, reducing potential for stalking	To increase deer control within woodlands as coupes are felled To participate in the local Deer Management Group
Soils	Generally heavy, wet soils, with rocky, drier areas	To match species to site, producing a diverse forest Wetter soils generally suited to spruce, drier areas suited to pine and larch which will be introduced for landscape diversity
Windthrow risk	Generally low windthrow hazard, with some exposed sections	Allows a range of silvicultural prescriptions to be considered
Topography	Limits economical working and road construction Steep ground will limit choice of harvesting systems	To designate some areas as no intervention, adding to biodiversity To fell some areas of conifers, and restock with broadleaves consistent with the amenity working circle
Native woodlands and designed landscape	Limit productive area	To maintain and enhance these areas in the long term To soften effect of commercial woodland To consider extension to replace inappropriate conifers To restore ancient woodland sites
Watercourses	Major gullies limit access	Improve riparian zones by removing conifers and extending native woodlands
Landscape prominence	Limit form and size of felling coupes	Will encourage felling over a long time period, phasing timber production and hence income
Poor internal access	Will contribute to some areas being uneconomical to work	To develop road construction and realignment proposals along with felling coupe planning

Source: Cawdor Forestry Limited

*Figure 10.24
a) The map and
b) a view of
constraints and
opportunities for
Dochfour.
Source: Cawdor
Forestry Limited.*

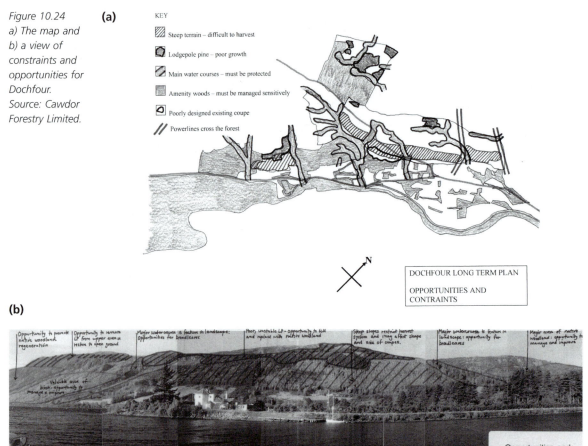

(a)

(b)

- Remove awkward shapes and retain diversity
- Concentrate commercial timber production in the most economic area.

Dochfour Hill/Blackfold
This area already possesses a significant expanse of native woodland that blends naturally into the landscape.

The design strategy is to:

- Fell the upper Blackfold area, to remove unsatisfactory lodgepole pine
- Extend native woodland by removing small areas of inaccessible commercial conifers and restocking with native broadleaves
- Fell commercial conifers over a long time period, and restock with suitable commercial species.

This will:

- Maintain and improve diversity by retaining, expanding and managing semi-natural woodland and other habitats
- Maintain landscape values and enhance views up into Blackfold by reshaping margins
- Concentrate commercial timber production in the most economic area.

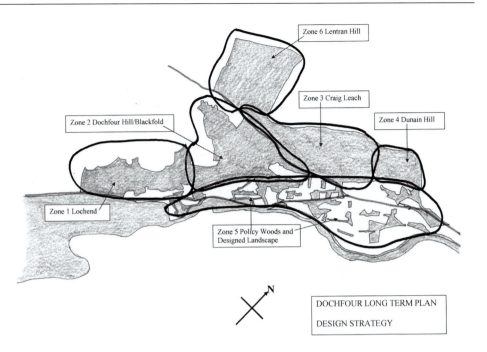

Zone 6 Lentran Hill

Zone 3 Craig Leach

Zone 2 Dochfour Hill/Blackfold

Zone 4 Dunain Hill

Zone 1 Lochend

Zone 5 Policy Woods and
Designed Landscape

DOCHFOUR LONG TERM PLAN

DESIGN STRATEGY

*Figure 10.25
Map of the
design strategy
for Dochfour.
Source: Cawdor
Forestry Limited.*

Craig Leach
This area provides the majority of the timber production potential. It has some steep, difficult slopes, forms an important backdrop to the policy landscape and forms a dominant skyline from many viewpoints around the area.

The design strategy is to:

- Introduce a pattern of felling coupes that reflects the shape and scale of the landscape
- Phase felling over a long time period
- Improve the shape of the existing felling coupe
- Fell conifers from the ancient woodland site and encourage native woodland regeneration
- Manage selected areas on the lower steeper slopes by continuous-cover silviculture, to maintain their current landscape diversity
- Restock with a mixture of species which will contain a significant proportion of larch for landscape diversity
- Introduce broadleaves and open space for biodiversity.

This will:

- Produce timber by felling at or close to the economic optimum
- Diversify landscape structure and enable poorly designed species and coupe shapes to be improved
- Protect the ancient woodland site
- Manage the skyline by phasing the felling
- Retain the option to increase the proportion of the area managed by a continuous-cover system
- Introduce greater structure and composition to assist biodiversity
- Retain the transitional links to the policy woods of the designed landscape
- Retain a mature element in the landscape.

Dunain Hill
This very prominent, steep hill is compromised by the power lines and the radio transmitter.
The design strategy is to:

- Retain a large proportion of the steep eastern slopes, which are difficult to work, and manage them by continuous cover
- Fell the commercial areas in coupes designed to reflect the shape and scale of the landform
- Restock with a species mix that retains the current diversity
- Redesign the most prominent shapes along the powerlines.

This will:

- Produce timber by felling at or close to the economic optimum
- Improve the landscape structure and correct some problems
- Manage the skyline
- Retain a mature element in the landscape and contribute to biodiversity values.

The designed landscape
These woods, clumps, avenues and single trees form the structure of the designed landscape.
The design strategy is to:

- Manage by various types of continuous-cover silviculture
- Retain the diverse species mix, including exotic conifers
- Plant isolated single trees to maintain the policy woodlands.

This will:

- Retain and conserve the layout and composition
- Retain and enhance landscape diversity
- Protect and enhance biodiversity values.

Lentran Hill
This area has low landscape sensitivity; however it is of high biodiversity value, due to the varied woodland structure at the southern end.
The design strategy is to:

- Continue to manage the northern end as commercial forest
- Fell using well-designed coupes
- Restock to provide greater species diversity, better matched to the site.

This will:

- Restructure the forest
- Retain and enhance biodiversity values
- Continue to produce timber
- Contribute to landscape diversity as part of a wider forest landscape.

Riparian zones
Running through most of the zones is a linear pattern of stream valleys.
The design strategy is to:

- Open up the stream sides
- Encourage the development of a native woodland structure and open space by planting and natural regeneration.

This will:

- Protect and enhance water quality
- Provide linear biodiversity corridors
- Enhance the landscape.

Felling and restocking

The sketch design stage comprises two parts, the felling design followed by the restocking design.

Felling design
The felling plan has been drawn up to reflect the stand structure, landform and the subsequent needs of restocking, and issues such as the removal of uneconomic areas.
    The coupe pattern includes:

- A series of well-scaled caps that crown the hilltop summits, which will be felled in sequence so that the integrity of the skyline is retained
- Mid-sized coupes on the side slopes to be clearfelled, and some larger coupes to be managed by continuous cover
- Smaller coupes and long-term retentions along the lower slopes
- Premature felling of poor-quality stands.

    The timing of the felling coupes has taken into account the need to fell as close to the economic optimum as possible while achieving a minimum of two-metres height differences between

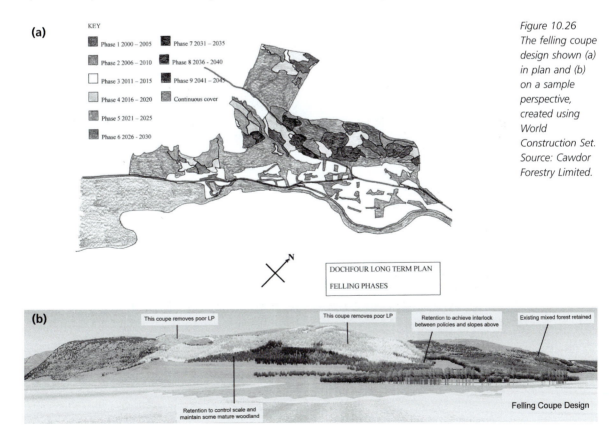

Figure 10.26
The felling coupe design shown (a) in plan and (b) on a sample perspective, created using World Construction Set. Source: Cawdor Forestry Limited.

adjacent coupes. The coupes are phased in five-yearly intervals, but it is recognised that two-metres height is unlikely to be reached within this period except on the best sites. Therefore, the phasing plan shows that each coupe, with one or two exceptions, is never adjacent to a phase immediately before or after it.

Attention has also been paid to the internal views obtained from along the minor public road that passes through the forest on the slopes of Craig Leach and the Great Glen Way.

The felling is phased over nine phases, that is a total of 45 years. This will add to the age diversity within the woodlands. The later phases are predominantly Scots pine, which have a longer rotation, and if possible will be managed by continuous cover rather than clearfelling, adding further diversity.

Restocking design

The restocking design is based on the structure provided by the felling plan, since the coupe shapes are sympathetic to the topography in shape and scale. A range of species will be used, according to site. The restocking species will contain a high proportion of larch and Scots pine either pure or in mixture, as well as Sitka spruce on the most productive areas. Within coupes, species will be allocated according to microsite, and will include up to 30% minor species, to add landscape and ecological diversity. For example, Sitka spruce areas will contain up to 30% Scots

*Figure 10.27
The restocking
design shown (a)
in plan and (b)
on a sample
perspective,
prepared using
World
Construction Set.
Source: Cawdor
Forestry Limited.*

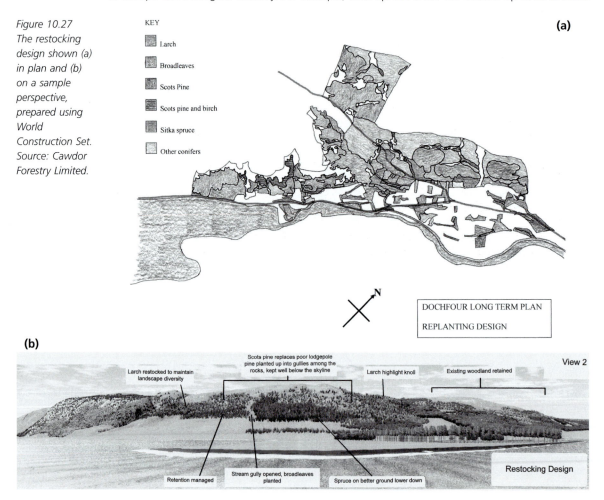

View at end of Second
Phase
Phase 1 Coupes Restocked
Phase 2 Coupes Still Bare

Figure 10.28
View of the felling progress as it should appear after 10 years of plan implementation. Source: Cawdor Forestry Limited.

pine and larch, etc. This, together with the new open areas, broadleaves, continuous-cover stands and existing deigned landscape and other woods, will produce an extremely diverse forest in both composition and structure.

Conclusion

Plantations probably stand to benefit more from the landscape design process than any other forest type. To a great extent this is because they are cultural artefacts, are simple in structure and have the potential to create negative impact on many resources unless they are carefully thought out. The jury is still out, and likely will be for some time as to whether plantation forests are truly sustainable over the very long term. The authors believe that plantation forests can be an important component in the larger goal of forest conservation by helping to take the pressure off remnant natural forests to be converted into fibre farms. We also believe that plantations can be more sustainable if managers pay close attention to the soil, and adopt the design strategies we have described.

The methods and techniques of forest landscape design were largely developed for application to plantation forests because these were seen by many people as some of the most unattractive additions to the landscape. It is likely that of the sum total of forest ecosystem designs undertaken around the world, the vast majority have been for plantation forests. The case studies described in this chapter are just two representative examples and display a degree of maturity in the application of the process as applied to plantation forests in the UK. What they lack, however, is a significant ecological analysis, so important in the managed natural forests that are prevalent elsewhere, and analysis that introduces more complexity and more challenges for the designer.

Afterword

One of our colleagues who reviewed a draft of this book noted that, if designing sustainable forest landscapes is such a good idea (as he agrees it is), then why is it not a more widespread practice?

This is a question that has troubled both of us for years. Other than the projects we have completed, plus those of a few close colleagues, there are few examples of projects completed along the lines described in this book. So we need to grapple with the question: What are the main impediments to the growth and development of the sustainable forest design method?

1. Rural and wild landscapes are seldom "designed"

As noted in the Introduction, there is a natural scepticism about the idea that rural, semi-wild and wild landscapes can or should be deliberately "designed". The UK is the only nation that has experienced a long history of landscape architects applying their skills to rural landscape problems. The profession of landscape architecture basically developed in the British Isles. Following the agricultural revolution and the Enclosures that consolidated landed estates and expanded the wealth and control of the landed gentry came opportunities for landscape gardener/architects, such as William Kent, Lancelot "Capability" Brown and Humphrey Repton, to reshape large swathes of the English countryside into an idealised park-like appearance. This tradition, while it has had ups and downs over the centuries, established a unique historic role for landscape architects in Great Britain. Thus, when the British Forestry Commission ran into problems with its programme to expand conifer plantations (largely based on aesthetic objections), it was natural for them to hire a prominent landscape architect, Dame Sylvia Crowe, to suggest solutions rooted in good design. She was building on a long tradition, and had only to apply her skills to the modern issues that presented themselves. After some initial resistance by foresters the discipline has become established and no one thinks it unusual to design a forest.

However, outside Great Britain there is little cultural tradition of design being applied to large tracts of rural or wild landscapes. In most of the developed world, the profession of landscape architecture has largely been confined to urban and suburban settings. Parks, golf courses, college campuses, housing developments and plazas make up the bulk of commissions. In the rural and wild hinterlands, landscape architects who work for public forestry and park agencies such as the US Forest Service and National Park Service have mostly been restricted to recreation site design, trail planning and aesthetic impact mitigation. The latter includes integrating scenic conservation within a framework of forest planning, but rarely includes proactive design that integrates multiple issues (see the comparison in Chapter 3). In the non-English speaking world (Central Europe,

Scandinavia, Russia) the forestry tradition is still strongly focused on site-scale silviculture, and landscape architecture and design are not tools that are used very often.

Rural, semi-wild and wild landscapes are seen as products of cultural and/or natural processes that are incremental, evolutionary and cumulative, not as first conceived by creative minds. One is reminded of the shock expressed by Arthur Dent in the renowned book *A Hitchhiker's Guide to the Galaxy* when he learns that Slartibartfast, a planet architect from Magrathia, designed Earth. "Are you trying to tell me," said Arthur, slowly and with control, "that you originally . . . *made* the Earth?" "Oh yes," said Slartibartfast. "Did you ever go to a place . . . I think it was called Norway? That was one of mine. We won an award you know. Lovely crinkly edges." Large-scale forest design, across tens of thousands of hectares, approaches the Slartibartfast hubris test. To many it just seems too fantastic even to be attempted.

This lack of awareness of the potential that forest design has to offer may be one reason why public stakeholders and communities do not typically advocate "design" in forestry. For the most part, they either advocate more intensive forest management (in communities that benefit from wood production) or they advocate less logging or complete preservation. Less logging can simply mean leaving more trees standing for wildlife, watershed and aesthetic benefits; it need not mean better design. Preservation means drawing lines around areas to keep foresters out.

Simply put, to most of the people interested in the fate of forests, they are seen as either fibre farms or "natural", and in either case they do not need to be designed. The former is a type of agriculture, and the design is simply functional. The latter is by definition self-organising, and needs no creative interference.

2. Forestry is catching on, but slowly

Foresters are not a naturally receptive audience to the idea of design. They pride themselves on being hard-headed, practical people who make rational decisions based on an analysis of economics, engineering and forest stand types. They use this knowledge to push for the efficient production of a maximum amount of timber, and the landscape pattern that results appears more by default than by deliberate design. For many years, foresters have assumed that if they did a good job managing productive stands, everything else would take care of itself (wildlife, water and aesthetics). Slowly, as values have shifted in favour of increased conservation, foresters have turned to various specialists and modified their practices to mitigate negative impacts. This has been reactive, not proactive.

An exception to this is the silvicultural specialist. Silviculturists understand the core idea of design, that objectives applied to the land drive its form, and that the spatial mosaic developed from the accumulated effect of many stand-level actions. However, they do not tend to think of the landscape pattern first and then the stand as an element of that pattern. Silviculturists simply need to learn to apply their skills to larger scales, adopt "outside-in" analytical techniques, and accept a more complex set of goals to adapt themselves to the forest design process.

Forestry is a very proud profession, once highly respected but increasingly under assault from many quarters for its slowness to adjust to changing values. Communities who depend on production forestry complain that foresters have failed them by reducing harvest rates or by companies having mechanised production, resulting in fewer employment opportunities. Environmental activists have derided and humiliated foresters with local and international campaigns against clearcutting. Science-based colleagues from other disciplines (such as ecology, biology, soil science and hydrology) have taken foresters to task for apparently ignoring resources that are not contributing to timber production. We landscape architects have also parried with foresters over aesthetics for decades. However, foresters are also very good at doing what is asked of them and many love forests, trees, nature and the landscape and they jump at the chance to move away from repetitive industrialised practices to something more challenging and rewarding from a professional and personal perspective. If they are asked to manage for multiple objectives and to take on some form of design, with training, support and encouragement, many of them rise to the challenge and produce excellent results.

At the political or policy level the focus in adapting forestry to new demands has often been to establish or adjust rules rather than to embrace a design approach. These rules include riparian protection, buffers around nests, road engineering, green or snag tree retention and aesthetic rules for scale and shape. Environmental activists push for ever stricter rules that all aim to leave more trees standing longer, while forest managers push for easier ones that allow more to be harvested sooner. Both groups seem to share an interest in having clear rules that take subjectivity and guesswork as much as possible out of management. It is simply easier to follow rules than to think, and neither side really trusts independent judgement. Rule-based forestry would appear to eliminate the need for proactive design, and also seems to take away risks. Rules allow everyone to make what they think are accurate calculations; of habitat conserved or wood fibre that goes to the mill. This is of course not true. Risks remain, rules constantly change and independent judgement is always necessary because every landscape and every forest differs from others in important ways, making inflexible rules problematic. Rules cannot be applied the same way in every landscape and this is their main downfall. They are easy for auditors and compliance monitors to check on but frequently ruin the landscape. An example is the limit on clearcut sizes found in many jurisdictions. Environmental groups often see size of cutblocks as being an issue – smaller must be better than bigger. However, natural disturbances may occur at a range of scales, some very large and others very small. This range often reflects the landscape. If an upper limit is applied then all cutblocks will tend to be planned to be of the maximum size and so the natural variability will be lost, the relationship of scale to the landscape will be lost (some cutblocks may appear too big for small-scale landscapes and far too small for large-scale ones) and because more cutblocks in total will be needed for a given level of harvest the landscape will become more fragmented. The rules can be checked and the foresters found to be in compliance but the results will not meet the requirements of sustainable forestry.

Thus, when a handful of landscape architects dare to step up with a fresh idea for solving multiple problems, of course the beleaguered foresters are sceptical. Who are we to tell them how to do their jobs? Design is something they are unfamiliar with, and they are naturally contemptuous of landscape architects who are viewed as only concerned with the aesthetics or the visual appearance of the forest, rather than as integrated thinkers and planners who work from a long tradition of design. They naturally assume that landscape design is another restriction on timber harvest. However, once some forest design is undertaken and either the cut level stays the same or actually rises (as a result of removing the problem of overlapping and duplicating restrictions on the cut) and in addition the public are reassured about the way the planning is being carried out and how the multiple objectives are being met, these sceptical foresters are soon converted. Some of the projects described in the case studies started life as pilot projects and made many converts.

3. Natural scientists do not trust or understand design

As previously noted, professionals from a number of natural science fields have become increasingly involved in forest management policies and decisions. Fish and wildlife biologists, plant ecologists, soil scientists, hydrologists, geologists and others have carved out important niches, particularly in public land forestry. In the USA, this process resulted largely from policies put in place in the 1970s, including the National Environmental Policy Act (NEPA) and the National Forest Management Act (NFMA) (see Chapter 7). These laws explicitly called for greater balance and professional diversification in forestry. Each discipline views the forest landscape through a different and unique lens, and all contribute important ideas about how forests work and what they are good for.

Also, each science-based discipline misunderstands or mistrusts "design" as much as the foresters do. Scientists think differently from designers. They come from a strongly left-brained and analytical tradition, notwithstanding the conceptual, creative breakthroughs that often drive science forward.

Their professional imperative is to reduce uncertainty, and the accepted method is to pose hypotheses, then undertake rigorous and replicable research or experiments to disprove or validate them. Isolation of variables is an important aspect of much research. Designers can appear cavalier about uncertainty, and as a rule are perfectly happy to offer imperfect solutions to any problem posed to them. To a scientist, uncertainty is a reason for not moving ahead until a problem can be further investigated. To a designer uncertainty is a given, and while there is no perfect solution there are clearly some that are better than others, at least based on the current state of knowledge.

We recall a joint session of landscape architects and ecologists, debating the merits of integrated forest design when it was first unveiled as an integrated concept in the early 1990s. A prominent ecologist raised this objection: *We simply do not know enough about how forests work to redesign them*. Our response was immediate. *If we do not know enough, then we should stop logging them immediately and wait until we have perfect knowledge*. A good quip perhaps, but it did not allay the fears expressed.

Some scientists view forest design as a useful way to conduct management experiments over large areas. But because these are experiments, they wish to narrow the range of research questions and would prefer to squeeze subjective ideas out of the picture altogether. They worry that designers tend to over-simplify complex ecological systems, and that the design process moves ecosystem management away from a science-based decision-making process. Design relies on simplified rules for spatial configuration (i.e. fitting vegetation patches to landforms, use of corridors or stepping stones for connectivity, creating natural shapes with feathered edges, etc.) and offers few opportunities for validating whether these rules actually work as intended.

Many scientists seem to see forest design as far too subjective, in that it relies too much on the knowledge or bias of whoever happens to be in the room when the sketches are made. The group problem-solving techniques (discussed in Chapter 4) are embedded in social, not natural, science. Group processes may make everyone feel good, but could easily result in bad decision making with respect to ecology.

However, the multi-disciplinary aspect of forest management also causes many problems because there is frequently little integration between specialists – typically each specialist produces a plan for conserving his or her own interest, such as a particular bird, fish or mammal. These plans frequently conflict with one another because they are based on species, not landscapes. One of the ways the Nancy Diaz approach was used to resolve these conflicts was by integrating them at the landscape level with the applied analysis (presented in Chapter 2). Subsequent forest design projects using multi-disciplinary project teams, especially the pilot projects, enabled these specialists to see how their work contributed to a greater whole and how design and the concept of the desired future conditions could not only satisfy the requirements of key species but of all species.

4. Management inertia and resistance

Designing forests represents a step into the unknown, and most people fear the unknown. They are caught up in the messy world of implementing imperfect and incomplete policies handed down from above. There are few rewards for innovation, yet high risks in misspending taxpayers' or their clients' money. In both American and Canadian public forests, there has been insufficient attention and commitment to the middle scale, or tactical level of forest planning (at which forest design operates). Managers are normally expected to go directly from strategic planning (land allocations and harvest estimates that lack spatial focus) to operational activities (road building, logging, restoration).

Uncertainty increases with time, and forest design by necessity forces managers to anticipate 50–200 years into the future. Given the rapid changes in social, demographic, economic and scientific views of forests over the past three decades, it is understandable that long-range commitments are shunned.

The public involvement aspect of forest design can also worry many forest managers. The idea of meaningful consultation or shared decision making puts them off. They worry that expectations will be raised, only to be dashed later as reality sets in. Then there is the uncertainty factor. The design process is a search for solutions, and it could lead to uncharted territory. Once again, the pilot projects with public participation demonstrated to some of these sceptical managers that better results are obtained and that suspicions about foresters' intentions held by the public can be removed. Far from handing power to the public, after a good experience with participation, the public hands the power back and waits expectantly for the plan to be implemented – including the logging!

Budgets for public forestry have been reduced over the years. Forest operations, particularly in the productive conifer forests of British Columbia and the Pacific Northwest of the USA, were historically financed by timber receipts paid by private companies who had logging rights (Canada) or purchased timber on a sale-by-sale basis (USA). As long as the timber cut remained high, there was little need to innovate, since promotions were easy to come by. As environmental problems mounted, the search for new solutions was unleashed, but not soon enough to forestall a rapid rise of environmental restrictions and an equally rapid decline of revenue. This fatal combination thwarted the initial burst of innovation. New methods, like the forest design approach, were generated or applied too late to save the day within that particular political–economic cycle.

Private forest managers would seem to have more incentive to forge ahead with forest design, particularly those seeking third-party certification for sustainable practices. But they have also been slow to take up the baton for many reasons. Firstly, private land owners do not like to share detailed information about their properties, since this can lead to a competitive disadvantage or hostile takeover bids (of publicly traded companies) based on high timber inventories that can be liquidated to pay off debts.

As is the case with public land managers, private foresters also like to keep the public at arms length (especially in the USA), and they certainly do not want to cede any decision making, so they avoid public consultation processes. Private forest lands held by publicly traded companies are subject to being sold off to another company with a different ethos. Why invest in a 50-year or longer plan only to walk away after five or ten years? In the UK, private owners have long been supported by public subsidies so that the availability of grants to prepare long-term forest plans has proved a valuable incentive. Public consultation is an accepted part of the process. However, in the UK, government intervention in private land management is more accepted than in the USA, for example.

The future of sustainable forest landscape design

So much for the reasons why forest design has had a slow take-up in various places. In spite of them all, we are optimistic that this is a process that will catch on sooner or later. Firstly, it is a natural evolutionary step within the field of forest and natural resource management. Burton *et al.*, in their comprehensive volume on management of the boreal forest, traced eight stages in the history of forestry:

1   *Subsistence management*: low population, light use
2   *Exploitation*: higher population, over-use, unrestricted harvest
3   *Regulation*: top-down restrictions on conversion and harvest rates
4   *Sustained yield timber production*: silviculture, even flow
5   *Multiple use*: recognition of other values, i.e. game and clean water
6   *Site specific*: use of ecosystem classification to improve effectiveness
7   *Biocentric*: the addition of values that are not utilitarian
8   *Landscape design*: strategic forest management over space and time that takes account of all values.

When prominent forest scientists and educators who are not landscape architects embrace the concept of design applied to forestry, a wider adoption of the practice cannot be far behind.

Secondly, we believe that forest design will increase because it works, and the growing (albeit slowly) number of successful case studies will build momentum, mostly "from below" for wider emulation and adaptation. Although the number of case studies is smaller than we would like, their distribution is impressive, including the USA, Canada and Europe.

Thirdly, there is also an emerging convergence of thought on the value of design in the service of conservation. Conservation biologists increasingly use the term "design" to discuss development of biological reserves. The corridor-matrix-patch landscape ecology paradigm developed by Forman and Godron is being adapted to land-use planning in and around urban areas, thus increasing its exposure to the design professions. Ecological restoration practitioners are also finding that design is an important tool in repairing damaged ecosystems.

Fourthly, the need is substantial. The World Commission on Forests and Sustainable Development points out that two-thirds of the world's terrestrial species live in forests, yet less than 10% of the world's forest land is protected in reserves. Only a little more than half of the original forest land 8,000 years ago is still in forest cover, much of this highly degraded or altered to plantations. This means the vast majority of remaining forests are subject to active management, much still primary forest and the rest already impacted by past exploitation. Thus the potential canvas and need for well-designed forest landscapes is simply enormous. The World Commission recognises the need for "implementing integrated planning and management approaches at the landscape level" to be a key tool in the development of sustainable forestry.

All of these factors suggest that wider adoption of forest design is just around the corner. The main competing approach appears to be forestry based on ecosystem models. What this seems to mean is direct emulation of disturbance ecology in terms of space (size), legacy retention (green trees, snags, downed wood), shape or configuration (coves, lobes, islands, ragged edges) and timing between harvests. We believe that while forestry based on ecosystem models is essential, the most promising application employs a design framework to help organise the spatial patterns. This is simply because no ecosystem model can capture the variability in a given landscape, and design techniques can also incorporate social dynamics.

In order for forest design to gain wider use, there are several challenges that need to be addressed by forest researchers, educators and managers.

Firstly, there needs to be a greater recognition that strategic-level planning alone (zoning) cannot supply all the answers spatially, and that forest design offers something more. We believe we have shown what this is and that the case studies demonstrate design as more than simply a new theory that has yet to be proved in practice.

Secondly, forestry and related natural resource students and teaching staff need to be educated and trained in understanding this process, at least to a level where they can know when it is going to be useful, as well as knowing when to ask and how to work with professional design help. We hope that the material in this book is useful for supporting training programmes and for guiding those interested in the process so that they can teach themselves.

Thirdly, there needs to be more forest landscape designers based within public forest services, in companies or in private practice, also educated and trained in this process so that the more complex and larger-scale projects can be solved to a high professional standard.

Fourthly, forest managers need to get more comfortable and confident with respect to implementing design results. This takes some experience but we hope that the examples and tools provided in this book will help instill some confidence that designs can be implemented within an operational framework that includes adaptive management.

Finally, forest managers will need to weigh up the cost of design versus the cost of not designing. If they only view the cost side of the ledger, and fail to consider the benefits, they will fail to recognise what can be saved over the long term. Our experience is that while the design process may cost more than less robust planning, this is more than compensated for by savings down the line, especially in the costs of fighting protestors or legal objections to plans. Timber volumes need not be reduced as a consequence of design. In some circumstances they may be increased by demonstrating that good design can enable logging to proceed in more visible landscapes or sensitive ecosystems without damaging them. To the extent that good design presents a viable third way, it can offer a new option between over-exploitation and preservation.

We hope therefore that while the process has only been applied in a limited number of locations this book will bring the benefits to a far wider audience and that it will enthuse foresters to develop sustainable forest planning using design as a tool in those many locations where it has much to offer.

Designing Sustainable Forest Landscapes is a new idea that builds on many good ideas of the past. It should be viewed as part of a wave of innovations that has been put forward to help societies cope with emerging challenges in resource use and conservation. There is no going back to simpler times, where forests are only seen as repositories of timber, waiting for the axe. In going forward, we hope that this book and the ideas it contains prove helpful in the wider effort to conserve, and still use, our world's forests.

Bibliography

Designing Sustainable Forest Landscapes is a book as much for planners, designers and managers as for students and teachers. In order to make it more readable and less academic we decided not to place references in the body of the text, although many publications and authors are mentioned there. Instead we have collected references of all the publications used in the text and also further reading on the subject and arranged it by chapter. Readers may find that the same publication appears in the references to more than one chapter. This is because it is easier to find out what is important for a particular subject and often the same reference is useful for different chapters, especially those that are key works lying behind the theory and practice. There tend to be more references for Section One because this is where the main theory is presented. The other two sections, being concerned with methods and application, are less fully referenced because much less is written about these subjects.

Introduction

Bechmann, R. (1990) *Trees and Man: The Forest in the Middle Ages*, translated from the French by K. Dunham, New York: Paragon House.

Boyd, R. (1999) *Indians, Fire, and the Land in the Pacific Northwest*, Corvallis, OR: Oregon State University Press.

Cotta, H. (1903) "Cotta's Preface", *Forestry Quarterly*, **1**: 1902–1903.

Diaz, N. and Apostol, D. (1992) *Forest Landscape Analysis and Design: A Process for Developing and Implementing Land Management Objectives for Landscape Patterns*, Portland, OR: USDA Forest Service, Pacific Northwest Region.

Egler, F. (1986) Commentary: "Physics envy," *Ecology Bulletin of the Ecological Society of America,* **67**: 233–5.

Forest Renewal BC, *Managing Your Woodland*, on-line publication available at: www.woodlot.bc.ca/swp/myw/index.htm.

Forman, R.T.T. (1995) *Land Mosaics: The Ecology of Landscapes and Regions*, Cambridge: Cambridge University Press.

Forman, R.T.T. and Godron, M. (1986) *Landscape Ecology*, New York: John Wiley.

Karr, J.R. (2002) "What from Ecology is Relevant to Design and Planning?", in B.R. Johnson and K. Hill (eds) *Ecology and Design*, Washington, DC: Island Press.

Keiter, R.B. and Boyce, M.S. (1991) *The Greater Yellowstone Ecosystem, Redefining America's Wilderness Heritage*, New Haven, CT: Yale University Press.

Maser, C. (1988) *The Redesigned Forest*, San Pedro, CA: R.& E. Miles Publishers.

Moeur, M., Spies, Thomas A., Hemstrom, M., *et al.* (2005) *Northwest Forest Plan – The First 10 Years (1994–2003): Status and Trend of Late-Successional and Old-Growth Forest*. Gen. Tech. Rep. PNW-GTR-646. Portland, Pacific Northwest Research Station: US Department of Agriculture, Forest Service.

Noss, R.F. and Cooperrider, A.Y. (1994) *Saving Nature's Legacy: Protecting and Restoring Biodiversity*, Washington, DC: Island Press.

Peck, S. (1998) *Planning for Biodiversity: Issues and Examples*. Washington, DC: Island Press.

Perlin, J. (1989) *A Forest Journey: The Role of Wood in the Development of Civilization* (1st edn), New York: W.W. Norton & Company Inc.

Ryle, G. (1969) *Forest Service. The First 45 Years of the Forestry Commission of Great Britain*. Newton Abbot: David and Charles.

United Nations (1992) *Report of the United Nations Conference on Environment and Development Annex I Rio Declaration on Environment and Development*. New York: UN.

United Nations (1992) *Report of the United Nations Conference on Environment and Development Annex III Non-Legally Binding Authoritative Statement of Principles for a Global Consensus on the Management, Conservation and Sustainable Development of all Types of Forests*, New York: UN.

United Nations (1992) "Convention on Biological Diversity", on-line publication available at: www.biodiv.org/convention/default.shtml.

United Nations (2000) "Framework Convention on Climate Change (Kyoto Protocol)", on-line publication available at: http://unfccc.int/cop6_2/index-4.html.

World Commission on Environment and Development (1987) *Our Common Future*, Oxford: Oxford University Press.

Chapter 1

Apostol, D. and Sinclair, M. (eds) (2006) *Restoring the Pacific Northwest*, Washington, DC: Society for Ecological Restoration/Island Press.

Bechmann, R. (1990) *Trees and Man: The Forest in the Middle Ages*, translated from the French by K. Dunham, New York: Paragon House.

Boyd, R. (1999) *Indians, Fire, and the Land In the Pacific Northwest*, Corvallis, OR: Oregon State University Press.

Cotta, H. (1903) "Cotta's Preface", *Forestry Quarterly*, **1**: 1902–1903.

Forestry Authority (1994) *Forest Landscape Design Guidelines* (2nd edn), London: HMSO.

Forestry Commission of Tasmania (1990) *A Manual for Forest Landscape Management*, Hobart: Forestry Commission of Tasmania.

Forestry Commission (1991) *Community Woodland Design Guidelines*, London: HMSO.

Forestry Commission (1992) *Lowland Landscape Design Guidelines*, London: HMSO.

Forestry Commission (2004) *UK Forestry Standard* (2nd edn), Edinburgh: Forestry Commission.

Forestry Stewardship Council *see* www.fscus.org/.

Fries, C., Carlsson, M., Dahlin, B., Lämås, T. and Sallnäs, O. (1999) "A Review of Conceptual Landscape Planning Models for Multiobjective Forestry in Sweden", *Canadian Journal of Forest Research*, **28**: 159–67.

Harris, L.D. and Miller, K.R. (1984) *The Fragmented Forest: Island Biogeography Theory and the Preservation of Biotic Diversity*, Chicago, IL: University of Chicago Press.

Kohm, K.A. and Franklin, J.F. (1997) *Creating a Forestry for the 21st Century*, Washington, DC: Island Press.

Lindenmayer, D.B. and Franklin, J.F. (2002) *Conserving Forest Biodiversity: A Comprehensive Multi-scaled Approach*. Washington, DC: Island Press.

Lewis, H.T. (1973) *Patterns of Indian Burning in California: Ecology and Ethnohistory*, Ramona: Ballena Press.

Macarthur, R.H. and Wilson, E.O. (2001) *The Theory of Island Biogeography*. Princeton University Press.

Marsh, G.P. (1864) *Man and Nature*, Georgetown, ON: UNI Presses.

Montreal Process (1999) "Criteria and Indicators" (2nd edn), on-line publication available at: www.mpci.org/criteria_e.html.

Noss, R.F. and Cooperrider, A.Y. (1994) *Saving Nature's Legacy: Protecting and Restoring Biodiversity*, Washington, DC: Island Press.

Pan-European Forestry Certification Council (2005) *Technical Document*, Paris: PEFCC.

Platt, R.H. (1996) *Land Use and Society: Geography, Law, and Public Policy*, Washington, DC: Island Press.

Province of British Columbia (1981) *Forest Landscape Handbook*, Victoria, BC: Ministry of Forests.

Steen, H.K. (1976) *The US Forest Service: A History*, Baltimore, MD: University of Washington Press.

The Wildlands Project: www.wildlandsprojectrevealed.org/htm/summary.htm.

UKWAS (2000) *UK Woodland Assurance Standard for the UK Woodland Assurance Scheme*, Edinburgh: UKWAS. On-line, available at: www.ukwas.org.uk/standard/ukwas_archive/index.html.

US Dept of Agriculture Forest Service (1995) *Landscape Aesthetics: A Handbook for Scenery Management*, Washington, DC: USDA Forest Service.

Watkins, C. (1998) *European Woods and Forests: Studies in Cultural History*, Wallingford: CAB International.

Chapter 2

Bell, S. (ed.) (2003) *The Potential for Applied Landscape Ecology to Forest Design Planning*, Edinburgh: Forestry Commission.

Bell, S. and Nikodemus, O. (2001) "Landscape Ecological Planning in Latvia", unpublished report of a workshop held in association with the State Forest Service, Riga, Latvia, October, 2000.

Burton, P.J., Messier, C., Smith, D. and Adamowicz, W.L. (2003) *Towards Sustainable Management of the Boreal Forest*, Ottawa, ON: National Research Council of Canada, pp. 433–80.

Diaz, N. and Apostol, D. (1992) *Forest Landscape Analysis and Design: A Process for Developing and Implementing Land Management Objectives for Landscape Patterns*, Portland, OR: USDA Forest Service, Pacific Northwest Region.

Dramstad, W.E., Olsen, J.D. and Forman, R.T.T (1996) *Landscape Ecology Principles in Landscape Architecture and Land Use Planning*, Washington, DC: Island Press.

Forman, R.T.T. (1995) *Land Mosaics: The Ecology of Landscapes and Regions*, Cambridge: Cambridge University Press.

Forman, R.T.T. and Godron, M. (1986) *Landscape Ecology*, New York: John Wiley.

Karr, J.R. (2002) "What from Ecology is Relevant to Design and Planning?", in B.R. Johnson and K. Hill (eds) *Ecology and Design*, Washington, DC: Island Press.

Kimmins, J.P. (1997) *Forest Ecology* (2nd edn), Upper Saddle River, NJ: Prentice Hall.

Kohm, K.A. and. Franklin, J.F. (1997) *Creating a Forestry for the 21st Century*, Washington, DC: Island Press.

Lindenmayer, D.B. and Franklin, J.F. (2003) *Conserving Forest Biodiversity*, Washington, DC: Island Press.

Macarthur, R.H. and Wilson, E.O. (reprinted 2001) *The Theory of Island Biogeography*, Princeton, NJ: Princeton University Press.

Marsh, G.P. (1864) *Man and Nature*, Georgetown, ON: UNI Press.

Maser, C. (1988) *The Redesigned Forest*, San Pedro, CA: R. & E. Miles Publishers.

Mellen, K., Huff, M. and Hagestedt, R. (1995) "HABSCAPES: Reference Manual and User's Guide", unpublished manuscript, US Forest Service.

Mills, L.S., Soule, M.E. and Doak, D.F. (1993) "The Keystone-species Concept in Ecology and Conservation", *Bioscience,* **43**(4): 219–24.

Noss, R.F. and Cooperrider, A.Y. (1994) *Saving Nature's Legacy: Protecting and Restoring Biodiversity*, Washington, DC: Island Press.

Perry, D. (1994) *Forest Ecosystems*, Baltimore, MD: Johns Hopkins University Press.

Ripple, W.J. and Beschta, R.L. (2003) "Wolf Reintroduction, Predation Risk, and Cottonwood Recovery in Yellowstone National Park", *Forest Ecology and Management,* **184**: 299–313.

The Wildlands Project: "Room to Roam", www.wildlandsproject.org/roomtoroam/endangered/.

Chapter 3

Anstey, C., Thompson, S. and Nichols, K. (1982) *Creative Forestry*, Wellington: New Zealand Forest Service.

Aspinall, P., Hill, A., Stuart-Murray, J. *et al.* (2000) 'Eye Movements and Visual Forces in the Landscape', *Scottish Forestry,* **54**(3): 133–42.

Bell, S. (1994) *Visual Landscape Design Training Manual*, Victoria, BC: Ministry of Forests of British Columbia Recreation Branch.

Bell, S. (1998) *Forest Design Planning: A Guide to Good Practice*, Edinburgh: Forestry Commission.

Bell, S. (1998) *The Landscape Value of Farm Woodlands*, Information Note 13, Edinburgh: Forestry Commission.

Bell, S. (1998) "Woodland in the Landscape", in M.A. Atherden and R.A. Butlin (eds) *The Landscape: Past and Future Perspectives*, York: The Place Research Centre, University College of Ripon and York St John.

Bell, S. (1999) "Plantation Management for Landscapes in Britain", *International Forestry Review*, **1**(3).

Bell, S. (1999) *Landscape: Pattern, Perception and Process*, London: E. & F.N. Spon.

Bell, S. (2001) *Landscape Pattern, Perception and Visualization in the Visual Management of Forests*, Landscape and Urban Planning.

Bell, S. (2004) *Elements of Visual Design in the Landscape* (2nd edn), London: E. & F.N. Spon.

Bell, S. (2004) "Visual Resource Management Approaches", in J. Burley, J. Evans and J.A. Youngquist (eds) *Encyclopaedia of Forest Sciences*, London: Elsevier.

Bell, S. and Murray, T. (2002) *Forest Ecosystem Design: Progress Report on Development and Implementation in the Maritimes*, Information Report M-X-215E, Fredericton: Canadian Forest Service, Atlantic Forestry Centre.

Bell, S. and Nikodemus, O. (2000) *Forest Landscape Planning and Design*, Riga: State Forest Service of Latvia (in Latvian).

Bell, S. and Shepherd, N. (1999) "Integrating Forest Design Guidance and Landscape Character", in M.Usher (ed.) *Landscape Character: Perspectives on Management and Change*, Edinburgh: Scottish Natural Heritage.

Berleant, A. (1992) *The Aesthetics of Environment*, Philadelphia, PA: Temple University Press.

British Columbia Ministry of Forests and Range (2006). *The Public Response to Harvest Practices in British Columbia at the Landscape and Stand Level*, Victoria, BC: Forest Practices Branch.

British Columbia Ministry of Forests (1996) *Clearcutting and Visual Quality: A Public Perception Study*, Victoria, BC: Recreation Branch.

Burke, E. (1958) *A Philosophical Enquiry into the Origins of Our Ideas of the Sublime and the Beautiful*, edited by J.T. Boulton, London: Routledge and Kegan Paul.

Crowe, S. (1978) "The Landscape of Forests and Woods," *Forestry Commission Handbook* 44, London: HMSO.

Entec (UK) (1996) "The Landscape Value of Farm Woodlands", unpublished report to MAFF/Forestry Commission.

Entec (UK) (1997) "Valuing Landscape Improvements to British Forests", unpublished report to Forestry Commission.

Forestry Authority (1994) *Forest Landscape Design Guidelines* (2nd edn), London: HMSO.

Forestry Commission of Tasmania (1990) *A Manual for Forest Landscape Management*, Hobart: Forestry Commission of Tasmania.

Forestry Commission (1991) *Community Woodland Design Guidelines*, London: HMSO.

Forestry Commission (1992) *Lowland Landscape Design Guidelines*, London: HMSO.

Forestry Commission (2000) "Monitoring the Landscape Quality of Upland Conifer Forests", unpublished report, Edinburgh: Forestry Commission.

Gobster, P.H. (1995) "Aldo Leopold's Ecological Aesthetic: Integrating Aesthetic and Biodiversity Values", *Journal of Forestry*, February: 6–10.

Gustavsson, R. and Ingelog, T. (1994) *Det Nya Landskapet* (*The New Landscape*), Jönköping: Skogstyrelsen (in Swedish).

Hauninen, E., Oulasemaa, K. and Salpakivi-Salomaa, P. (1998) *Forest Landscape Management*, Helsinki: Forestry Development Centre Tapio.

Kaplan, S. (1988) "Perception and Landscape: Conception and Misconception", in J.L. Nasar (ed.), *Environmental Aesthetics*, Cambridge: Cambridge University Press.

Kohler, W. (1947) *Gestalt Psychology: An Introduction to Modern Concepts in Psychology*, New York: Liveright Publishing.

Landscape Institute and Institute for Environmental Assessment (2002) *Guidelines for Landscape and Visual Impact Assessment* (2nd edn), London: Sponpress.

Lee, T.R. (2001) 'Perceptions, Attitudes and Preferences in Forests and Woodlands', Technical Paper 18, Forestry Commission.

Litton, R.B. (1968) *Forest Landscape Description and Inventories*, USDA Forest Service Research Paper PSW-49, Washington, DC: USDA.

Lucas, O.W.R. (1991) *The Design of Forest Landscapes*, Oxford: Oxford University Press.

Ministry of Forests, Forest Practices Branch (2001) *Visual Impact Assessment Guidebook* (2nd edn), Victoria, BC: Ministry of Forests, Forest Practices Branch.

Province of British Columbia (1981) *Forest Landscape Handbook*, Victoria, BC: Ministry of Forests.

Rackham, O. (1986) *The History of the Countryside*, London: Dent.

Sheppard, S.R.J. and Harshaw, H.W. (eds) (2001) *Forests and Landscapes: Linking Ecology, Sustainability and Aesthetics*, Wallingford: CABI Publishing.

Swanwick, C. (2002) *Landscape Character Assessment: Guidelines for England and Wales*, Countryside Agency and Scottish Natural Heritage.

US Dept of Agriculture Forest Service (1970) *National Forest Landscape Management*, Vol. 1, Washington, DC: USDA Forest Service.

US Dept of Agriculture Forest Service (1995) *Landscape Aesthetics: a Handbook for Scenery Management*, Washington, DC: USDA Forest Service.

Chapter 4

Apostol, D., Sinclair, M. and Johnson, B. (2000) "Design Your Own Watershed: Top-Down Meets Bottom-Up in the Little Applegate Valley", in R.G. D'Eon, A.E. Ferguson and J.F. Johnson (eds) *Ecosystem Management of Forested Landscapes: Directions and Implementation*, Vancouver, BC: UBC Press, pp. 45–50.

Arnstein, S.R. (1969) "A Ladder of Citizen Participation", *Journal of the American Planning Association*, **35**(4): 216–24.

Bell, S. and Komulainen, M. (2000) *Crossplan: Integrated, Participatory Landscape Planning as a Tool for Rural Development*, Oulu: University of Oulu.

Bell, S. (ed.) (2003) *Crossplan: Integrated, Participatory Landscape Planning as a Tool for Rural Development*, Edinburgh: Forestry Commission.

Countryside Commission (1998) *Participatory Action in the Countryside – a Literature Review*, CCWP07, Cheltenham: Countryside Commission.

Democracy Network (1999) *Guidance on Enhancing Public Participation. Modern Local Government Series*, London: Department of Environment, Transport and the Regions.

Derounian, J.G. (1998) *Effective Working with Rural Communities*, London: Packhard Publishing.

Francis, D. and Henderson, P. (1992) *Working with Rural Communities*, London: Macmillan.

Hislop, M. (2004) *Involving People in Forestry: A Toolbox for Public Involvement in Forest and Woodland Planning*, Edinburgh: Forestry Commission.

Kaufman, H. (1960) *The Forest Ranger: A Study in Administrative Behavior*, Baltimore, MD: Johns Hopkins University Press.

Shannon, M. and Antypas, A.R. (1997) "Open Institutions: Uncertainty and Ambiguity in 21st Century Forestry", in J.W. Thomas, M. Hunter, F. Cubbage *et al. Creating a Forestry for the 21st Century: The Science of Ecosystem Management*, Washington, DC: Island Press, Chapter 28, pp. 437–45.

Chapter 5

Bell, S. (1994) *Visual Landscape Design Training Manual*, Victoria, BC: Ministry of Forests of British Columbia Recreation Branch.

Bell, S. (1998) *Forest Design Planning: A Guide to Good Practice*, Edinburgh: Forestry Authority.

Hammond, H. (1991) *Seeing the Forest Among the Trees: The Case for Holistic Forest Use*, Vancouver, Canada: Polestar Press.

Harrington, S. (1999) *Giving the Land a Voice: Mapping Our Home Places* (2nd edn), Salt Spring Island, BC: Salt Spring Island Community Service Society.

Johnson, D.H. and O'Neil, T.H. (2001) *Wildlife Habitat Relationships in Oregon and Washington*, Corvallis, OR: Oregon State University Press.

Mellen, K., Huff, M. and Hagestedt, R. (1995) "HABSCAPES: Reference Manual and User's Guide", unpublished manuscript, US Forest Service.

Chapter 6

Bell, S. (1994) *Visual Landscape Design Training Manual*, Victoria, BC: Ministry of Forests of British Columbia Recreation Branch.

Bell, S. (1998) *Forest Design Planning: A Guide to Good Practice*, Edinburgh: Forestry Authority.

Diaz, N. and Apostol, D. (1992) *Forest Landscape Analysis and Design: A Process for Developing and Implementing Land Management Objectives for Landscape Patterns*, Portland, OR: USDA Forest Service, Pacific Northwest Region.

Lucas, O.W.R. (1991) *The Design of Forest Landscapes*, Oxford: Oxford University Press.

Chapter 7

British Columbia Ministry of Forests (2001) *Visual Impact Assessment Guidebook* (2nd edn), Victoria, BC.

Landscape Institute and Institute for Environmental Assessment (2002) *Guidelines for Landscape and Visual Impact Assessment* (2nd edn), London: Sponpress.

Lindenmayer, D.B. and Franklin, J.F. (2003) *Conserving Forest Biodiversity*, Washington, DC: Island Press.

Mellen, K., Huff, M. and Hagestedt, R. (1995) "HABSCAPES: Reference Manual and User's Guide", unpublished manuscript, US Forest Service.

Noss, R.F. and Cooperrider, A.Y. (1994) *Saving Nature's Legacy: Protecting and Restoring Biodiversity*, Washington, DC: Island Press.

Chapter 8

Anderson, Kat M. (1996) "Tending the Wilderness", *Restoration and Management Notes*, **14**(2): 154–66.

Apostol, Dean and Buursma, Ed (1995) "Watershed Analysis for the Collowash and Hot Springs Fork Rivers", unpublished report, Mt Hood National Forest, Sandy, OR: USDA Forest Service.

Bell, S. (ed.) (2003) *The Potential for Applied Landscape Ecology to Forest Design Planning*, Edinburgh: Forestry Commission.

Diaz, N. and Apostol, D. (1996) *Forest Landscape Analysis and Design: A Process for Developing and Implementing Land Management Objectives for Landscape Patterns*, Portland, OR: USDA Forest Service, Pacific Northwest Region.

Forest Ecosystem Management Assessment Team (1993) *Forest Ecosystem Management: An Ecological, Economic, and Social Assessment*, Portland, OR: USDA Forest Service.

Luoma, J.R. (2000) *The Hidden Forest: The Biography of an Ecosystem*, New York: Henry Holt and Company.

Priester, K. (1995) "Words into Action", unpublished report by the Rogue Institute for Ecology and Economy.

Society for Ecological Restoration International and the Science & Policy Working Group (2002) *The SER Primer on Ecological Restoration*, available on-line at: www.ser.org/content/ecological_restoration_primer.asp.

Chapter 9

Bell, S. (1996) "Vuokatti Forest Landscape Design", unpublished report for Metsakeskus Tapio, Helsinki.

Bell, S. and Murray, T. (2002) *Forest Ecosystem Design: Progress Report on Development and Implementation in the Maritimes*, Information Report M-X-215E, Fredericton: Canadian Forest Service, Atlantic Forestry Centre.

Diaz, N. and Apostol, D. (1992) *Forest Landscape Analysis and Design: A Process for Developing and Implementing Land Management Objectives for Landscape Patterns*, Portland, OR: USDA Forest Service, Pacific Northwest Region.

Forest Ecosystem Management Assessment Team (1993) *Forest Ecosystem Management: An Ecological, Economic, and Social Assessment*, USDA Forest Service.

Kohm, K.A. and. Franklin, J.F. (1997) *Creating a Forestry for the 21st Century*, Washington, DC: Island Press.

Lindenmayer, D.B. and Franklin, J.F. (2003) *Conserving Forest Biodiversity*, Washington, DC: Island Press.

Mellen, K., Huff, M. and Hagestedt, R. (1995) "HABSCAPES: Reference Manual and User's Guide", unpublished manuscript, US Forest Service.

Morrison, M.L. (2002) *Wildlife Restoration*, Washington, DC: Island Press, p. 156.

Murray, T. and Singleton, J. (1995) "Sutherland Brook Forest Ecosystem Design Pilot Project", Fredericton: unpublished report of Canadian Forestry Service Maritimes.

Perry, D. (1994) *Forest Ecosystems*, Baltimore, MD: Johns Hopkins University Press.

Slocan Forest Products (1995) "Bonanza Face Ecosystem Design", unpublished report.

Thomson, A.J., Goodenough, D.G., Archibald, R., *et al.* (1996) 'Landscape Management and Biodiversity: Automating the Design of Forest Ecosystem Networks', *AI Applications*, **10**(3): 57–65. British Columbia Ministry of Forests Research Branch, available at: www.for.gov.bc.ca/hre/pubs/pubs/1129.htm.

Chapter 10

Bell, S. (1995) "New Woodlands in the Landscape", in R. Ferris Kaan (ed.)*The Ecology of Woodland Creation*, Chichester: John Wiley and Sons.

Bell, S. (1998) *Forest Design Plans: A Guide for Good Practice*, Edinburgh: Forestry Authority.

Bell, S. (2000) "Agroforestry in the Landscape", in M. Hislop and J. Claridge (eds), *Agroforestry in the UK*, Forestry Commission Bulletin 122, Edinburgh: Forestry Commission.

Bell S. and Heath, G. (2000) *Nant-yr Hwch Long Term Forest Design Plan: An Example of Good Design from the Private Sector*, Practice Note 10, Edinburgh: Forestry Commission.

Cawdor Forestry (2000) "Dochfour Long-term Forest Plan", unpublished report, Ardersier, Inverness: Cawdor Forestry.

Forestry Authority (1994) *Forest Landscape Design Guidelines* (2nd edn), London: HMSO.

Forestry Commission Scotland (2006) *The Creation of Small Woodlands on Farms*, Edinburgh: Forestry Commission.

Harris, L.D. and Miller, K.R. (1984) *The Fragmented Forest: Island Biogeography Theory and the Preservation of Biotic Diversity*, Chicago, IL: University of Chicago Press.

Lucas, O.W.R. (1991) *The Design of Forest Landscapes*, Oxford: Oxford University Press.

Maser, C. (1988) *The Redesigned Forest*, San Pedro, CA: R. & E. Miles Publishers.

Noss, R.F. and Cooperrider, A.Y. (1994) *Saving Nature's Legacy: Protecting and Restoring Biodiversity*, Washington, DC: Island Press.

Perry, D. (1994) *Forest Ecosystems*, Baltimore, MD: Johns Hopkins University Press.

Woodland Trust (2005) "Victory Wood: Proposals for the Creation of a New Woodland to Commemorate HMS *Victory* at the Battle of Trafalgar", unpublished report.

Afterword

Adams, D. (1979) *A Hitchhikers Guide to the Galaxy*, London: Pan Books.

Andison, D.W. (2003) "Tactical Forest Planning and Landscape Design", in P.J. Burton, C. Messier, D. Smith and L.A. Wiktor, *Towards Sustainable Management of the Boreal Forest*, Ottawa, ON: Canada,National Research Council of Canada, pp. 433–80.

Apostol, D. and Sinclair, M. (eds) (2006) *Restoring the Pacific Northwest: The Art and Science of Ecological Restoration in Cascadia*, Washington, DC: Island Press.

Forman, R.T.T. and Godron, M. (1986) *Landscape Ecology*, New York: John Wiley.

Hoskins, W.G. (1955) *The Making of the English Landscape*, London: Penguin Books.

Krishnaswamy, A. and Hanson, A. (1999) *Summary Report World Commission on Forests and Sustainable Development*, Winnipeg: International Institute of Sustainable Development.

Lindenmayer, D.B. and Franklin, J. (2002) *Conserving Forest Biodiversity: A Comprehensive Multiscaled Approach*, Washington, DC: Island Press.

United Nations (1992) *Report of the United Nations Conference on Environment and Development Annex III, Non-Legally Binding Authoritative Statement of Principles for a Global Consensus on the Management, Conservation and Sustainable Development of all Types of Forests*, New York: UN.

Index